Structural Studies, Repairs and Maintenance of Heritage Architecture XVII

&

Earthquake Resistant Engineering Structures XIII

WITPRESS

WIT Press publishes leading books in Science and Technology.
Visit our website for new and current list of titles.
www.witpress.com

WIT eLibrary

Home of the Transactions of the Wessex Institute.
The WIT eLibrary provides the international scientific community with immediate and
permanent access to individual papers presented at WIT conferences.
Visit the WIT eLibrary at www.witpress.com.

SEVENTEENTH INTERNATIONAL CONFERENCE ON
STRUCTURAL REPAIRS AND MAINTENANCE OF HERITAGE ARCHITECTURE
&
THIRTEENTH INTERNATIONAL CONFERENCE ON EARTHQUAKE RESISTANT
ENGINEERING STRUCTURES

STREMAH XVII & ERES XIII

CONFERENCE CHAIRMAN

Santiago Hernandez
University of A Coruña, Spain
Member of WIT Board of Directors

LOCAL ORGANISER

Guido Marseglia
University of Seville, Spain

INTERNATIONAL SCIENTIFIC ADVISORY COMMITTEE

FOR STREMAH XVII

Tawfiq Abuhantash
Khalid Al Saud
Jaroslaw Bakowski
Luisa Basset
Maria Teresa Broseta Palanca
Rene Brueckner
Robert Cerny
Chi-Jen Chen
Dan Constantinescu
Carlos Cuadra
Mariella De Fino
Fernanda De Maio
Khalid El Harrouni
Fabio Fatiguso
Lukas Fiala
Antonio Galiano Garrigos

Josip Galic
Fabio Garzia
Luis Fernando Guerrero Baca
Antonella Guida
Kabila Hmood
Bashir Kazimee
Jiri Madera
George Manos
Hazel Nash
Cordelia Osasona
Antonello Pagliuca
Carlos Rubio-Bellido
Nilufer Saglar Onay
Wael Shaheen
Miroslav Sykora
Claudia Vitone

FOR ERES XIII

Sudha Arlikatti
Francisco Javier Baeza
Raul Campos
Abel Diaz
Maria Favvata
Salvatore Grasso
Yasser Ibrahim
Salvador Ivorra
Yoshihiro Kimura
Jens-Uwe Klugel
Marios Phocas
Maria Cristina Porcu
Ivan Roselli
Michal Sejnoha
Stavros Syngellakis
Alexandros Tsonos
Humberto Varum

ORGANISED BY

Wessex Institute, UK

SPONSORED BY

WIT Transactions on the Built Environment

WIT Transactions

Wessex Institute
Ashurst Lodge, Ashurst
Southampton SO40 7AA, UK

Structural Studies, Repairs and Maintenance of Heritage Architecture XVII

&

Earthquake Resistant Engineering Structures XIII

EDITORS

Santiago Hernández
University of A Coruña, Spain
Member of WIT Board of Directors

Guido Marseglia
University of Seville, Spain

WITPRESS Southampton, Boston

Editors:

Santiago Hernández
University of A Coruña, Spain
Member of WIT Board of Directors

Guido Marseglia
University of Seville, Spain

Published by

WIT Press
Ashurst Lodge, Ashurst, Southampton, SO40 7AA, UK
Tel: 44 (0) 238 029 3223; Fax: 44 (0) 238 029 2853
E-Mail: witpress@witpress.com
http://www.witpress.com

For USA, Canada and Mexico

Computational Mechanics International Inc
25 Bridge Street, Billerica, MA 01821, USA
Tel: 978 667 5841; Fax: 978 667 7582
E-Mail: info@compmech.com
http://www.witpress.com

British Library Cataloguing-in-Publication Data

A Catalogue record for this book is available
from the British Library

ISBN: 978-1-78466-429-9
eISBN: 978-1-78466-430-5

ISSN: 1746-4498 (print)
ISSN: 1743-3509 (on-line)

Contents

Part I – Structural Studies, Repairs and Maintenance of Heritage Architecture XVII

Section 1: Heritage architecture and historical aspects

Giuseppe Vaccaro: An "experimenter" of a constructional modernity
Antonello Pagliuca, Pier Pasquale Trausi & Donato Gallo PI-5

Building renewal: The centrality of eventual occupancy in design decisions
Cordelia O. Osasona PI-15

Deliberations on conservation of built heritage: Paying homage to a
historical past through architectural education, learning and research
Achilles Ahimbisibwe & Anthony K. Wako PI-29

Archive research as a diagnostic and cognitive investigative method
of memory of the multistratified urban built heritage:
A case of urban archeology
*Lucrezia Longhitano, Santi M. Cascone, Stefano Cascone
& Giuseppe A. Longhitano* PI-39

Urban planning and its impact on the city of Hebron, Palestine
Wael Shaheen PI-51

Evolution of the traditional Turkish house
Tamara Kelly PI-63

Woodworking trade at the turn of the 19th and 20th centuries and
its protection
Klara Kroftova & Lukas Hejny PI-77

Section 2: Maintenance and conservation issues

Towards an autonomous management maintenance model applied to
a heritage building: The case of Hernando Colón College, Universidad
de Sevilla, Spain
Miguel León-Muñoz, David Bienvenido-Huertas & Carlos Rubio-Bellido PI-89

Optimization of hydrated lime putties and lime mortars using nopal pectin for conservation of cultural heritage
Angélica Pérez Ramos, Jose Luz González Chávez, Luis Fernando Guerrero Baca, Miguel Ángel Sánchez Espinosa & Anaí Chiken Soriano PI-101

Section 3: Monitoring and damage detection

Monitoring and study process of gothic buildings
Sergio Coll-Pla, Guillem Mateu, Agustí Costa Jover & Jaume Roset-Calzada PI-115

Indoor corrosivity in Klementinum Baroque library hall, Prague
Katerina Kreislova, Pavlina Fialova & Tereza Bohackova PI-123

Simplified procedure for structural integrity evaluation of complex walls through signal energy analysis based on vibrational data
Ivano Aldreghetti, Giosuè Boscato, Mauro Fiorin, Lorenzo Massaria & Vincenzo Scafuri PI-133

Section 4: Vulnerability assessment

Novel risk assessment methodology for cultural heritage sites
Fabio Garzia PI-149

Built heritage in the 2020 earthquakes in Zagreb and Petrinja, Croatia: Experience and consequences
Josip Galić, Davor Andrić, Lucija Stepinac & Hrvoje Vukić PI-161

Digital survey and parametric 3D modelling for the vulnerability assessment of masonry heritage: The Basilica of San Domenico in Siena, Italy
Angelo Massafra, Carlo Costantino, Davide Prati, Giorgia Predari & Riccardo Gulli PI-173

Towards a sustainable *re-construction* method for seismic-prone heritage settlements of Gujarat, India, based on advanced recording technologies
Bernadette Devilat, Jigna Desai, Rohit Jigyasu, Mohamed Gamal Abdelmonem, Felipe Lanuza & Mrudula Mane PI-185

Section 5: Social, cultural and economic aspects

Role historic assets can play in reviving the retail high street: A case study of Derby's retail high street
Sarah Ball, David Higgins & Hazel Ann Nash PI-201

Capital Model Prison: A political tool for government power
Weiqi Chu PI-215

Remains of the beauty: A definition of beauty in the built space
Silvana Kühtz PI-229

Adaptive reuse heritage buildings addressing sustainability potentials:
Analytical case studies in Sharjah, United Arab Emirates
Iman Ibrahim & Fatma Eltarabishi PI-237

Heritage sites: The problem of economic, social and cultural valuation
Roni Demirbag, Vivienne Saverimuttu, Yiyang Liu & Sajan Cyril PI-249

Section 6: Industrial heritage

Urban regeneration of industrial sites:
Between heritage preservation and gentrification
Rafaela Simonato Citron PI-263

Postcolonial industrial heritage in north Africa:
Investigations and insights into the city of Casablanca, Morocco
Chaima Seddiki, Jeremy Cenci & Isabelle de Smet PI-275

Research of the industrial heritage category and spatial density
distribution in the Walloon Region, Belgium, and northeast China
Jiazhen Zhang, Jeremy Cenci, Vincent Becue & Sesil Koutra PI-285

Section 7: Learning from the past

Tradition and innovation in the scenery city's architectures:
The impact of Filippo Juvarra in Carlos Mardel's 1733 plan for
Lisbon's riverfront – A water-city proposed design for the envisioned
"Rome of the Occident"
Arménio da Conceição Lopes & Carlos Jorge Henriques Ferreira PI-297

Social role of the wall:
The domestic vernacular architecture of south India
Anjali Sadanand, Ramasamy Veeranasamy Nagarajan
& Monsingh Devadoss PI-311

Towards developing an ecological tourism settlement in Siwa Oasis,
Egypt: Case study of Babenshal eco-lodge
Ola Ali Bayoumi, Mohamed Abdelall & Mohamed Anwar Fikry PI-327

Crisis or opportunity: Looking at the past for the resilience of settlements
Gulız Ozorhon & Ilker Fatıf Ozorhon PI-337

Author index PI-349

Part II – Earthquake Resistant Engineering Structures XIII

Fragility assessment of the inter-story pounding risk between adjacent
reinforced concrete structures based on probabilistic seismic demand
models
Maria G. Flenga & Maria J. Favvata PII-3

Proposal of in situ parameters for the assessment of physical
vulnerability to seismic events: A Peruvian case study
Luis Izquierdo-Horna & Andrea Galván PII-15

Strength balance of steel damper columns and surrounding beams
in reinforced concrete frames
Kenji Fujii & Mizuki Kato PII-25

Proposal of territorial parameters for seismic hazard assessment
in Pisco, Peru
Luis Izquierdo-Horna & Patricia Aranibar PII-37

Structural monitoring for seismic damage evaluation: A case study
*Stefano Anastasia, Pedro Poveda-Martínez, Benjamín Torres-Gorriz
& Salvador Ivorra-Chorro* PII-47

Examining the adequacy of separation gaps between adjacent buildings
under near-field and far-field earthquakes
Yazan Jaradat, Harry Far & Ali Saleh PII-59

Author index PII-73

Part I

Structural Studies, Repairs and Maintenance of Heritage Architecture XVII

WIT *eLibrary*

Home of the Transactions of the Wessex Institute.
Papers published in this volume are archived in the WIT eLibrary in volume 203 of
WIT Transactions on the Built Environment (ISSN 1743-3509).
The WIT electronic-library provides the international scientific community with immediate and
permanent access to individual papers presented at WIT conferences.
Visit the WIT eLibrary at www.witpress.com.

Preface

This volume comprises selected contributions to STREMAH 2021, the 17th International Conference on Studies, Repairs and Maintenance of Heritage Architecture, topics that are becoming increasingly important in modern society. The initial venue for the conference was the city of Rome, which boasts a large number of heritage constructions spanning several historical times, but the situation generated by the COVID-19 pandemic prevented the planned in-person event, which instead transformed into an online forum.

The meeting has taken place on a regular basis since the first conference started in Florence (1989) and continued in Seville (1991), Bath (1993), Crete (1995), San Sebastian (1997), Dresden (1999), Bologna (2001), Halkidiki (2003), Malta (2005),Prague (2007), Tallinn (2009),Tuscany (2011), the New Forest, UK ,home of the Wessex Institute (2013), A Coruña, Spain (2015), Alicante, Spain (2017) and Seville, Spain (2019).

The importance of retaining built cultural heritage cannot be overstated. Rapid development and inappropriate conservation techniques are threatening many unique heritage sites in various parts of the world. This conference aimed to provide the knowledge required to formulate regulatory policies to ensure effective ways of preserving architectural heritage.

The papers included in the book address a vast variety of topics. Some deal with policies on the reuse of historical buildings and the role of public officials, architects and other actors involved in the debate.

Papers on historical archeology are also included and describe the problem generated by multi-stratified built heritage in urban areas where the level of knowledge of each historical stratum is different.

The topic of structural reinforcement of heritage buildings is addressed in many papers, with several examples from wood, masonry or steel frameworks. The methods to alleviate and solve structural pathologies is present in other contributions.

Modern digital techniques for monitoring and surveying construction damage and the evaluation of the damage to indoor halls are also described in the book. Research devoted to investigate the materials used and historical construction methods to solve structural defects are also in the book.

The examples presented in the book belong to different historical times, from the Middle Ages to the 20th century, and include locations from Europe, Asia and Africa. The method of the construction is very diverse, with religious, industrial and military purposes.

Some contributions describing strategies for the regeneration of urban heritage aimed to preserve its value and the potential for creation of ecological tourism and prevention of gentrification can also be found.

This conference brings together contributions from scientists, architects, engineers and restoration experts from all over the world dealing with different aspects of heritage buildings.

Papers contained in the volumes in the series have been published in both paper and digital format, and widely distributed around the world. They are also permanently archived in the Wessex Institute eLibrary where they are available to the international community (https://www.witpress.com/elibrary).

The editors are also grateful to all authors for their excellent papers and to the reviewers for their help in ensuring the quality of this issue. They also want to acknowledge the contribution of the members of the International Scientific Advisory Committee (ISAC).

The Editors
Ashurst Lodge, 2021

SECTION 1
HERITAGE ARCHITECTURE
AND HISTORICAL ASPECTS

GIUSEPPE VACCARO: AN "EXPERIMENTER" OF A CONSTRUCTIONAL MODERNITY

ANTONELLO PAGLIUCA, PIER PASQUALE TRAUSI & DONATO GALLO
Dipartimento delle Culture Europee e del Mediterraneo, Università degli Studi della Basilicata, Italy

ABSTRACT

In the first half of the 20th century throughout Europe, on the rubble produced by the two World Wars, new cultural ferments arise as a "critical" response to the socio-political events that affected Europe. In particular, in the Italy of the early 20th century there was a series of changes that affected not only architecture but every aspect of community life. The most important one was the introduction of the industrial process also in the construction sector favoured the search for innovative architectural solutions that led to the definition of a style called 'Modern': architecture must communicate clarity and knowledge, in order to achieve greater utility through the use of materials and construction systems to combine beauty and functionality of the building (see B. Taut). In the wide panorama of the most important figures of Modern Italian architects, the architect Giuseppe Vaccaro distinguished himself particularly in his different (often daring) architectural experiments for the ability to combine the typical instances of the typically Italian construction tradition with the new avantgardes constructive and technological of the beginning of the century. In fact, different times Giuseppe Vaccaro was collaborator of architect Marcello Piacentini and forefather of Modern architectural culture in Italy, he was the author, in particular, of one of the most important buildings of the Modern: the "Palazzo delle Poste" in Napoli. The building, designed together with Gino Franzi between 1933–1936, is close by the ancient Cloister of Monteoliveto; in particular, the main façade, articulated along a hyperbolic line, becomes an occasion for architectural and urban renewal of the district of Charity, in the centre of Napoli. The monolithic nature of this architecture and the typological, technological, constructive and of material characteristics become an instrument through which the architect tells of a "new Italy" that experiments innovations and avant-gardes of materials and constructive aspects, as result of experiments of Italian industries of the early 20th century. The research, part of a broader study of 20th-century architecture, emphasizes, therefore, the need for a historical-analytical cognitive approach as a tool for the conservation the importance of each "value" of these architectures as an expression of a "Made in Italy" style.

Keywords: '900 architecture, construction, technological characterization, modern movement.

1 INTRODUCTION AND METHODOLODY

In Italy the 20th century was a period in which the industrial processes contributed to the creation of a new architectural type. This period, in fact, constitutes one of the most interesting phase in the cultural and architectural Italian history. Today, after more than a century, these buildings represent an historical heritage to be preserved and protected.

Therefore, these architectures today constitute a very "fragile" heritage and an in-depth knowledge is required to design their recovery.

Therefore it is important to know:

- the typological aspects (which are the evolution of the classical compositional theory);
- the social aspects (in particular the collaboration between industry and architecture for the realization of building components); and
- aspects related to materials and construction techniques (the technological elements represent an experimental idea of the Modern culture of the 20th century, contributing to create a new housing idea and social types).

WIT Transactions on The Built Environment, Vol 203, © 2021 WIT Press
www.witpress.com, ISSN 1743-3509 (on-line)
doi:10.2495/STR210011

In order to plan a recovery intervention, therefore, it is necessary to analyze the archive documentation ("indirect" sources) and realize a survey on site ("direct" sources), in order to increase the level of knowledge on the architectures, especially regarding the technological and construction aspects and the materials characterization.

Through the analysis of a case study of rationalist architecture (the "Palazzo delle Poste e dei Telegrafi" in Naples, designed by the architects Giuseppe Vaccaro and Gino Franzi), the research aims to experiment a methodology that starting from the above said analysis sources (i.e. technical manuals, brochures of industrial products, architecture and engineering journals), examining the detail of materials, the construction systems, led up to the creation of an "axonometric section," as a useful instrument to analyze the relationship between architecture and constructional system.

2 RATIONALISM IN ARCHITECTURE

In the first half of the 20th century, the South Italy was involved in a "restructuring project" carried out by the Government to create the image of a "new Italy" [1]. In fact, in 1925 the Government gave an important sign of relaunching the economy of different cities; among them Naples, one of the most important centers in the South; there, the Liberty modernism was soon supplanted by monumentalism carried out by the group of Gustavo Giovannoni, in particular by some Neapolitan academics such as Roberto Pane, Marcello Canino, Camillo Guerra, etc. [2].

Thus, Neapolitan architecture begins to share the directives of the group of "academic architects," who work according to the canons of classical architecture, which already characterized the historic city of Naples [3].

A series of urban interventions and buildings are designed that change the layout of some parts of the city, as the demolition of a stratified area in the district "Carità." Here a new political and financial center was built with a series of important institutional buildings, such as the "Regia Questura," the "Casa del Mutilato," the "Palazzo della Provincia" and the "Palazzo delle Poste e dei Telegrafi." This last building was the subject of a competition announcement that was won (after a series of variations) by Giuseppe Vaccaro and Gino Franzi. Compared to the headquarters of the Banco di Napoli (in the nearby via Toledo), built between 1936 and 1939 by Marcello Piacentini, the "Palazzo delle Poste" has a very interesting architectural quality, closer to rationalism than to neoclassical monumentalism.

For the architects was difficult to design the building in an interclosed space in the historic center; in fact, the neighboring architectures limited the project by imposing the plan (irregular), the height of the building (equal to surrounding ones) and the inclusion in the project of a Renaissance loggia [4].

The complexity of the building is highlighted in the formal choice but also in the technological and constructive ones, with the inclusion of a stone cladding system that attempts to clarify the troubled "question of stone cladding envelope" [5], through the study of different types of anchors that create a new formal idea of buildings.

"A first sketch of the façade, designed with impetus in a moment of disheveled inspiration, was then completely revised, checked, strengthened and this building, which even had the daring and imagination of the architect, but which fell ahead of an examination of environmental values, has turned into a healthy and environmental form. And no less Modern for this reason: indeed, precisely for this reason, more Modern than ever" [6].

3 GIUSEPPE VACCARO:
THE "PALAZZO DELLE POSTE E TELEGRAFI" IN NAPOLI

The "Palazzo delle Poste e dei Telegrafi" takes up the concept of "artwork" applied to Italian rationalism buildings that aims at defining every detail at different scales, from the architectural one to the interior design one, with a refined control of finishes and materials.

The envelope cladding is entirely realized in marble (taking up the formalism of the Roman academic school) while the semicircular staircase and the central pillar (located in the center of the monumental entrance) are realized in black granite (called "Diorite") and gray marble caming from Valle Strona.

Particularly interesting is the line that characterizes the external elevation, a hyperbola with maximum curvature at the center; it allows to solve the design problem at the corners and to emphasize the external entrance to the building. This building represents the right compromise between the rational-avant-garde instinct and the academic obligations imposed by the Regime.

The "Palazzo delle Poste e dei Telegrafi," in fact, became an important reference for modernity, also published in the same architectural magazine called "Architettura" directed, at that time, by Marcello Piacentini; he also was the coordinator of the Faculty of Architecture in Rome after Gustavo Giovannoni.

3.1 Construction and industry

The architect Giuseppe Vaccaro distinguished himself for his various architectural experiments and for his ability to combine the demands of the construction tradition with the new construction and technological avant-gardes of 20th century.

The "Palazzo delle Poste e dei Telegrafi," in fact, collects numerous constructive instances of the Modern that are inserted in the (growing) Italian building industry and that leads to a renewed "construction culture" and new stylistic experiments, innovative materials and the coding of construction technologies more suitable for general trend of "machine civilization" [7].

3.1.1 Main and secondary structure

The structure, designed by Eng. Francesco Fiacchi, is made of reinforced concrete with "multiple frames" resting on a foundation realized with a slab of reinforced concrete (50 cm thick) and stiffening beams (1 m × 2 m in section) (Fig. 1). The foundation is divided into ten elements by expansion joints, on which the pillars of the "Mannesmann" typology weigh down; those pillars were made with the "beton frétte" technique which involved the confinement in the pillar of a helical reinforcement, capable of significantly increasing the resistance to compression of the column.

The masonries are realized in perforated brick for the internal partitions and solid bricks for the external envelope. In addition, there are also masonries realized using the glass block technique to ensure the lighting of internal rooms.

The floors are a solid slab type in reinforced concrete and of the mixed type concrete-brick, as well.

The false ceilings are made with the construction system of the "Graticcio Stauss," a flexible mat "consisting of an orthogonal mesh of 20 mm on each side made with iron wire (diameter 1 mm) which at the intersection has polyhedral clay crosses, cooked at high temperature with a special process" [8]. The mat is easily adaptable to any flat or curved geometric surface and provides excellent support for the subsequent plastering phase.

Figure 1: Historical image of building site. *(Source: Archivio Giuseppe Vaccaro.)*

3.1.2 Envelope system

The above said iconic hyperbolic shape of façade of "Palazzo delle Poste e dei Telegrafi" represents one of the most interesting part of design of stone cladding envelope, realized (as said before) in reinforced concrete and with an innovative technological systems (with the experimentation of the new system of thin coatings in stone elements). The cladding materials are: black diorite called "Baveno" (cladding in 5 cm thick sheets) for the lower part of the façade and the central pillars; marble called "Vallestrona" (cladding in 3 cm thick slabs) for the remaining part of the façade (Fig. 2). "The façade is not intended as a delimitation surface between external and internal architecture: the first is the natural continuation of the second" [9].

The windows are realized in stainless steel (produced by the Bombelli Company of Milan and installed by the Cassinelli & Guercini Company of Rome) whose execution is "of exceptional preciousness both in the up and down gates that characterize the façade and in the large glass walls, made up of from 38 large unbreakable slabs 4.50 m wide, 75 cm high and 1 cm thick, framed by a metal structure; the vertical uprights are invisible because they are embedded in the marble" (Fig. 3) [10].

3.1.3 Coatings and finishes

The interior is also embellished with marbles such as that called "Rosso Monte Amiata," "Vitulano" marble, "Bellona" stone, "Trani" and "Verde del Brennero" stone.

The stone materials, worked by local artisans, are combined with ceramic and glassy artificial materials, coming from the wise Italian experimental ingenuity and synonymous of constructive progress and industrialization of production processes.

Some of them are the "Desagnat" and the "Fontanit" (both produced by the Società Anonima Luigi Fontana of Milan) that are applied in the form of mosaic tiles for covering large surfaces [11].

WIT Transactions on The Built Environment, Vol 203, © 2021 WIT Press
www.witpress.com, ISSN 1743-3509 (on-line)

Figure 2: Historical image of stone cladding envelope of façade. *(Source: Archivio Giuseppe Vaccaro.)*

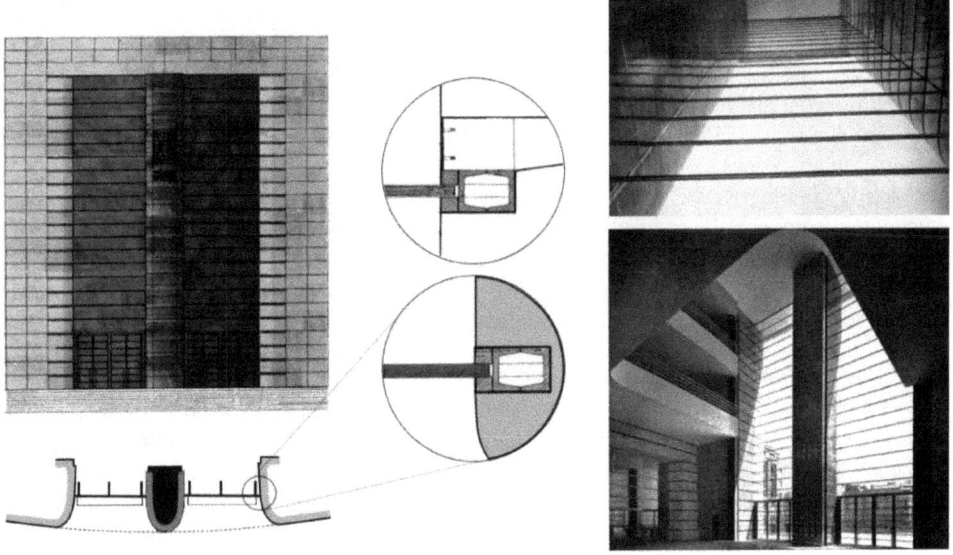

Figure 3: Stainless steel windows with the vertical uprights (details).

The "Desagnat," in particular, is "made up of colored glass mosaic tiles crossed in the two orthogonal directions by hollows forming many small mosaic tiles joined to a support of highly resistant fabric mat" [8], [12] and used to embellish the entrance shelter [13] (Fig. 4).

Figure 4: Detail of the shelter covered in "Desagnat" glass with (on the left) and the hall (on the right) with the vaulted surface in "Desagnat" glass and the "Mannesmann" columns also covered in "Desagnat" glass. *(Source: "Architettura: rivista del Sindacato Nazionale Fascista Architetti".)*

There are several applications of reinforced glass blocks, too; they constitute a very widespread construction system in Italian rationalist architecture through the marketing of numerous diffusers designed by the "Società Anonima of Saint Gobain, Chauny and Cirey."

In fact, there are numerous types of diffusers (Nevada, Novalux, Prismalith, Duralux, Clarilux, Cupolux, Discolith, Isocalor, Luxfer, Opalina, Planilux, Quadralith, etc.) also used in the Palazzo delle Poste and Telegrafi, especially in the envelope (i.e. entrance vestibule, the cylindrical glass block wall of boardroom) [14] (Fig. 5).

The flooring are realized using precious glassy mosaic, marble (in the reception rooms) and Linoleum with different colors (brown, blue, black marble effect, dark gray, Pompeian red, etc.) in all offices, corridors and wainscots (Fig. 6).

Finally, as regards the architraves and window jambs are finished with white calcium stucco, while the cantilever roof roofing (with reinforced concrete structure) has the external finish with plaster called "Duralbo" "produced by the "Società Istriana dei Cementi" which represents the only artificial Portland cement manufactured in Italy and obtained from the "clinkerization" process of kaolin, marl in special rotary kilns" [8].

3.1.4 The technological equipment

"The central heating system is based on radiators, with partial application of convectors and heating system by air. The building is equipped with pneumatic post systems, mechanical belt conveyors for posting (the letters set arrive directly on the sorting tables), a system of belt conveyors and clamps in the room of telegraph equipment, electric clocks to central command in main offices" [9]. This set of systems, with attention to every detail, even from an aesthetic point of view, constitutes the most complete and new idea implemented in the growing Modern architecture.

Figure 5: Detail of the walls made using the reinforced glass block construction system in the entrance vestibule (on the left), of the Provincial Bank room (above, on the right) and of the staircase enclosed in the cylindrical glass block wall (below, on the right). *(Source: Architettura: rivista del Sindacato Nazionale Fascista Architetti.)*

Figure 6: The floorings in Linoleum (details).

3.1.5 Interior design

Most important is also the elegant interior design, studied by the designers down to the smallest detail with a careful choice of both traditional and innovative materials (i.e. Albes, Buxus, etc.) which undoubtedly contribute to confer nobility to the simple architectural volumes, transformed them in a "monumental expression," where "no superfluous elements and no excessive dimensions have been adopted to render this expression rhetorical, which instead it was entrusted only to the unitary composition of the elements and, therefore, of the forms, and to the expressive value of the relationships" (Fig. 7) [9].

Figure 7: Interior design elements realized with traditional and innovative materials.

4 CONCLUSIONS

This building, like others built in the same period, represent a very "fragile" heritage [15]. To operate a recovery intervention on these buildings is necessary to know all its architectural, technological and constructive elements [16].

As for G. Vaccaro's building, many other architectures hide much more innovative construction systems behind natural stone claddings envelopes.

In fact, it's enough to analyze the structure framed in reinforced concrete (with the innovative "Mannesmann" columns), the experimental use of the anchoring system of the cladding in very thin solid marble and the new materials made in Italy.

All these elements have contributed to transform and change the Italian cultural trend and its traditional building practice.

The "Palazzo delle Poste e dei Telegrafi" – as said above – a monument of Italian Rationalism, represents one "of the first significant episodes in the context of a construction problem that would have assumed considerable importance in the history of Modern construction in Italy" [17].

Through the analysis of this case study, therefore, it is possible to highlight that the knowledge of technology and construction systems is crucial for a suitable design of recovery interventions. In the same way, the used system of the "axonometric section" (Fig. 8) is a useful tool for the designer and technician as it allows to show, using a single drawing, the relationship between technique and architecture. This system of representation highlights the technological innovation of these structures, otherwise hidden behind envelope materials (i.e. natural stone, bricks, etc.), too; this approach allows to plan recovery interventions on this cultural heritage, through a "scientifically correct" method of knowledge and representation [18].

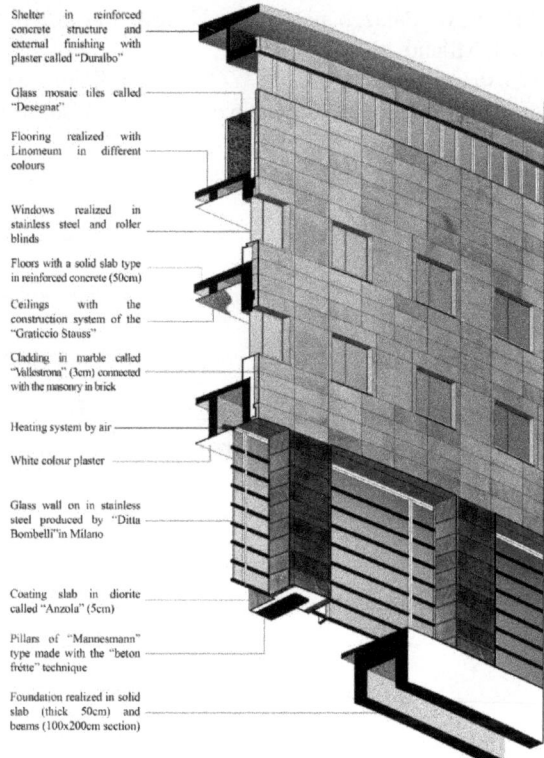

Figure 8: Information model of Palazzo delle Poste e Telegrafi in Napoli.

REFERENCES

[1]   De Seta, C., *La civiltà architettonica in Italia. 1900–1944 arte e architettura*, Clean: Napoli, 2019.

[2]   Frampton, K., *Storia dell'architettura moderna*, Zanichelli: Bologna, 1982.

[3]   Ciucci, G. & Muratore, G., *Storia dell'Architettura Italiana. Il Primo Novecento*, Electa: Roma, 2004.

[4]   Cislaghi, P., *Il rione Carità*, Electa: Napoli, 1998.

[5]   Poretti, S., *Modernismi Italiani. Architettura e costruzione nel Novecento*, Gangemi: Roma, p. 24, 2008.

[6]   Opere di Giuseppe Vaccaro. *Architettura. Rivista del Sindacato Nazionale Fascista Architetti*, Fascicolo X: Milano, 1932.

[7]   Gallo, D., Modernità italiana nel sistema involucro. *Quaderni di tecnologia dell'architettura. Sistemi tecnologici e costruttivi del "Movimento Moderno"*, ed. A. Pagliuca, Favia: Bari, pp. 10–11, 2020.

[8]   Pagliuca, A., *Materiali Made in Italy: Avanguardia Italiana nell'Industria delle Costruzioni del Primo '900*, Gangemi: Roma, pp. 525–541, 2019.

[9]   Vaccaro, G., Edificio per le Poste e Telegrafi di Napoli. *Architettura. Rivista del Sindacato Nazionale Fascista Architetti*, vol. 8, Fascicolo VIII: Milano, pp. 353–391, 1936.

[10]   Vitellozzi, A., Il nuovo Palazzo postale di Napoli. *Periodico trimetrale "Edilizia Moderna"*, vol. 23, Milano, pp. 1–9, 1936.

[11]   Griffini, E., *La Costruzione Razionale della Casa*, IV Edizione, Hoepli: Milano, 1948.

[12]   Zorzi, L., *Intonachi, Pavimenti, Rivestimenti Nella Moderna Edilizia*, Edizioni Tecniche Utilitarie: Bologna, p. 167, 1935.

[13]   Griffini, E., *Costruzione Razionale della Casa – I nuovi materiali*, Hoepli: Milano, 1932.

[14]   Dal Falco, F., *Stili del Razionalismo, Anatomia di Quattordici Opere di Architettura*, Gangemi Editori, 2002.

[15]   Iori, T., *Il Cemento Armato in Italia, Dalle Origini Alla Seconda Guerra Mondiale*, Edilstampa: Roma, 2001.

[16]   Carbonara, G., Il restauro del moderno. *Trattato di Restauro Architettonico*, UTET: Torino, 1996.

[17]   Poretti, S., La facciata del Palazzo delle Poste di Napoli e la questione dei rivestimenti lapidei nell'architettura italiana negli anni trenta. *Architettura e Costruzione*, nn. 84–85, pp. 28–37.

[18]   Poretti S., Modernismi Italiani. *Architettura e Costruzione nel Novecento*, Gangemi: Roma, p. 24, 2008.

BUILDING RENEWAL: THE CENTRALITY OF EVENTUAL OCCUPANCY IN DESIGN DECISIONS

CORDELIA O. OSASONA
Department of Architecture, Obafemi Awolowo University, Nigeria

ABSTRACT
Architectural renewal, otherwise termed "adaptive re-use", has long established its efficacy as a more effective way of justifying the continued existence of heritage structures, than mere restoration. Ile-Ife, though not particularly indicated in architectural heritage conservation in Nigeria, has its own share of such buildings worthy of conservation. Adopting a comparative case-study approach, two renewal projects in Ile-Ife are critiqued – with a view to highlighting the centrality of the role eventual tenancy played in their design proposals. In each case, an old family house was commissioned for renewal: one to be used, thereafter, only by blood relatives; the other, essentially to be let out to paying tenants. Specifically, Lokore House underwent renewal to give modern accommodation to members of an extended family; on the other hand, Agbedegbede House, was to be put up for rent (with a few rooms reserved for other extended family members). While the former was executed on a fairly liberal budget, the latter was to be "shoe-string". In articulating the latter mandate, prime consideration was given to communal spaces being perfunctory (since not carrying a need for family-bonding); maintaining as much of the existing building as was still structurally sound; ensuring ease of circulation in common spaces and specifying rugged finishes. Inflation that set in since the Lokore House project (2018–2019) has accounted for higher prices of similar materials for the Agbedegbede one (started in 2020 and still ongoing). This notwithstanding, the major constraint of making the intervention comparatively low-budget is being met. In both cases, client and architect remained constant. Despite this (and also the fact that the end-use was still residential, and in similar cultural and physical neighbourhoods), a comparative analysis shows that it was the issue of clearly distinguishable variance in eventual occupancy that dominated design decisions.
Keywords: Lokore, adaptive re-use, Agbedegbede, tenancy, building renewal.

1 INTRODUCTION

Despite the fast-paced nature of modern living (and the increasing premium placed on the value of land), the conservation of architectural monuments will forever be topical. This is so for a variety of reasons, which include: a fundamental human psychological need for anchorage with the past; visual relief, akin to serendipity, some such monuments provide; alternative accommodation (in places of urban residential stress), and the real economic benefits of architecture-based tourism [1], [2].

With these established bases for the phenomenon, architectural conservation (encompassing restoration and renewal) is a highly advanced practice in the West [3]. In much of Africa, this is not so; specifically, in Nigeria, the practice is virtually non-existent. Even the tentative steps taken by environmental-interest groups (such as LEGACY), have not only been met with indifference on the part of government; recently, a major undertaking by LEGACY was frustrated – and eventually demolished outright [4], [5]. In light of the prevalent governmental neglect, it should not be surprising that corporate organisations in Nigeria also, generally, do not see the need to get involved. Nevertheless, some such concerns – coupled with private individuals – by virtue of international exposure and financial wherewithal, have gradually seen it as their corporate social responsibility, on the one hand, and their family duty, on the other, to try and salvage specific buildings in the local landscape of perishing architectural heritage. LEGACY has even compiled a compendium of such [6].

 WIT Transactions on The Built Environment, Vol 203, © 2021 WIT Press
www.witpress.com, ISSN 1743-3509 (on-line)
doi:10.2495/STR210021

Stemming from this awakening, the Nigerian Institute of Architects (NIA) recently established a Heritage Architecture Committee, with the mandate to identify and recommend for conservation-restoration, heritage buildings worthy of such intervention in the six geo-political zones of the country. Expectedly, this is with a view to initiating (and superintending) such conservation exercises. Similarly – but less dramatically – in Ile-Ife, for instance, certain individuals have risen to the challenge of not just restoring decrepit family houses "for posterity"; the interest of the initiators of these schemes has majorly been that these houses also serve living members of their extended families, now, as comfortable, modern-day accommodation.

The present work highlights two projects in Ile-Ife, executed in the spirit of the latter orientation. The first (Lokore House) was the focus of a previous publication [7]. It features in the present discourse as a comparative case-study with Agbegbede House, the main focus of this work; this is with a view to emphasising the salient points of departure. The appropriateness of this approach was informed by the fact that original use, cultural/neighbourhood context, client and project architect are the same in both adaptive re-use schemes. Details of the essentials of the comparative analysis are set out in Table 1.

2 THE BACKGROUND

Both the Lokore (Odeyemi Family) House and the Agbedegbede House are landed property to which the sponsor of their renewal, Chief John Agboola Odeyemi, is tied by birth. The Lokore House was his paternal grandfather's home, while that at Agbedegbede was his maternal grandmother's. Chief Odeyemi himself is a well-travelled professional, fully conversant with what subsists in developed parts of the world, with respect to valuable architectural legacies. In 2018, based on his emotional attachment to the Lokore (Odeyemi Family) House – where he had spent a substantial part of his childhood – he sought architectural expertise in upgrading the property to a more modern residence. The beneficiaries were to be his sister (who was currently resident in the house) and other members of the extended family who would later join her.

In 2020, based on the success of the adaptive reuse intervention on the Lokore House, Chief Odeyemi decided to similarly sponsor renewal for the Agbedegbede House. (Significantly, on the occasion of the "unveiling" of the new-look Odeyemi Family House, the French Ambassador, Monsieur Jerome Pasquier, who was on an official visit to the Obafemi Awolowo University, decided to grace the occasion – having heard about the complete transformation that had overtaken the building; see Figs 1 and 2). In this new enterprise, Chief Odeyemi decided to maintain the services of the architect he had used previously – for reasons of continuity, as evidence of his confidence in services previously rendered, and as an overall endorsement. The project kicked off in September, and is still underway.

3 THE GENERAL MANDATE

While the Lokore project had been exclusively for the comfort of family, the new scheme at Agbedegbede was to, essentially, provide good quality rentable single-room, and "room-and-parlour" accommodation. On the Lokore (Odeyemi) project, the client had been content to invest whatever it would take to transform the building into a well-lit, well-aerated residence with modern conveniences and furnishing – to the extent to which the cultural background and general social orientations of the end-users could cope with such facilities. In essence, a fairly liberal budget. For Agbedegbede, there was to be no furniture, and all amenities were to be spartan – merely utilitarian and durable. A major restriction in fulfilling this latter mandate was that it was not expedient to bring down existing walls – at least not for

Table 1: Comparative analysis of interventions at Lokore and Agbedegbede.

Intervention	Lokore			Agbedegbede			Justification for difference
	Utilities	Finishes	Outcome	Utilities	Finishes	Outcome	
Modernization	1 bedroom, en-suite; 2 other communal bathrooms (with WCs); 1 communal kitchen; use of both conventional septic-tank and soakaway pits; also, well septic tank.	All floors tiled; bathroom walls tiled; kitchen walls partially tiled; all other interior walls gloss-painted.	Appropriate status accorded en-suite master bedroom. Conventional family privacy guaranteed by provision of doors, generally.	2 communal toilets; 2 communal showers; separated conveniences; 1 communal kitchen; no doors into kitchen, and out into shower area; use of well septic tank.	All floors tiled; bathroom walls tiled; kitchen walls partially tiled; all other interior walls gloss-painted.	Subjugated privacy; limited to bedrooms, toilets and shower enclosures.	Owing to nature of proposed occupancy, no basis for undue celebration of privacy.
Water-access	Provided by a well.		Beneficial to residents and neighbours.	Provided by a well.		Beneficial to residents and neighbours.	–
Improved daylighting	Use of secondary roof, with clerestory windows.		Phenomenal improvement in interior lighting condition.	Introduction of skylight into roof configuration; supported by fanlight over door and light through uncovered doorways.		Phenomenal improvement in interior lighting condition.	Multi-pronged approach (giving up some privacy) adopted since configuration of overall complex prevented same approach.
Improved ventilation	Enlarged existing bedroom windows, and created new ones; use of glass louvre-blades to replace timber shutters.		Additional windows mostly high-level, guaranteeing cross-ventilation.	Enlarged existing windows, and created new one; use of glass louvre-blades to replace timber shutters.		Additional windows only to increase lighting in rooms. No high-level windows.	Since end-users are tenants, budget did not cater for drastic structural overhaul.
Retail outlet	None provided in the proposal.		–	Provided – for retail trade.		Frontage shop, with non-conflicting customer access.	Tenancy-occupancy should generate as much returns as possible.

Table 1: Continued.

Intervention	Lokore		Outcome	Agbedegbede		Outcome	Justification for difference
	Utilities	Finishes		Utilities	Finishes		
Common leisure spaces	Central hall, naturally well-lit.		Appropriately furnished to enhance regular family interaction.	Central hall, naturally well-lit.		Mostly general circulation space; non-personal, occasional space; provision of electrical points only.	Would-be tenants' activities more room- (and outdoor-) based. Also, different backgrounds, so no "family" meeting point.
Surrounding graves	Two located close to the house, pertaining to known relatives, upgraded by identification marble plaques and surround-tiling.		Complemented general facelift given to the environment. Accorded respect to the dead.	Flanking sites previously unidentified, slabbed over; those to the rear, left untouched. Family matriarch's well-celebrated: identification plaque and complementary tiling installed.		Slab deliberately steeply canted, to prevent abuse by walking over or lounging on – as had previously been the trend.	Despite upgrade, tenancy occupancy of refurbished house likely to invite greater influx of guests – so, greater abuse would have likely been inevitable.

architectural "correctness" or overt aesthetics. This could be deemed necessary, if doing so increased the overall accommodation provided.

Figure 1: (a) Hall at Lokore house, before renewal; and (b) After renewal; lighting facilitated by clerestory.

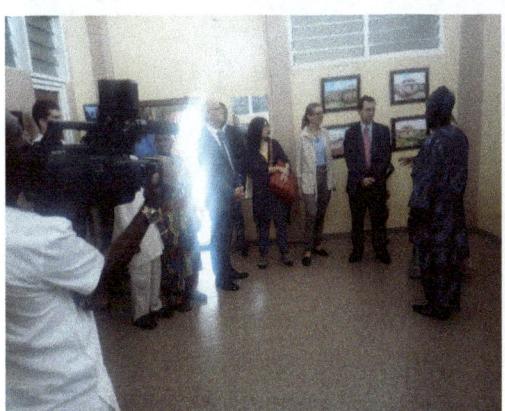 

Figure 2: (a) French Ambassador and entourage viewing transformed Hall at Lokore; and (b) Relaxing in the lounge there, after the tour (September 16, 2019).

4 CONSTRAINTS

In both cases, the buildings were of cob construction, and over eighty (80) years old. Coupled with this, was the fact that local building practice orientations were adopted (characterised by rule-of-thumb measurements, resulting in visible imprecision). Specifically, most walls were skewed – generally out of alignment with the 90° angle – while horizontal planes were commonly out of plumb. Also characterising the two buildings, was extended-family constraints on configuration: there was more than one dominant character to dictate the fate of the building. This imposed additional complications on physical extent (and actual handling) of the intervention. A notable feature to be offset in both schemes was the twilight-like darkness defining interior communal space within the buildings.

4.1 Peculiar constraints at *Agbedegbede*

Agbedegbede is actually the name of the street (On its part, "Lokore" refers to the neighbourhood in which the Odeyemi Family House is located) (Fig. 3). One of the peculiar difficulties encountered in the course of the adaptive reuse intervention at Agbedegbede was the fact that, in addition to implied extended-family complications, there was tacit hostility from those still in residence in the untouched wing of the house. Others were: gross dis-alignment of walls in the existing building configuration; restrictions for correction imposed by adjoining burial sites of dead family members; general constraining influence of built-up environmental context; neighbourhood hooliganism, and other constraints imposed by sizeable ruins of demolished adjacent structure.

Figure 3:   Relative locations of Agbedegbede (upper left, yellow) and Lokore (lower right, red). (*Source: Google Maps.*)

5 AGBEDEGBEDE HOUSE BEFORE RENEWAL

Situated less than a kilometre from the Odeyemi Family House at Lokore (as the crow flies), the Agbedegbede House is in the ancient heart of Ile-Ife; in fact, it is barely half a kilometre from the palace of the *Ooni*, the traditional monarch. (Interestingly, the traditional family rallying-point of the present ruler, the Giesi *akodi*, is directly opposite the house). As such, the environment is densely-populated with typical Ifes, steeped in the local culture – up to Lokore and beyond.

Despite the generally decrepit state of the building, it was still partially occupied. Similarly, the surroundings were still in use, though extremely un-complimentarily. On the north-western flank, a huge mound of essentially laterite debris (from the government demolition of a structurally-weakened and dangerous building), dominated the landscape. This formed a massive obstruction to the frontage of the rest of the extended-family accommodation, essentially screening it from the road (see Fig. 3). The north-eastern (rear) side of the building was practically a public latrine. The frontage of the building itself had virtually no setback (as even part of its plinth narrowly missed the rim of the gutter running in front of the house). Abutting this frontage is a grave of the matriarch of the home, Madam Jemilat Ayanlola Hassan-Mosaku; traditionally, this was just routinely slabbed over, with her name scrawled on it. Though perfunctory, it served the purpose of identification of location, and according it due recognition. It was common practice for this slab to serve as a safer way to connect with the entrance stairway, much of the lower part of which had been eroded away over time. In addition, it served as a relaxation area for neighbourhood layabouts (who routinely brought plastic seats from a drinking "joint" across the road, and ensconced themselves there). Fig. 4(a) and 4(b) show the frontage of the Agbedegbede House before the renewal exercise. In addition to the public sewer, there is a lamp-post at the front of the house – broken and non-functioning.

Figure 4:   (a) Frontage of Agbedegbede House before renewal; and (b) Close-up of eroded plinth and steps leading into the building, 2020.

6 WORK UNDERTAKEN AT AGBEDEGBEDE

In an attempt to renew the house at Agbedegbede (and adapt it for its new intended use) several things needed to be taken peculiarly into consideration. The existing rooms in the house were extremely dingy – most of them significantly less than 9 m². This informed an immediate resolve to increase room sizes (in spite of the obvious land constraint on the flank of the building indicated for extension). In addition, interior lighting needed to be improved – despite the issue of the barrier imposed by the other adjoining spaces that were not to be tampered with, and a tight budget. There were facilities that had not been previously provided, but non-negotiable, in a "modern building" (such as bathrooms and a kitchen), which needed to be accommodated.

6.1 The design

With constraints centred on available site area, general scope of intervention and a lean budget, it was obvious that the proposal would feature only necessary alterations. The design problem was seen as meeting the needs of the major target group (paying tenants), essentially within the confines of the existing building, and to meet modern exigencies. Several initial proposals had to be jettisoned – based on discoveries on site in the process of implementation. (Figs 1, 2 and 4 trace the progression). Reasons for this include the discovery of burial sites on one flank, and a septic tank (under the debris), on another.

At Lokore, the hall (constituting a concourse of sorts) was appropriated as a "bonding space" for family members (see Fig. 5). Though a similar space exists at Agbedegbede, it could not be treated the same way; paying tenants do not necessarily relate with such levels of intimacy. Other spaces of (inevitable) common use (e.g. the kitchen and conveniences) also called for peculiar spatial treatment. On their part, bedrooms are unequivocally places of great privacy and intimacy.

6.1.1 Spatial provisions

In the articulation of the final design (Fig. 4), six of the original seven rooms retain their erstwhile designations: two room-and-parlour suites, and two independent rooms. The seventh has been sub-divided to provide toilets for the house; a common wash-hand basin, located in the entrance lobby, serves the two toilets. Also, in a bid to ensure height clearance of the pipework related to toilet fixtures (with respect to the tombs behind the rear wall), a platform was created in the toilets.

Figure 5:   The common hall at Lokore; used for dining and general-family lounging (at 1.25 pm, on April 8, 2021).

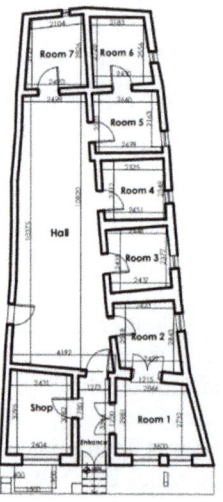

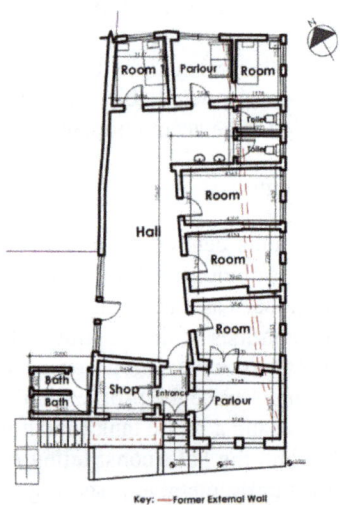

Figure 6: Existing floor layout. Figure 7: First proposal.

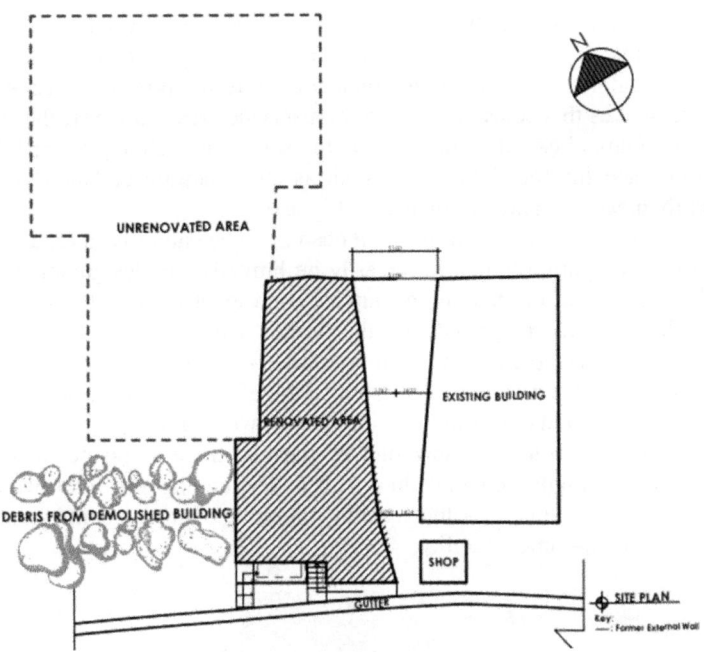

Figure 8: Agbedegbede site layout.

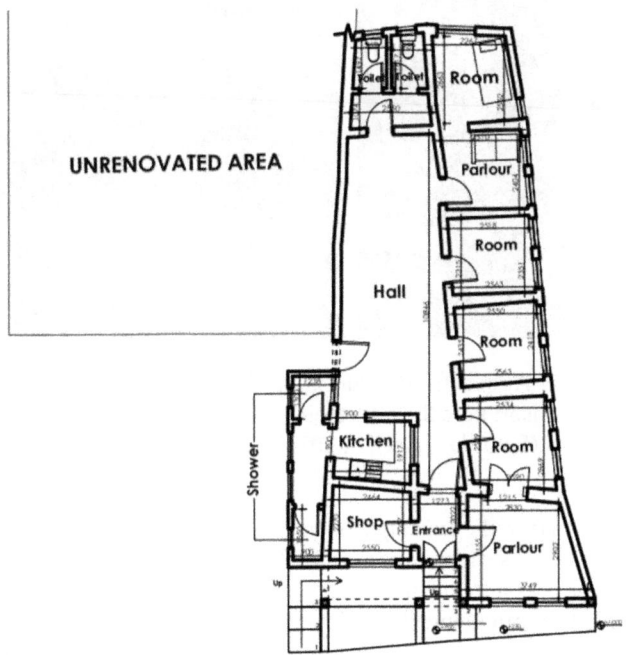

Figure 9:   Final proposal, showing separated conveniences: toilets at the rear, showers by
the kitchen.

The common hall is essentially circulation space. However, given the constrained nature of the private rooms, electrical power outlets (4) have been provided here. This is to allow those with refrigerators, freezers, etc. to connect them to the power-source at these points. This, therefore, makes this space also a common service area; however, the onus to secure their refrigeration units lies with tenants using the space for such appliances! Alternatively, the hall can be used for social functions such as child-naming ceremonies, and nominal extended-family meetings, among others (see Fig. 6).

In recognition of the sheer size of this hall needing to be naturally lit (and the fact that the area devoted to sky-lighting would necessarily be limited), the design sought to introduce additional day-lighting from alternative sources. High-level lighting (above the alternative door into the house) was integrated into the design. However, the major supplementary lighting was to be provided from the bathroom area – a new wing added to substantively provide this facility to the residents. To optimise the effectiveness of the fenestration here, doors (into the kitchen, and out from there into the showers) are totally eliminated. This was deemed acceptable as, from the viewpoint of tenants, there is no personal, sentimental attachment to the (communal) cooking area. This understanding also informed even the way shelving has been provided in the kitchen cabinet: nothing of any personal value will ever be actually stored in an all-comers facility.

Figure 10:   The hall at Agbedegbede, during the formal "opening ceremony". (a) Waiting for the sponsor; note the effectiveness of the skylight; and (b) Special prayers by clergymen (Saturday, April 3, 2021, 3.30 pm).

A shop had been part of the existing design. Strategically positioned at the frontage of the house, it had been a tailor's shop; however, the entrance had been through a side wall – after essentially gaining access into the house, first. As part of the modern-day clear resolution of circulation pattern, this arrangement has been changed (more so as the retail trade is more identified with a home-base, today, than services). As such, prospective customers are restricted to the front of the shop, getting sales attention through a window provided for such, and being physically confined to the veranda from which they interact with the vendor, by a metal grille (see Fig. 11).

6.1.2 Cost-control measures

The mandate to cut costs had been unequivocal. However, the local realities on ground (recession, attendant inflation resulting in galloping prices of building materials), may not have made strict adherence noticeable. While needless spending was certainly not indulged in, professional judgment also dictated the use of durable materials that would offset their initial capital outlay, over a reasonable part of the building's lifespan.

The strategy adopted was to retain whatever was still useful: door-frames, doors, roof-truss members, walls – crooked or otherwise. This has resulted in a motley arrangement of building elements and features. In keeping with the requirement (and to synchronise with the still-dominant existing roof of the rest of the building complex), aluminium sheet-cladding was specified – as opposed to the stone-coated, profiled steel (Gerard) sheets used at Lokore. The former corrugated zinc sheets had, over decades, rusted and assumed a homogeneous brown colour. As such, specifying brown aluminium sheets was clearly indicated.

In the spirit of long-term reduced maintenance costs (and similar to the situation at Lokore), gloss paints were specified for all interior spaces (with tiling in toilets, showers and parts of the kitchen). This was to pre-empt the need for frequent re-painting, based on the sure knowledge that in such communities, casually (and frequently) running hands over wall surfaces is a quite established habit.

With basic floor-tiling (using the sturdy, anti-slip, easy-to-clean tiles presently available on the market), cost-in-use considerations prevailed in the specification. Though not significantly inferior to those employed at Lokore, they are less classy. Generally, going for the same grade as the previous, would certainly have been at a significantly higher cost.

Comparing the expenditure deployed at Lokore to achieve the results obtained, with current prevailing prices would, doubtless, present a conservative mark-up of at least 60%. Even at that, the present scheme has not come drastically cheaper. As such, in absolute terms, the Agbedegbede renewal project has not been achieved within the envisaged "shoe-string budget" scope – for the reasons stated above.

6.1.3 Expansion of spaces

Figs 1, 2 and 4 show the progression in attempts at resolving this. Squaring off the right-hand side of the building from the frontage to the rear, automatically enlarged the rooms, six of the existing seven being in this wing. As such, this measure was immediately adopted in the (initial) design proposal. However, as soon as digging up the new foundations for this extension started, work was abruptly stopped by an extended family member. She claimed there were graves in the intervening space between this flank and the building adjacent to it – despite there being no visible evidence at ground level! This notwithstanding, the assertion was soon verified as being true. It resulted in this design being abandoned, and a return to the original constraining walls and undersized spaces. Nevertheless, alternative spaces were created, while some existing ones were re-assigned (see Fig. 4).

6.1.4 Improved interior lighting

At Lokore (and with a similar constraint on the renewal not to cover the whole existing building), the course of action to improve daylighting, had been fairly straightforward. However, here, additionally, the client was categorically averse to a general overhaul of the roof to effect this. This was to support his position on not spending "too much" on the project.

Rather than clerestory windows (employed at Lokore, by creating a secondary roof-system), this mandate was realised by the use of a skylight. Fig. 6 shows its efficacy.

7 COMPARATIVE ANALYSIS

Table 1 looks at specific features both Lokore and Agbedegbede renewal interventions have in common – and where they differ. It also sets out the rationale for these attributes, in light of the respective mandates and enabling designs.

7.1 Plumbing

As previously-stated, the Agbedegbede House is seriously cramped, in its physical deployment. In the northeast, neighbours' properties restrict any building activities there – more so, as the little intervening space is taken up with (actually celebrated) tombs. The south-eastern flank looked innocuously useful for expansion, though still constrained – until proved to be a no-go area by the existence of unmarked graves. In the southwest, the building virtually sits on the rim of the public sewer; in the northwest, debris prevents any meaningful activity there (in addition to the fact that the land in that area belongs to another family).

With respect to the issue of providing mechanical services to the building, this restriction on all sides has proved a big challenge. The conventional septic-tank/soakaway-pit arrangement could not be invoked; even inspection-chambers had to be ingeniously situated and crafted. However, recourse was made to the use of a combined "well" septic-tank-soakaway pit. The provision of water is currently underway on the northwest flank, close to the alternative entrance to the house; a well is being dug – after successful overtures to hitherto hostile members of the extended family. Since there is no public water utility in the area, this will be a community benefit – just like the one at Lokore is currently being.

Figure 11:   Unpainted north-western flank; ongoing well-works to the left.

Figure 12:   Tomb of matriarch, Madam Hassan-Mosaku, raked to discourage sitting.

Figure 13:   Access to the house, from the Agbedegbede road.

Figure 14: View from the rear.

Figure 15:   Front façade of Agbedegbede House, after renewal. The stairs to the left lead to the sales veranda; those on the right, lead to the main entrance of the building, April, 2021.

8 CONCLUSION

Both the Lokore and Agbedegbede house-renewal schemes had peculiar challenges, even though contexts and mandates were generally similar. The major point of departure was that, whereas the former had family comfort as a priority, the latter was more concerned with practical, utilitarian provisions for paying tenants. This fundamental difference not only dictated spending profile, but also conditioned general design inclinations in articulating the proposal. Figs 7–11 capture the essence of the new-look house at Agbedegbede Street.

As an architectural professional expedient, the purpose to which a building will ultimately be put invariably dictates design orientation. However, for schemes that are typologically the same, similar in environmental and cultural context, sponsored by the same client and handled by the same architect, very few issues should be able to produce significant differences in outcome. Two such obvious conditioners are budget and occupancy.

In the case-studies reviewed, it has been seen that these two factors have significantly inter-played to justify the noticeable differences in the end-results – all other factors having remained the same. Despite the injunction to cut costs as much as possible on the Agbedegbede House project, the predominant factor making the significant difference in design orientation, has been the type of occupancy proposed for the building. As such, it is reasonable to conclude that the nature of eventual occupancy in a residential building is cardinal in determining the proposed initial investment on the building, and the actual eventual design to fit the budget.

REFERENCES

[1]   Harvey, J., *Conservation of Buildings*, John Baker Ltd.: London, p. 18, 1972.
[2]   Osasona, C.O., Nigerian architectural conservation: A case for grassroots engagement for renewal. *International Journal of Heritage Architecture*, **1**(6), p. 15, 2017.
[3]   Hardy, M., *The Venice Charter Revisited: Modernism, Conservation and Tradition in the 21st Century*. Cambridge Scholars Publishing: Newcastle upon Tyne, 2008.
[4]   Osasona, C.O., Nigerian architectural conservation: A case for grassroots engagement for renewal. *International Journal of Heritage Architecture*, **1**(6), pp. 4–6, 2017.
[5]   Osasona, C.O., Architectural renewal: A rising dawn in Ile-Ife? *Proceedings of the 16th STREMAH Conference*, pp. 43, 56, 2019.
[6]   LEGACY, *Historical Sites of Nigeria*, LEGACY: Lagos, 1999.
[7]   Osasona, C.O., Architectural renewal: A rising dawn in Ile-Ife? *Proceedings of the 16th STREMAH Conference*, pp. 43–56, 2019.

DELIBERATIONS ON CONSERVATION OF BUILT HERITAGE: PAYING HOMAGE TO A HISTORICAL PAST THROUGH ARCHITECTURAL EDUCATION, LEARNING AND RESEARCH

ACHILLES AHIMBISIBWE & ANTHONY K. WAKO
Faculty of the Built Environment, Uganda Martyrs University, Uganda

ABSTRACT

Memorialization of Africa's architectural past continues to fade in time by way of exposure to the cruelty of natural or man-made forces, iconic buildings from as recent as the 1960s are torn down with little consideration of their heritage value. In 2010, controversy surrounding the demolition of Uganda's National Museum for a proposed high-rise redevelopment brought to light the blatant disregard for thematic value of Uganda's memorable architecture. Unfortunately, this was a lone survivor among a myriad of projects where developers show no sympathy, architects offer no guidance, and research efforts draw no attention to protect built heritage or safeguard rich historical narratives. Architecture Education should adopt to support participatory approaches that underpin the integration of revitalising heritage values. It is envisaged that through immersive design experiences students could gain a critical awareness of the realities, insight on regional success stories, an appreciation of limitations around conservation efforts, as well as a lasting memory of taking part in the design of integrated conservation projects. Faculty of the Built Environment (FBE), engenders learning activities appropriated with: conservation doctrines, community needs, sense of the cultural context of historic buildings. A pertinent concern during these undertakings was to initiate processes that instigate students' abilities to band together and work jointly with: students from two similar schools of thought at the faculty of Engineering and Built Environment (EBE), University of Cape Town (UCT) and School of Architecture and Design (SADE), Ardhi University (AU), in conjunction with African Architecture Matters, a non-profit consultancy firm working in fields of design planning, research and education. This paper thus seeks to highlight, from both socio-cultural inferences and historical perspectives, the significance of the learning opportunities for students, researchers and academics through adaptive reuse project of the oldest cinema in East Africa, the Majestic cinema.
Keywords: architectural education, built heritage, conservation, research, adaptive reuse, historic buildings.

1 INTRODUCTION

To be considered holistically, conservation of built heritage necessitates an integrated approach to its protection and enhancement. Professionals of the built environment across East Africa evidently place more attention to new developments and less to the possibility of restoration or adaptive reuse of historic buildings, a consequence that has partially stemmed from their educational background. Veritably, there has been a minimal collaboration between the architecture education sector vis-a-vis schools of architecture and practising professionals across Uganda. This divide widens a gap within which built heritage and conservation education would have thrived. Places and buildings around us have irreplaceable identities that are made up of social and cultural values that represent, embody and give significance to our societies [1]. Built heritage conservation education, from the onset, should entail a set of activities based on shared cultural values if sustainable solutions are to be realized [2], lest neglect or damage of historic buildings rooted on non-researched information. Heritage as a pedagogical resource for research and incorporated in teaching project-based work, allows students to understand it as symbolic, a physical and economic resource, platform for critical thinking and construction of new meanings [3]. This paper presents an appreciation of research-based design project as consideration for the thematic value of historic buildings.

Historians, from a social-cultural perspective and at an urban scale, have framed historic buildings in what Myers [4] coined as "Verandahs of Power" that drew inspiration from the physical verandahs of the House of Wonders *in Arabic Beit-al-Ajaib* (Fig. 1), an outstanding monument in Stone Town. The collapse of part of the House of Wonders in December 2020 was preceded by the collapse of the roof after heavy rainfall in November 2015. This was also preceded by the collapse of a corner section of the building (Fig. 2(a)) in 2012 leading to its closure.

(a)   (b)

Figure 1:   (a) House of Wonders [5]; and (b) House of Wonders in December 2020 [6].

(a)   (b)

Figure 2:   (a) Collapsed corner of the House of Wonders in 2012 [7]; and (b) The Uganda Museum [8].

The conundrum of the House of Wonders is relatable to the Uganda Museum (Fig. 2(b)) in that both buildings have been victims of neglect. The missed opportunities as a result of the limited partnership between the architectural education sector and the professionals have exacerbated this neglect. The case of disappearing heritage in East Africa and across sub-Saharan Africa is of little concern to a countless number of people. Related views that may apply to this situation can be observed in education, practice and society's perception of heritage. Inherent to this concern is the negative impact of socialization in architectural professional education in East Africa that overlooks society's desires and promotes sameness

[9]. Partly because "many of the important African heritage sites are not known […] or very few are declared World Heritage Sites" [10]. This narrative shall highlight principles of heritage conservation within architectural education that can be used as the basis to underscore the relevance of historic buildings. The rich historical past of East Africa with influences from Arabia, India, Persia, Portugal and in recent colonial times the western world, has left scattered gems worthy of restoration consideration. Teaching and learning in architecture education ought not to be limited to one single event but should be continuous throughout one's career. The scope and applicability of built heritage as a teaching tool should, therefore "[…] incorporate the past as a dynamic supply towards contemporary values" [3].

2 GLOBAL PERSPECTIVES ON TEACHING CONSERVATION AND HERITAGE IN ARCHITECTURAL EDUCATION

Incorporating heritage studies and conservation into architectural education is vital to the training of architecture students and later in their professional career. Needless to say, conservation is deeply engaged in the debate on sustainable practices in local contexts when our identities and control of our resources are progressively global [11]. From the beginning of the education career, various schools across the globe have introduced effective methodologies for teaching conservation education that may work and entail an approach through career structure [12]. Although there is a significant divide between practice and education, particularly in the field of architecture, in East Africa, "training for the purposes of executing tasks is a world apart from educating someone to develop critical thinking skills in how or why they would choose to execute a particular task" [11]. Further, the divide exists in gaps between different localised contexts, of which international organizations like UNESCO, ICCROM and ICOMOS have sought to bridge through various international methodologies and concepts, and challenged to rethink the meaning(s) of heritage and its values. Localizing such contexts demystifies the concept of scholarship in post-colonial societies across Africa as an invention of Western academia [13]. For architectural education to gradually contribute to the conservation profession, training should be carried out sequentially to integrate knowledge and understanding of the history of buildings [12] and, an introspective approach that portrays architecture education as philosophy as opposed to mere curriculum should be explored [14]. Application of methodologies and concepts, that apply to different types of buildings and are context-specific could be implemented. Architectural education then becomes paramount and provides a platform for re-interpreting any heritage evidence, re-evaluating history and reassessing places and values of any society as a whole.

3 FBE'S PEDAGOGICAL SCHEME AND PLATFORM FOR ARCHITECTURAL EDUCATION

Heritage and conservation-concerns are research areas that the FBE introduces to architecture students at an early stage. Historical issues are at the forefront of education and research at the faculty because of their relevance to the built environment, heritage and conservation. The research on heritage and conservation is not introduced as a chronological block course but as thematic design theories of historical and socio-cultural contexts of pre-19th-century human settlements across the globe. Hence, offering a grounded approach to historical studies that does not obscure readings of phenomenon [15]. In addition, these research studies emphasise the placement of historic places and buildings in their wider context that integrates sustainable development and their apprehension about the environment and climate responsiveness. Architecture, with its project-based education, varies from other traditional disciplines and does not imply a fundamental intellectual model [3]. As an outcome, the FBE has attempted to engage both learners and educators in a scheme of learning how to learn by socialization processes that are rooted in discursive environments whereby it is important to listen to students and appreciate their needs as they transition through their architecture education [16]. To further engage in the learning process and promote this research, the Field Experience travel exercise that was undertaken in 2020 by students proceeding to their 2nd year focussed on documenting buildings across East Africa from the 1950s/1960s. As the students get to their advanced stages of the Bachelor of Environmental Design degree and/or proceed to postgraduate studies of Master of Architecture (Professional), some key research areas that have gleaned towards heritage and conservation have emerged over the past three years. The studio project, discussed in this paper, was undertaken by third-year finalist students of the Bachelor of Environmental Design. The bachelors program is the first stage of a two-tier architecture program at the faculty. The Master of Architecture program, which must be undertaken by students that have completed the bachelors program or its equivalent, has an entire course unit dedicated to adaptive reuse and conservation of historic buildings. In the same regard, the faculty of Engineering and the Built Environment (EBE) at the University of Cape Town, who were collaborative partners with FBE, have an honours program with an emphasis on heritage research among other study areas. EBE engaged in the previous collaborative-adaptive-Reuse-workshop with FBE and the University of Rwanda in 2019 that realised the importance of working with stakeholders, highlighted meaning(s) for the local context and added value to the built heritage of Kigali. Both EBE and FBE later carried on with the design project of the Majestic Cinema virtually because the site visit to Zanzibar was cancelled in March 2020 due to the global pandemic. Working remotely also extended to African Architecture Matters, who provided background historical information, drawings, photographs and availed time for video conferences.

4 THE PROJECT

4.1 The building and its contribution to heritage and learning

The Majestic Cinema building (Fig. 3) is not listed, however its context (Stone Town) was designated as a UNESCO World Heritage site in 2000. To fulfill UNESCO's requirements of World Heritage management standards, and resources and skills of Zanzibaris, adaptive reuse of designated buildings started with research and design at the forefront, and laid the groundwork. The research on this project began with understanding the culture of the Zanzibaris because "[…] the significance of an inhabited World Heritage site, cannot be fully understood, safeguarded and developed without considering the interest, dreams and

Figure 3: Majestic Cinema building in 2011 [18].

priorities of the inhabitants" [17]. Design exploration for the Majestic Cinema project began as discussions within the studio environment between instructors and students, as well as amongst students. This is often done to break the hierarchies of power and promote networks of interaction so as to downgrade what Olweny [9] posited as "[...] instructors in a position of authority could inculcate their views and values onto students, often with little contention."

The studio project thematically focussed on revitalization of the Majestic Cinema as an inclusive and sustainable model in which the new use of the building would be optimised for Zanzibar's communities and its cultural operators. Stakeholders (organizations) who would provide financial support during the planning and construction of the project were: Hifadhi Zanzibar, a coalition between the private sector and government; Reclaim Women Space, empowers women to work together to take back public space; Sauti Za Busara Zanzibar, celebrates African music under African skies; Zanzibar International Film Festival, one of the largest annual cultural events in East Africa; and African Architecture Matters. The project was intended to be the first 'culture hub' in Zanzibar, a place for various communities to meet, engage in dialogue and collaborate beyond the prevalent public spaces. In an increasingly mobile world, the project highlighted and contributed to ongoing debates around community heritage in lieu of students from Uganda and South Africa, and the communities in Zanzibar.

The majestic Cinema, built in the 1920s, derives its significance within the socio-cultural context as the first cinema in East Africa. From its inception, the cinema played an important role on the island, as an inclusive community space for Europeans, Indian, Arabs and indigenes. Located about 120 m west of the Creek Road (Fig. 4) that structured the urban sphere by absolute oppositions: Stone Town or "the city proper" and Ng'ambo the "Other Side" [19], the cinema then became an anchor space between the often forwarded "dual city" model of the colonial city in Africa [20].

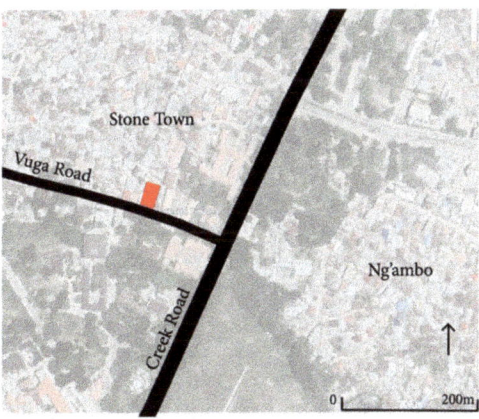

Figure 4: Location of Majestic Cinema. *(Source: Author.)*

4.2 Value addition through adaptive reuse

The redevelopment of the cinema into a self-sustaining Cultural Hub was chosen as a consequence of the myriad of cultures present in Zanzibar. The two student projects presented in this paper highlight the historic importance of the place and building's role, as an inclusive community space to Stone Town and the greater Zanzibar. Both projects aimed at; cultivating a sense of place through heritage ownership for the residents, achieving an integrated building program relevant to the stakeholders, and presented an operational management model strategy. As such, themes like *"From Ruin to Rebirth: The Awakening of the Majestic"* and *"An architecture design where the dimensions of time in space are manifested to the user,"* were generated by students as a guide to the design.

The historic ambience embedded within the Majestic Cinema's spaces was given immediate care through the building program by the first student, Elizabeth Nabagerekka's proposal shown in Fig. 5(a). As a way of bringing this dilapidated building back to life, the program's target users were from within a 500 m radius from the cinema and these included: students from the State University of Zanzibar neighbouring the site to the east; and a public space, the Victoria Gardens south of the site.

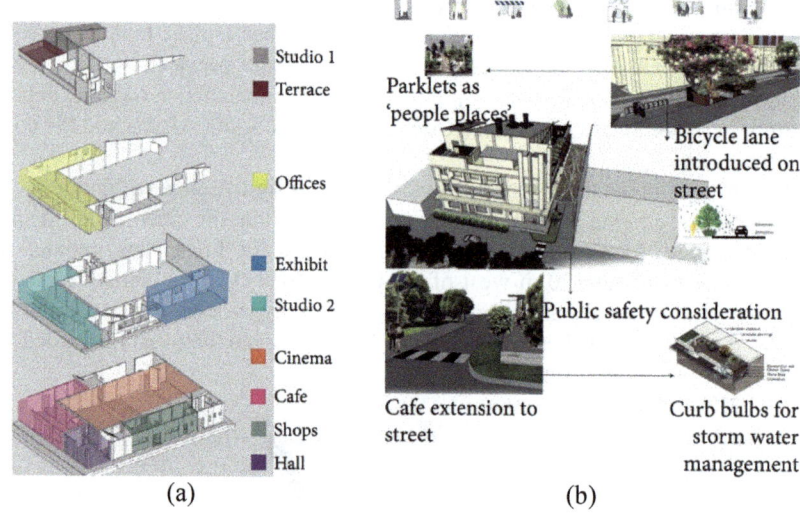

(a) (b)

Figure 5: Building program proposal by Elizabeth Nabagerekka.

Nabagerekka's design proposal sought to activate the streets (Fig. 5(b)) along Vuga Road and that adjacent to the university, and took advantage of the public park through activities such as open-air movie shows, public gatherings for cinema sensitization, public art and furniture. Activities such as these contribute to the revitalization of the cinema as a cultural hub through the space program and thus highlighting the contextual relationship.

The second student, Ronald Businge pursued an aesthetic approach to the revitalization of the cinema building in an attempt to capture its historic highlights as dimensions of time manifested to the users through material selection. The old and the new materials were represented by a palette (Fig. 6) that overtly merged them, and bridged the past, present and future anticipation. The cinema's role in this case became a seamless continuum with the bygone and a diverse multi-dimensional cinema spatial experience introduced at the roof top of the building (Fig. 7)

Businge's proposal used the form, facades and structure to further anchor old building elements into a *bold modern bone*: an attempt to maintain authenticity of the cinema - pay homage to the past; and cultural hub – showcase the future expression. The *bold modern bone*

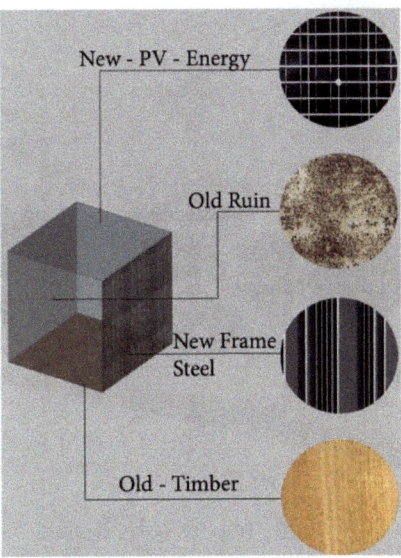

Figure 6: Interior ideas by Ronald Businge.

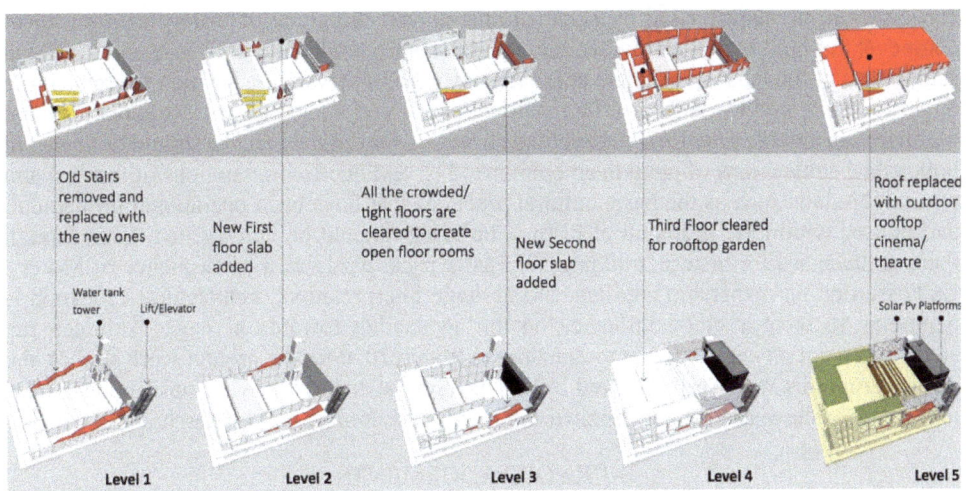

Figure 7: Old vs new interventions by Ronald Businge.

was represented by a steel structure that braced and encased part of the building (Fig. 8), an endeavour that resulted from engaging research at the onset of the assignment. Both projects showcased steps towards approaches to sustainability concerns such as social sustainability: participatory processes; and environmental sustainability: bioclimatic design. Such approaches are within FBE's goal to educate professionals who contribute to cultural and socio-economic development through participation in design, construction and interpretation of built environments.

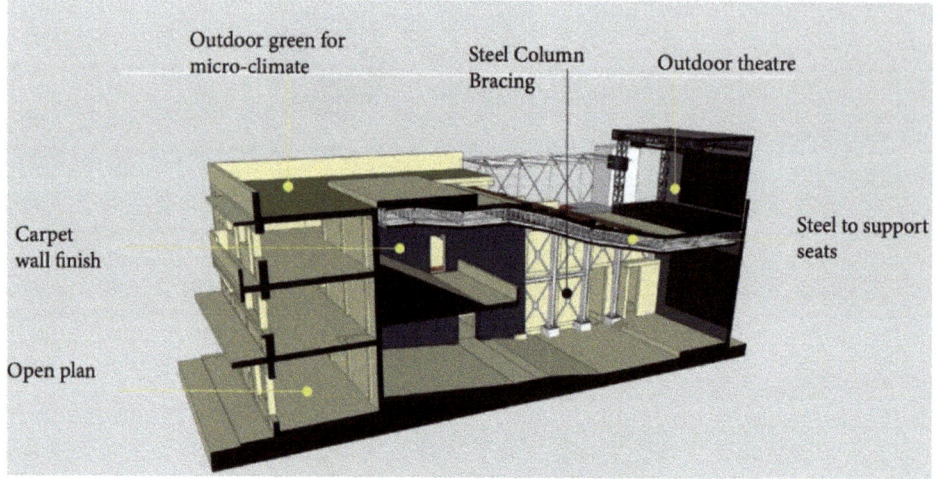

Figure 8: Section perspective by Ronald Businge.

5 CONCLUSION

Conservation of built heritage by paying homage to the past through architectural education can serve as a worthwhile role in the evolution of the built environment. Education, like conservation, is characterised by constant change and the study of history as a pertinent component of built heritage provides static platforms from which to draw inferences. Global standpoints of the most appropriate approach towards conservation education in architecture ought to be tailored towards specific historical narratives, socio-cultural extrapolations and highlight sustainable practices. Unravelling strategies for critical practice should be vested in both trivial subversions of prejudiced opinions [21], and questioning notions of the past and present. Notions such as the basic cultural practices that have been operational throughout, and housed within the lifeblood of historic buildings should be investigated. Even though there is disregard for historic buildings in East Africa, partly as a consequence of Moon's [22] assertion on ownership conflicts and heritage interpretation, architectural education is beginning to have a subtle influence on the awakening towards heritage. Although the students did not physically travel to Zanzibar in March 2020 for the ground-work studies, the design proposals were well received by the different stakeholders, with positive feedback, and are to be showcased at the Zanzibar International Film Festival in March 2021.

ACKNOWLEDGEMENTS

Seven students participated in the Zanzibar project, two of whom had their work included in the paper: Elizabeth Nabagerekka and Ronald Businge. We would like to acknowledge students from the FBE whose research areas have contributed to conservation and built heritage studies in the last five years. These study areas at undergraduate level included: *Conservation of Modernist Industrial Heritage in Kampala: Case Study of the Uganda Coffee Marketing Board (Diana Amanya Kacooni)* and *Heritage Significance of Late 19th and early 20th Century Buildings in Buganda Kingdom, Uganda (Gilbert Kafuuma and Esther Muhwezi);* and at postgraduate level they included: *Making a Case for Tropical Modernism as Built Heritage in Uganda (Joseph Kasimbi)* and *A Historical Study of Jinja, Uganda: A City Influenced by Industrial Developments During the Early 20th Century (Anthony K. Wako).*

REFERENCES

[1] Worthing, D. & Bond, S., *Managing Built Heritage: The Role of Cultural Significance*, John Wiley & Sons, 2008.

[2] Cody, J. & Fong, K., Built heritage conservation education. *Built Heritage*, **33**(3), pp. 265–274, 2007.

[3] Lapadula, M. & Quiroga, C., Heritage as a pedagogical resource and platform for exploration in architectural design education. *Journal of Architecture*, **17**(4), pp. 591–607, 2012.

[4] Myers, G.A., *Verandahs of Power: Colonialism and Space in Urban Africa*, Syracuse University Press, 2003.

[5] Iconic house of wonders collapses leaves Zanzibaris wondering about fate of cultural heritage. https://globalvoices.org/2020/12/28/iconic-house-of-wonders-collapse-leaves-zanzibaris-wondering-about-fate-of-cultural-heritage/. Accessed on: 21 Jan. 2021.

[6] Oman: Collapse of House of Wonders to be investigated. https://gulfnews.com/world/gulf/oman/oman-collapse-of-house-of-wonders-in-zanzibar-to-be-investigated-1.76124462. Accessed on: 21 Jan. 2021.

[7] World Monuments Fund. https://www.wmf.org/project/house-wonders-and-palace-museum. Accessed on: 10 Jan. 2021.

[8] Yamarella: An armchair travel blog that takes you around the world. https://yamarella.wordpress.com/2017/08/08/visiting-uganda/. Accessed on: 20 Jan. 2021.

[9] Olweny, M., Socialisation in architecture education: A view from East Africa education and training. *Education+Training*, **59**(2), pp. 188–200, 2017.

[10] Rüther, H., An African heritage database the virtual preservation of Africa's past. *International Archives of the Photogrammetry, Remote Sensing and Spatial Information Sciences*, **34**(Pt 6), p. W6, 2002.

[11] Cody, J. & Fong, K., Built heritage conservation education. *Built Heritage*, **33**(3), pp. 265–274, 2007.

[12] Jokilehto, J., An international perspective to conservation education. *Built Heritage Conservation Education*, **33**(3), pp. 275–286, 2007.

[13] Marschall, S., The heritage of post-colonial societies. *The Ashgate Research Companion to Heritage and Identity*, eds B. Graham & P. Howard, Ashgate Publishing: Hampshire, pp. 347–364, 2008.

[14] Embaby, M., Heritage conservation and architectural education: An education methodology for design studios. *HBRC Journal*, **10**(3) pp. 339–350, 2014.

[15] Arnold, D. (ed.), *Reading Architecture History*, Routledge: London, pp. 1–14, 2002.

[16] Olweny, M., Listen without prejudice: The design studio as a discursive environment (or) Helping students learn in architecture education. Presented at *Association of Collegiate Schools of Architecture 108th Annual Meeting*, San Diego, 2020.

[17] Mathisen, B., East Africa world heritage network and stakeholder priorities. *International Journal of Heritage Studies*, **18**(3), pp. 332–338, 2012.

[18] Tanzania's art deco ruins the Majestic Cinema, inspires restoration campaign. *The Guardian*. https://www.theguardian.com/world/2011/jun/02/zanzibar-mjaestic-cinema-restoration-campaign. Accessed on 31 Dec. 2020.

[19] Bissell, W.C., *Urban Design, Chaos, and Colonial Power in Zanzibar*, Indiana University Press, 2011.

[20] Beeckmans, L., Editing the African City: Reading planning in Africa from a comparative perspective. *Planning Perspectives*, **28**(4), pp. 615–627, 2013.

[21] Wigglesworth, S., Critical practice. *The Journal of Architecture*, **10**(3), pp. 335–346, 2005. https://doi.org/10.1080/13602360500162238.

[22] Moon, K., Ownership conflicts and heritage interpretation in Uganda and Tanzania. *8th Annual International Symposium in Heritage Interpretation*, US/ICOMOS, 2005.

ARCHIVE RESEARCH AS A DIAGNOSTIC AND COGNITIVE INVESTIGATIVE METHOD OF MEMORY OF THE MULTISTRATIFIED URBAN BUILT HERITAGE: A CASE OF URBAN ARCHEOLOGY

LUCREZIA LONGHITANO, SANTI M. CASCONE,
STEFANO CASCONE & GIUSEPPE A. LONGHITANO
University of Catania, Italy

ABSTRACT

The urban contexts that we live today are often the result of a stratification spontaneously followed over time, incorporating buildings in use in ancient times but abandoned as a result of the historical and cultural evolution of the city. This situation is very frequent in many historical centers, where archaeological buildings are located below the modern urban plan. An exemplary case of this situation is given by the Roman amphitheater of Catania. It was covered and used as a substructure for the new plan steal and for the buildings rebuilt after the earthquake of 1693 that destroyed the city. These reconversions of ancient structures cause several conservative problems, often difficult to understand and which must be investigated in an interdisciplinary way to understand their origins and solve them. The complexity of urban systems requires an inevitable multidisciplinary approach that combines archival analysis, historical, archaeological analysis, diagnostics, architectural, plant engineering and urban. This research aims to show, in particular, how the part of historical analysis of archival type can be a useful diagnostic survey tool to be applied in multi-layered contexts, through the example of the Roman amphitheater of Catania which today is in a precarious state of conservation and accessibility that requires intervention. This work was carried out by studying the administrative documentation of the 19th century produced by responsables for the protection of the historical buildings of Catania, kept at the State Archive of Catania, and allowed to understand the conservative events of the heritage architectures and to deepen the knowledge of the causes of the degradation that today the monument undergoes, demonstrating how such problems have a long history never radically resolved.

Keywords: urban stratification, archive research, history of conservation, diagnostic.

1 INTRODUCTION

Today, the archaeological and historical building heritage is part of a complex and stratified urban system. The natural growth of cities has led to the creation of urban archaeology contexts that affect the conservation of the oldest heritage. By natural overlap and development, the oldest heritage is incorporated and subjected to the construction of subsequent eras. Examples are the Roman Marcello's theater surmounted by the latter stratified building, the Crypta Balby, the Domizian's stadium whose shape is legible from the buildings surrounding piazza Navona, the temple of Athena in Ortigia incorporated in the construction of the Cathedral and many others. An urban context offering several examples of these stratifications is the city of Catania, founded by Greeks and reconstructed by Romans. After an uninterrupted development until the medieval phase, it underwent an important break due to the earthquake in 1693 that devastated the city. From this moment, the reconstruction restarted with a new planning incorporating the old buildings, e.g. the Roman ruins such as the Greek-Roman theater, the odeon, the Roman bath named Terme Achilliane and the amphitheater, with priority assigned to the new late-baroque urban plant, existing until today [1].

WIT Transactions on The Built Environment, Vol 203, © 2021 WIT Press
www.witpress.com, ISSN 1743-3509 (on-line)
doi:10.2495/STR210041

Focusing on the case of the amphitheater, that is subject of this research, it is incorporated in the urban stratification following the earthquake of 1693, in the area of Piazza Stesicoro (Fig. 1). Among the original 56 sectors, only 22 remain, divided between an uncovered area, visible from the square and set up in 1906, and a hypogea area, underlying modern structures.

Figure 1:   Piazza Stesicoro overlooking the ruins of the amphitheater, north-west view. *(Source: Poloregionale.net.)*

The hypogeal area has problems due to the unnatural location, and these are: structural, being the ruins used as substructures, and conservative, due to the continuous infiltrations from the upper floors. This situation is compromising the use of the monument and it would be necessary to plan cognitive survey of the ancient structures and the upper urban stratification, with multidisciplinary approaches, in order to prevent it from getting worse. In fact, it is necessary to investigate the conditions and causes of degradation through environmental and material analysis, monitoring performance and historical–archaeological studies to understand the phases, transformations, possible damages and restorations, without neglecting the study of the relationship, over time, between the amphitheater and the upper layers, trying to understand the sequence, evolution and plan systems of the latter construction. It is important that any intervention is aimed to the full enjoyment of the monument, supported by a careful basis of studies that goes back in time to understand not only the conservation history of the building but also to discover information on the development of degradation phenomena and their causes [2]. During the course of this study, in addition to the direct approach to the building, an important support is given by the study of historical and archival documentation. In fact, it is interesting to show how archival research could add information on the conservative history and degradation, representing a useful diagnostic tool.

2 PRESENTATION OF THE CASE STUDY OF URBAN ARCHEOLOGY
The multi-layered context of Catania joins two important "urban layers" of the city: a Roman time layer, represented by the amphitheater, and the later baroque layer, which was built after the earthquake in 1693. The area where the amphitheater was built during the imperial era is a sloping soil, at the base of the Montevergine hill which was the northern boundary of the Roman city lapped by the ancient walls. This part of the city was important for the Regis

gate, in which came close to the Via Pompeia (connecting Catania to Messina), and for this it was enhanced by the Emperor Hadrian [3]. Although starting from the 11th century the religious and political core of the city moved to the south, with the construction of the Cathedral, the northern area remained an important area due to the presence of the church of St. Agata la Vetere, saint of the city [4]. After the earthquake in 1693, the reconstruction of new urban area, incorporating the rest of the amphitheater, the adjacent 16th-century walls and the rest of the medieval city, was directed by the Duke of Camastra G. Lanza [5].

This area is part of the Piazza Stesicoro, between Via dei Cappuccini to the North, Via Penninello and Via Cerami to the South, Via Alessandro Manzoni to the East and Via Gallo to the West (Fig. 2).

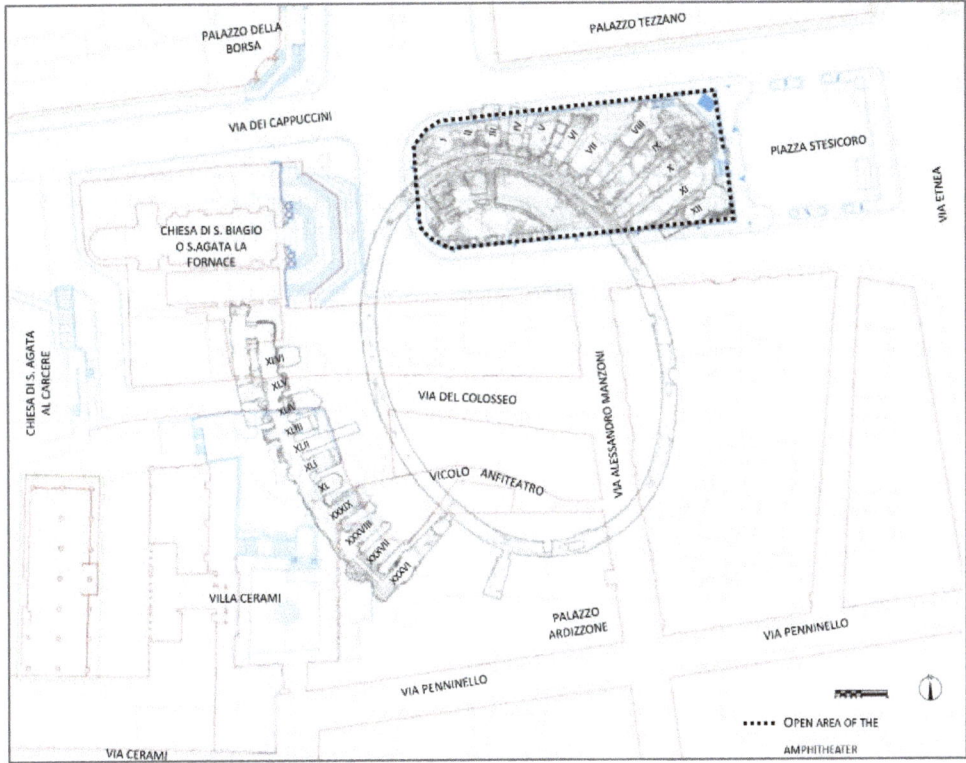

Figure 2:   Framing of the amphitheater. *(Source: Malfitana, Mazzaglia 2018, graphic reworking by L. Longhitano.)*

It is not clear if this area is the result of urban planning, or it is more likely a spontaneously system established as result of urban growth [6]. However, the attention to this area grew at the beginning of the 18th century, when started an expansion northward. The gentry was the first encouraging construction in this area, commissioning many buildings such as: Palazzo Tezzano, Palazzo del Toscano and, above all, the Villa of the Prince of Cerami (or Villa Cerami), provided with a roof garden supported by the amphitheater arches and by part of the 16th-century bastions [7]. At the same time, the Benedictine order intervened by rebuilding the church of St. Agata la Vetere and started the churches of St. Agata al Carcere and St. Biagio or St. Agata la Fornace, continuing to exploit the walls and the vaults of the amphitheater [8] (Fig. 3). Therefore, the medieval and Roman ruins were totally hidden and used as foundations. The re-discovery of the amphitheater, as consequence of the new urban plan, started progressively from the mid-1700s when Ignazio Paternò Castello, Prince of Biscari, began the excavation of the external corridor and the western façade of the structure, facing the hill Montevergine. These activities were consistent with the 18th century idea of rediscovering antiquities [9]. In 1841, the excavations started under the supervision of Francesco Saverio Cavallari and Domenico Lo Faso Pietrasanta. In this phase, a sector was dug to connect the middle ambulatory with the internal one, and the two main entrances, located on the major and minor axis of the arena, were discovered. In 1875, walls and portions of the lava stone cornice were found under some houses in Via Penninello. The last phase of discovery of the building was in 1904–1906 when Filadelfo Fichera freed the cavea, the wall of the podium, the corridor adjacent to the podium below the cavea and a "precinct" separating the arena from the base of the stairs. In this phase, in line with a desire for urban enhancement through ancient memories, the Piazza Stesicoro square was partly opened, creating the view to the monument that exist today. In the next years, new portions were not added to those excavated, but the interventions and studies continued. In the 1970s and 1980s, there were the restoration and consolidation of the amphitheater's arches in the south-western area, located below the Villa Cerami. In 1996 a campaign of interdisciplinary studies was carried out, consisting in an in-depth analysis of archaeological excavations, geologic surveys and shoring; in 1997, in Via Penninello, some archaeological parts emerged, following works, which led to new investigations. The latest interventions and studies were carried out in 2006, with the participation of the Germanic Archaeological Institute in Rome [10].

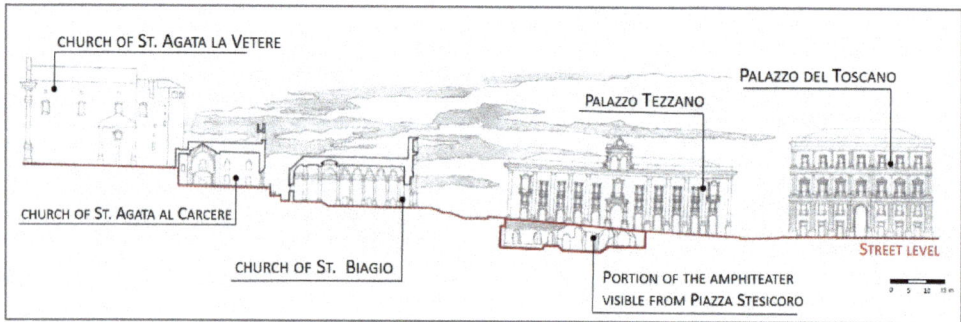

Figure 3:   Section of the Piazza Stesicoro. *(Source: Quaderni dell'Istituto di disegno, 2, 1967, graphic reworking by L. Longhitano.)*

3 THE STATE OF THE MONUMENT: CONSERVATIVE PROBLEMS

Today, the amphitheater has a different condition from the original one. It was designed as an "above-ground" structure, however, today it is located underground and it is used as foundation and support of numerous structures, areas and houses characterized by important architectural value (Fig. 4). Over time, this unnatural condition has led to many problems, especially since the 18th century, i.e. structural collapses that required the construction of both concrete and temporary shoring, and degradation phenomena. In particular, walking through the ambulatory, there is the presence of biological colonization, efflorescence, perennial infiltrations and drips, which don't allow a pleasant and real enjoyment of the monument, as well as acting negatively on the Roman vaults and walls. The current condition is not easy to investigate, because the amphitheater is a system made up of different parts in heterogeneous conditions and subject to a complex stratification. In light of this situation, studies and projects have been ongoing for many years with the aim of securing and enhancing the underground portion [11].

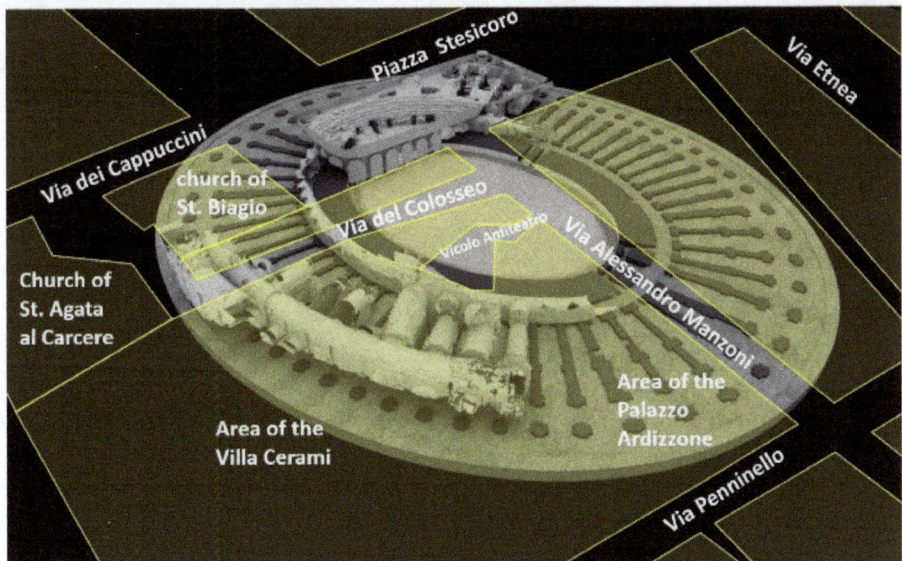

Figure 4: Surviving parts of the amphitheater with the new building in yellow. *(Graphic reworking, by L. Longhitano, of laser scanner made by ISPC-CNR of Catania.)*

Recent studies [12] have deepened the analysis of the XXXVI surviving sectors, (see Fig. 2), located in the underground part, that is the only intact sector and connection point between the internal and external ambulatory, even if it is seriously compromised by the decay phenomena. The analysis carried out allowed to understand the main manifested degradation phenomena such as deposits, efflorescence and various types of encrustations.

These encrustations, following the X-ray diffractometry analysis, have been distinguished into three types: chalky encrustations (Type 1) in all walls of the sector (conditioned by the rising and lateral infiltrations of water crossing the higher layers) and carbonate encrustations (Type 2–Type 3) derived from the dissolution of the mortar binder of the Roman vaults in *opus caementicium* (lime mortar and stones), due to the infiltrations (Fig. 5).

Therefore, the main causes of degradation are infiltrations, however, it is not easy to understand the origin. It requires an in-depth study of the context through the integration of different approaches, applied to the entire building, because the precarious condition of XXXVI sector is not isolated but similar scenarios were found in other parts of the structure.

Considering this situation, this study proceeded by carrying out an analysis of the evolution of the upper urban context through the study of historical cartography which demonstrated how the current system was organized in the 18th century and was not changed.

Especially, the authors made a detailed study of the archival documentation concerning the historical and artistic heritage of the city of Catania and, in particular, concerning the amphitheater, in order to find more information on conservative history of the building as detailed in the following paragraph [13].

 WIT Transactions on The Built Environment, Vol 203, © 2021 WIT Press
www.witpress.com, ISSN 1743-3509 (on-line)

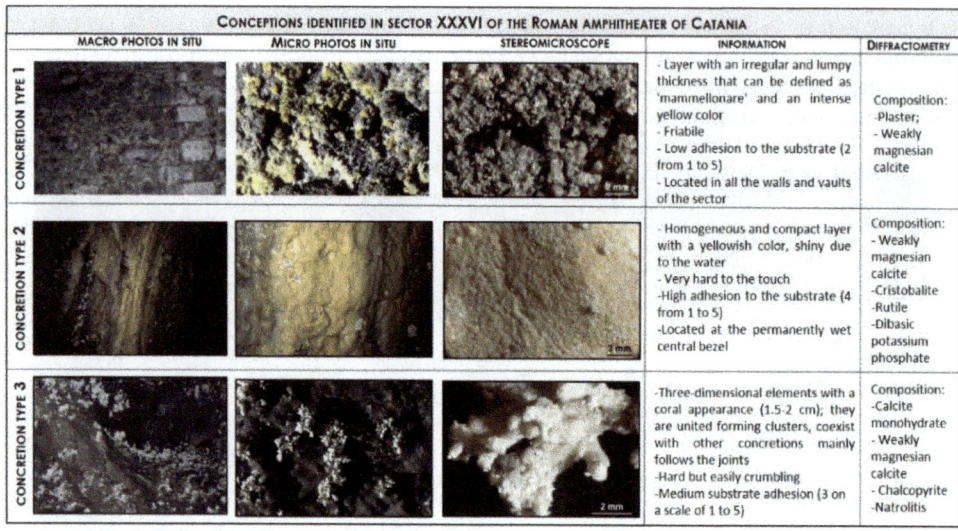

CONCEPTIONS IDENTIFIED IN SECTOR XXXVI OF THE ROMAN AMPHITHEATER OF CATANIA				
MACRO PHOTOS IN SITU	MICRO PHOTOS IN SITU	STEREOMICROSCOPE	INFORMATION	DIFFRACTOMETRY
CONCRETION TYPE 1			- Layer with an irregular and lumpy thickness that can be defined as 'mammellonare' and an intense yellow color - Friable - Low adhesion to the substrate (2 from 1 to 5) - Located in all the walls and vaults of the sector	Composition: -Plaster; - Weakly magnesian calcite
CONCRETION TYPE 2			- Homogeneous and compact layer with a yellowish color, shiny due to the water - Very hard to the touch -High adhesion to the substrate (4 from 1 to 5) -Located at the permanently wet central bezel	Composition: - Weakly magnesian calcite -Cristobalite -Rutile -Dibasic potassium phosphate
CONCRETION TYPE 3			-Three-dimensional elements with a coral appearance (1.5-2 cm); they are united forming clusters, coexist with other concretions mainly follows the joints -Hard but easily crumbling -Medium substrate adhesion (3 on a scale of 1 to 5)	Composition: -Calcite monohydrate - Weakly magnesian calcite - Chalcopyrite -Natrolitis

Figure 5: Different types of concretions found in sector XXXVI of the Roman amphitheater of Catania.

4 RESEARCH METHODOLOGY: ARCHIVE ANALYSIS

In order to expand the knowledge framework on the history of the amphitheater conservation, the archival documentation preserved in the State Archives of Catania was examined. It dates from the early 19th century, when Sicily prepared new control systems of the antiquities as well as being it was reorganized with a new division into seven Intendencies [14].

Especially, the documentation analyzed belongs to the sections of the Intendenza Borbonica and of the Prefettura of Catania. Concerning the protection policy history implemented in Sicily, in place of older management system of the assets governed by local nobles, appointed royal guardians (called Regi custodi) started since 1778, a Commission (Commissione di Antichità e belle arti di Sicilia) was imposed from 1827 in order to create a more centralized system. By 1830 local commissions were added, as well as collegial bodies, destined to control the peripheral areas [15]. In particular, in Catania, from 1829, a Deputation of Antiquities was also set to ascertain the state of the monuments in Catania considered particularly disfigured [16]. In general, from the 19th century, there was a greater interest in the control and care of assets, which was followed by an increase in exchanges of letters between central and local government. In fact, more information on monuments in Catania can be obtained from letters between various administrative bodies such as: the Intendenza of the Vallo di Catania, the Patrizio of Catania, the Police Station, the Catania's Deputation of Antiquities and from the proceedings of this Deputation.

In particular, for example, from the acts of the sessions in 1830, it is possible learning that the remaining parts of the amphitheater, subject to inspections and discussion, consisted of: ruins, corridors, cavea and part of the portico (Intendenza Borbonica, envelope 643, cc. 587r). This research carried out an examination on the latter documentation.

5 RESULTS AND INFORMATION OBTAINED FROM THE RESEARCH

It was found that the damage in the amphitheater due to the upper urban layer has characterized the building life for several centuries and, therefore, is not related to the current state. The problems are clearly delineated as early as the 19th century and can be attributed to the infiltration, to the use of the building as a landfill and to the lack of a sewerage system to dispose the rainwater that flow from the street level to the ambulatories of the amphitheater placed below. The oldest report is dated 1832, when the caretaker of the monuments urged the authorities to take action to resolve a series of inconveniences due to the use of the amphitheater as a rubbish area by the buildings placed around and above. In particular, it is reported that the sacristans of the parish of S. Agata al Carcere (located near the western area of the amphitheater) continuously threw their rubbish into the "place of antiquity", as well as fellow local merchants. In particular, the master Carmelo Pulvirenti is mentions because the waste water of his workshop was discharged into the internal ambulatory of the amphitheater (Intendenza Borbonica, envelope 643, cc. 695–696). A few years later, in 1838, the same caretaker wrote to the Superintendent of the province to denounce how in the amphitheater there were problems caused by "falling water and other rubbish", this one from the garden adjacent to the amphitheater, owned by Mr. Cappellano Cosentino. It is therefore urged to intervene since this had never been done, causing damage not only to the amphitheater itself but also to visitors (Intendenza Borbonica, envelope 643, cc. 609). In February 1844, the architect Mario Musumeci was designated to carry out emergency work in the ancient buildings of Catania, the architect testified in a report that the most issue to manage is the relationship between the archaeological building and the street upper level. To face the situation, the architect suggested building a circuit walls to avoid use it as a landfill (Intendenza Borbonica, envelope 643, cc. 678–683). However, in the restoration works to be done, dated to 1854, there are still infiltrations for both ambulatories (Intendenza Borbonica, envelope 643, cc. 554, 556). This report documents that from the vault of the external ambulatory were coming "continuous leaks of water". This water occurred from the garden upstairs belonging to the Prince of Cerami. The same problems also appear on the vault of the internal ambulatory, located below the Mr. Carlo Ardizzone's house and precisely the cause would be the running water used by owners. As interventions, it was suggested, in these old documents, to "expurgate and trim the aqueduct and bring water elsewhere" and bringing the fallen material in one of the surviving compartments.

As some letters from the middle of the 19th century show, expropriating and demolishing the houses that damaged the archaeological ruins was proposed among the available solutions. These problems were discussed for both the amphitheater and the theater. In fact, another important building in Catania, the theater was in the same conditions due to the upper urban layer. The Commission began performing a preliminary estimation on property value and on the compensation to be paid to owners, but these analyses were carried out only for the buildings built on top of the theater and not for the amphitheater. The architectural value of the buildings built above the amphitheater hindered the demolition of the upper house.

These buildings are the Villa Cerami and Professor Ardizzone's house, both emblematic buildings of late baroque architecture in Catania, on which important local architects intervened, such as Giovan Battista Vaccarini and Francesco Battaglia [17], as well as the important churches dedicated to the patron saint of the city. The documents revealed many causes against the owners of the demolished houses above and near the theater, while none for the houses near the amphitheater. As the years passed, even in the aftermath of the unification of Italy in 1861, the problems were not solved, on the contrary they were continued amplified by the administrative changes and centralized decision to detriment of local force.

In 1860, a Central Commission of antiquities in Palermo has been reconfirmed with total jurisdiction over archaeological excavations and restorations but which could avail itself, where necessary, of the advice of provincial commissions. However, communication between the center and the periphery had many problems and consequently the Commission's action had trouble, and it was accused of negligence. The inability of the administrative bodies in charge of monument's protection is demonstrated by the existence of problems and the long intervention times, in fact, the Prefect of Catania emphasized that the amphitheater still has problems due to swamping and stagnation of water that flow together. In 1875, in letters discussing the findings in the southern part of the amphitheater incorporated by the houses in Via Penninello, it was notified as beyond the usual drains, in a corridor of the amphitheater (probably below the Ardizzone's properties) disposable water drainage wells were opened (Prefettura, list 14, envelope 258). It was a recurring practice until today. More useful is the documentation drawn up by the Inspector of Monuments Carmelo Sciuto Patti, strongly interested in the protection of Catania's assets and on several occasions he documents the situation [18]. In 1881, Patti urged the municipal administration to intervene urgently the monument after he was being informed about the damage and inconveniences during last many years. These problems concerned the internal ambulatory subject to infiltrations due to the water drained from a sewer in the ascent of the Capuchins below the bell tower of the church of S. Agata la Fornace, where the rainwater converge from Via Botte dell'Acqua and from various aqueducts. The part of the city in question, being at the basis of the Montevergine hill, is sloping morphology towards Piazza Stesicoro, thus favoring a natural channeling towards the amphitheater. There were also problems in other points. In particular, in the portion of the external ambulatory, placed below the Professor Ardizzone's house and the Villa Cerami, where the "unclean waters of the wash houses" flow into.

It is reported that the Prince built on his own initiative, a canal that collected water from the various roofs to bring them right into the amphitheater (Prefettura, envelope 258, series I, list 14). Also, great of interest is the letter written by Sciuto Patti in 1885 where the latter locates and carefully explains the damage in the amphitheater, the related causes and interventions. There are two drawbacks. The first was the infiltration of water, especially in winter times, in the vaults of the external ambulatory, making it uncomfortable to transit, especially due to the stagnated humidity in the clay soil. The second drawback was the humidity involving the internal ambulatory all year round. Regarding the causes, for the Inspector, rains beating on the various houses then flow into the amphitheater. In particular, the house of the Professor Giovanni Ardizzone heirs is denounced (relative of the mentioned Carlo) because dirty water continues draining and is absorbed by the soil and handed down to the vaults of the underlying archaeological ruin. Therefore, such problems are to be found in the South-eastern area of the building. As important intervention, Sciuto Patti proposes to channel all water into special channels to dispose it in a large adjacent area outside the building. The causes of infiltrations in the internal ambulatory are traced back to the poor or "poorly maintained" construction of the municipal aqueducts, under the paving of Via Cappuccini, in the north of Piazza Stesicoro. Regarding the problems, the municipal administrators were repeatedly urged to intervene, but it never happened. Also, the Inspector suggested to repair and to create a disposal system inside the building by opening a well in the most depressed part of the corridor, in order to free it by exploiting the natural internal slope of the soil (Prefettura, section I, list 14, envelope 257). Despite the great interest shown, there was not an effective solution to the problems and this was demonstrated when in 1899 there was still discharges and continuous use of the amphitheater as a garbage dump. These drains always refer to the above buildings and specifically to the Villa Cerami and the chaplaincy of the church of S. Agata la Fornace (Prefettura, section I, list 33, envelope 59).

In 1913, after the liberation of the visible portion in Piazza Stesicoro, Sciuto Patti underlines how the problems still persisted and they were still studying the most appropriate way to radically solve the problems of infiltrations, a solution that perhaps still today it was not found. Fig. 6 shows all the possible areas subject to the problems documented by the archival documentations analyzed.

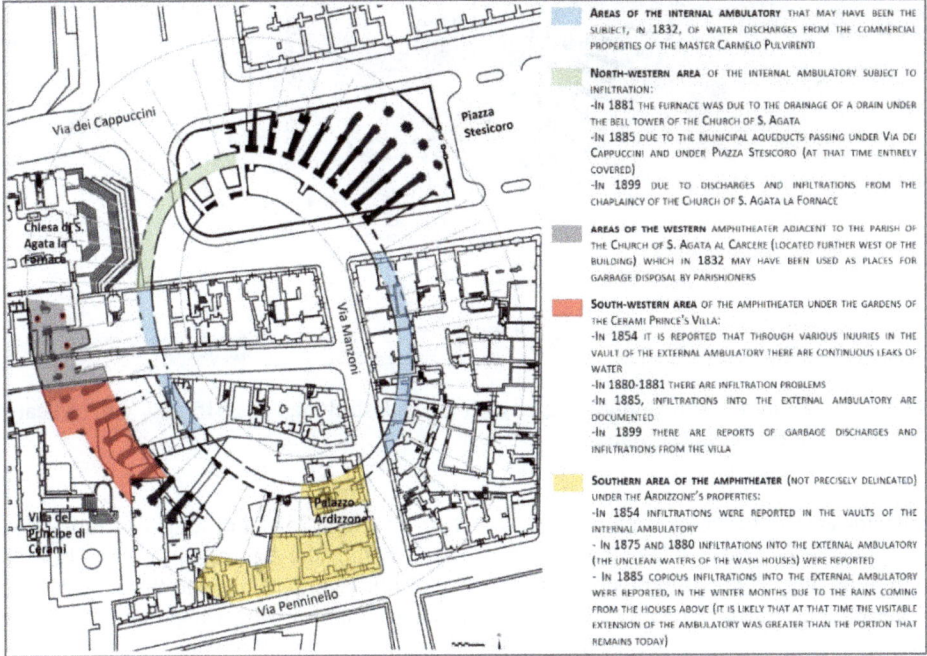

Figure 6:   Hypothesis of identification of the areas of the amphitheater subject to discharges and infiltrations cited in the archive documentation.

6 COMPARISON BETWEEN URBAN AND ARCHIVAL DATA

At this point, an interesting topic is the attribution of the houses to the owners mentioned by the documents and find out with which current buildings they coincide. From the most updated cadastral surveys of 2015, it is possible to distinguish the lots and buildings weighing on the hypogeal area of the amphitheater (Fig. 7).

Building n. 4112 facing onto Via Manzoni at the corner with Via Penninello is recognizable, as it is the 18th century palace of the Ardizzone family mentioned in the papers several times. Even today, the underlying portion of the amphitheater is the one that presents the greatest problems both in terms of static and infiltration, in accordance with what the archive papers documented several times for the past centuries. With the same ease, the area of the Villa Cerami (n. 32639) is identified, today it is the seat of the University of Catania, Department of Law, and the terrace with garden is entirely supported by the South-Western sectors and by the external ambulatory of the amphitheater. These parts of the buildings are subjecting to perennial infiltrations and consequent manifest alterations, even in this case the problems mentioned by the documents are still present today. The more-doubts remain for the identification of the other houses. Clues can be drawn from the sources documenting in 1875 the discovery of parts of the ruin in to the house belonging to a certain Mr. Puglisi and in the adjacent house of the Nicosia family, both built after 1693 in the street Via Penninello,

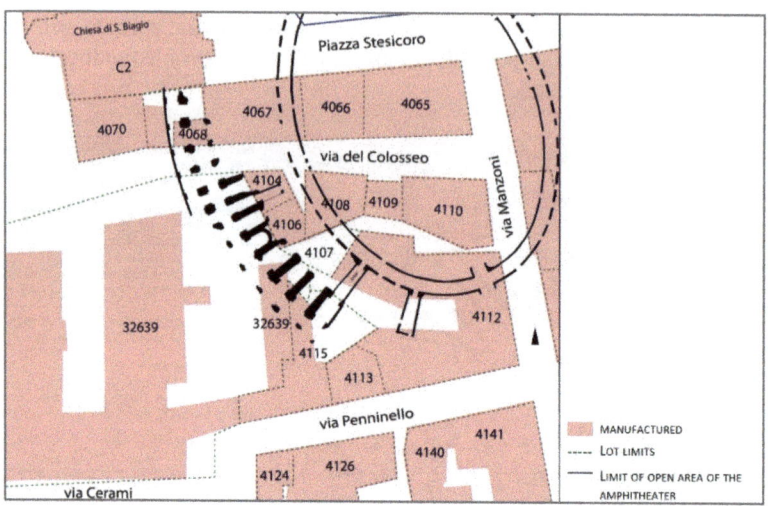

Figure 7:   Framing of the amphitheater with the cadastral map of 2015 and with indication of the buildings and lots.

above the amphitheater (Prefettura, series I, list 14, envelope 258). For logic of stratification, it is possible to consider that these properties fall within the lots n. 4115 and 4113. Not negligible, then, is the aforementioned existence of other gardens, next to the amphitheater, in addition to those of the Cerami. In particular, that of Mr. Cappellano Cosentino (Intendenza Borbonica, envelope 643, cc. 609) from which the fall of water and "garbage" is denounced in 1838 and the citation, in 1885, among the houses competing against, that of the knight Nigo (Prefettura, series I, list 14, envelope 257). The attribution to the lots for the latter cases remains more doubtful. However, it is not excluded that the knight's house may be located in the remaining blocks above the amphitheater (lots n. 4014, 4106, 4108), while the garden could coincide with lot n. 4107, as well as the only green area in the area.

7 CONCLUSIONS

The archival sources relating to the historical and artistic heritage of the city of Catania and in particular to the amphitheater, clearly show how it is possible to trace a real "history of the deterioration of the amphitheater" connected to the upper urban area, coming to note how the problems that today we are faced with have actually been there for centuries [19].

Reasoning and summarizing, the areas of the amphitheater most subject to problems were the western and southern areas during the 19th century and continue to be so today. About the intervention proposals that are drawn from the documents, both active water diversion interventions by building new channels and passive interventions were suggested: closing of the parts, plugging of the entrances. At present, it is not easy to understand which ones have actually been made. Certainly, however, still at the beginning of the 19th century the problems were present so it is possible to hypothesize an actual lack of interventions up to that period. Certainly, an important fact dates back to 1870 when a document talk about the construction of a municipal road sewer in the area where the amphitheater is located, probably part of those located under Via dei Cappuccini and Piazza Stesicoro. However, this does not seem to have been designed to solve the present situations because, rather, in 1875 with private houses caused as many problems of infiltration to the ancient structure that lay under. Despite the interest in the intervention shown by the institutional managers, there is a lack of timeliness and indeed a predominance of the abusive private action. We can remember the interventions carried out on the vaults of the amphitheater by the Prince of Cerami, the uncontrolled construction of canals that convey waters and the use of the ancient rooms as places of unloading. These practices are also found in other archaeological sites in Catania; as demonstrated by the construction of a latrine in a house built above the second corridor of the Roman theater (Prefettura, list 33, envelope 59). What, in conclusion, the archival research shows is that the problems with which we have to face today have already characterized the life of the amphitheater for centuries. These problems must obviously be traced back to the articulated stratification system that unites a building born to be above ground to an urban and plant context that spontaneously stratifies above it and incorporates it following a culture that does not care about antiquities but is projected towards reconstruction. To approach the problems, in addition to a study that combines the archeology of architecture with the analysis of urban transformations, an area to be investigated to trace the causes of the problems certainly concerns the systems of the inhabited areas and the urban area above the amphitheater. On this theme, from the examination of the sources, a precise provision is not deduced, at most we note an illicit exploitation of the underlying amphitheater with direct drains. Private individuals used to build gathering canals which they unloaded in illegally arranged points with, as we have seen, negative consequences. The drains, such as those of the Ardizzone's house, still present today, or of the sacristies, often exploited voids and pockets of the land according to a "disposable" system widespread in the city and seen both in the amphitheater and in the theater. This condition is very likely to still be active today, just as it is likely that the systems built in a careless way have never been modified or diverted from the archaeological area. Today the conditions of the plant system of the city are very complex [20], without going into much detail on the theme, it is difficult to understand how the individual building complexes that affect the area of the amphitheater are set up. However, it is indisputable that this situation is as in general the development over time of an urban system that incorporates an older one, lead to consequences that are not easy to investigate. In addition, to the losses caused by underground discharges in the concurrence of the causes, the external rainwater drainage systems and the burials that lean against the building, even more difficult to manage than plant engineering, should not be underestimated. Any restoration or large-scale intervention in urban terms, therefore aimed at blocking the ongoing process, cannot exclude an accurate analysis of all these situations.

ACKNOWLEDGEMENT
We would like to thank the *Ministero dell'Istruzione, dell'Università e della Ricerca* who financed the project *EWAS – An early WArning System for Cultural Heritage –* PON ARS01_00926 PNR2015-2020.

REFERENCES
[1] Correnti, S., *Catania attraverso i secoli, nei testi e nelle immagini*, Stamperia Valdonega: Verona, 1966.
[2] Cascone, S.M., Longhitano, G.A., Longhitano, L. & Tomasello, N., Dallo studio archeologico delle malte storiche alla prospettabile progettazione delle malte da restauro attraverso una procedura di metodo. Case study: le Terme Achilliane di Catania. *Conference Proceeding in Colloqui.AT.e 2020 – Nuovi orizzonti per l'architettura sostenibile (Catania, 9–11 dicembre 2020)*, 2020.

[3]   Tomasello, F., La viabilità suburbana in età imperiale. *Tra lava e mare, Contributi all'archaiologhia di Catania, Edizioni private e varie*, eds M.G. Branciforti & V. La Rosa, Le Nove Muse Editrice: Catania, pp. 289–318, 2010.

[4]   Privitera, S., Lo sviluppo urbano di Catania dalla fondazione dell'apoikia alla fine del V secolo d.C. *Catania. L'identità urbana dall'antichità al settecento*, ed. L. Scalisi, Sanfilippo: Catania, pp. 37–71, 2009.

[5]   Boschi, E. & Guidoboni, E., *Catania. Terremoti e lave dal mondo antico alla fine del Novecento*, Editrice Compositori: Roma, 2001.

[6]   Dato, G., *La città di Catania*, Officina Edizioni: Catania, 1983.

[7]   Basile, F. & Magnano di San Lio, E., *Orti e Giardini dell'aristocrazia catanese*, Sicania: Messina, 1996.

[8]   Cantone, U., La Chiesa di S. Agata al Carcere a Catania. *Quaderni dell'Istituto di disegno 2,* Tip, dell'Universita: Catania, pp. 215–240, 1964–1965.

[9]   Bonaventura, V., *La Sicilia al tempo del Grand Tour*, GBM: Messina, 2009.

[10]   Beste, H., Becker, F. & Spigo, U., Studio e rilievo sull'anfiteatro romano di Catania. Rapporto preliminare sul rilievo archeologico – Recente campagna di indagini. *Bullettino dell'Istituto Archeologico Germanico*, **113**, pp. 595–613, 2007.

[11]   Malfitana, D. & Mazzaglia, A., Archeologia globale a Catania. Nuove prospettive dall'integrazione di ricerca archeologica e tecnologie ICT. Nuovi dati sull'anfiteatro romano. *Studi e materiali 1, Dipartimento culture e società – sezione beni culturali area archeologia Università di Palermo*, Palermo, pp. 327–353, 2018.

[12]   Longhitano, L., L'anfiteatro romano di catania settore XXXVI. Ricerca storica, stratigrafia degli elevati, prima diagnostica per la verifica delle condizioni di conservazione. Thesis, Politecnico di Milano, 2018–2019.

[13]   Pelagatti, P., Dalla Commissione Antichità e Belle Arti di Sicilia (CABAS) alla amministrazione delle Belle Arti nella Sicilia post-unitaria. Rottura e continuità amministrativa. *Mélanges de l'école française de Rome Année*, pp. 599–621, 2001.

[14]   Iozzia, A.M., Documenti dell'Archivio di Stato di Catania per la storia dell'archeologia catanese. 1743–1932. *Catania antica nuove prospettive di ricerca*, ed. F. Nicoletti, Grafica Saturnia: Siracusa, pp. 673–720, 2015.

[15]   Oteri, A.M., Tutela dei monumenti antichi e trasformazioni urbane a Catania, 1779–1949. *Aree Archeologiche e centri storici. Costruzione dei Parchi archeologici e processi di trasformazione urbana*, ed. G.P. Treccani, FrancoAngeli: Milano, pp. 153–186, 2011.

[16]   Muscolino, F., Il principe di Biscari e il principe di Torremuzza «i due Dioscuri della passione antiquaria settecentesca». *LANX*, **21**, pp. 1–40, 2015.

[17]   Ottorino Russo, A.G., *Catania e il suo Settecento*, Tringale Editore: Catania, 1984.

[18]   Sciuto Patti, S., I recenti restauri dei Monumenti antichi di Catania, l'Odeon, l'Anfiteatro romano, il Teatro greco, il Foro sotto l'ispettorato dell'Ing. S. Sciuto-Patti. *Archivio Storico per la Sicilia orientale*, X, I, Catania, pp. 312–315, 1913.

[19]   Cascone, S.M. & Longhitano, L., L'importanza della ricerca d'archivio per un'analisi dello stato di fatto degli edifici storici e delle cause dei fenomeni di degrado: il caso dell'anfiteatro romano di Catania. *Simposio Internazionale-Assemblea dei soci-REUSO* (30 ottobre 2020), 2020.

[20]   Murabito, L., Lo scandalo della Rete Fognaria di Catania: emergenza per gli scarichi a mare. SUD PRESS giornalismo d'inchiesta, online. www.sudpress.it. Accessed on: 11 Feb. 2021.

URBAN PLANNING AND ITS IMPACT
ON THE CITY OF HEBRON, PALESTINE

WAEL SHAHEEN
Department of Civil and Architectural Engineering, Palestine Polytechnic University, Palestine

ABSTRACT

Due to its history of more than 5,000 years, Hebron is considered one of the most ancient cities in the world. It has been occupied several times throughout history, by Romans, Byzantines, Ottomans, Mamluks, the British Mandate and finally the Israeli Occupation. All these people marked the urban planning of the town in many ways, until the present day. There is no doubt that Hebron City suffers from severe weaknesses related to urban planning, because of the failure of structural plans that aren't keeping up with the rapid development that the city has been going through for the past few years. Actually, if we took a moment to observe some Arab cities like Beirut and Amman, we would realize they're all in the same situation; all of them lack a proper, well-planned layout. On the contrary, very old cities such as Al-Basra in Iraq, had very special environmental considerations still valid nowadays, but there are other examples and models of sustainability standards within the planning of the Arab cities. However, we can still find some global obstacles facing the Arab World regarding their planning potentials. Palestine consists of many cities, Hebron is one of the biggest cities in terms of total area and population density. Unfortunately, it lacks an effective structural planning to fulfill the city's continuous unstoppable development, due to various circumstances affecting the planning process in the city, which has influenced the land use. Moreover, there is a losing control over areas inside and around the city because of the Israeli Occupation. Due to the huge importance of urban planning and the human ultimate need for it in all different sectors, this research aims to explain and clarify the definition of the space and incorporate its cultural values and characteristics. The factors affecting urban planning will be clarified, as well as explaining how to create a different reality that develops the city and preserves its heritage, history, resources and sectors.

Keywords: planning, law, rehabilitation, economy, development, tourism, culture.

1 INTRODUCTION

Hebron is considered one of the most important Palestinian cities, especially from a religious and historical point of view because it dates back to around 5,000 years ago [1]. Hebron Governorate contains more than 650 archaeological and historical sites, among which the site of the Tomb of the Patriarchs and the remains of ancient churches, as well as The Monastery of the Holy Trinity, located in the west of the city, highly visited by tourists from abroad. Furthermore, we find more than 170 mosques, of which more than 50 are old mosques and others are newly constructed.

Geographically, Hebron is located 36 km from the city of Jerusalem and Bethlehem [2].

The city center was affected after the establishment of outposts following Israeli occupation in 1967, which led to the displacement of residents from the heart of the city to outside. Most of the migration from the city center occurred after the settlers' assault on the Ibrahimi Mosque in 1994, and this affected it commercially, as all the commercial markets were in the old center, and it represented one of the greatest commercial centers in the West Bank.

Furthermore, after the signing of the Oslo Accords in 1993, the Palestinian Authority took control of part of the town and opened the door, without any planned strategy to import products from China, Turkey and elsewhere, which had a negative impact, and led to the deterioration of economic life and negatively affected most of the cities, of the entire West Bank. So, factories were closed, especially factories making shoes and clothes which were

WIT Transactions on The Built Environment, Vol 203, © 2021 WIT Press
www.witpress.com, ISSN 1743-3509 (on-line)
doi:10.2495/STR210051

prosperous in the city, and life deteriorated and consequently this led to poverty. Besides, without structural plans random urbanization appeared, which was not related to the old architecture [3], as well as visual distortion, traffic congestion, and the spread of vendors in the streets and on its sidewalks because of the lack of city markets (Fig. 1).

Figure 1: Inside the old city of Hebron.

The author would like to shed light on some broad outlines of urban planning in the city to contribute to reducing the damage, even with a simple thing, and to reduce the negative effects that the city experiences and strive to find determinants for development and improvement.

2 HISTORICAL STAGES OF PALESTINIAN CITIES AND LAND-USE PLANNING

Historical stages of Palestinian cities, especially last century history, greatly affected the land use:

1) The first stage: the stage of the Ottoman Empire 1517 AD–1917 AD.
2) The second stage: the period of the British Mandate 1917 AD–1948 AD.
3) The third stage: the annexation of the West Bank to Jordan, and the Gaza Strip to Egypt, 1948 AD–1967 AD.
4) The fourth stage: the stage of the Israeli occupation 1948 AD–1994 AD.
5) The fifth stage: the stage of the Palestinian Authority from 1994 to today.

These last phases had great impact on Hebron for its management and planning, as everyone who came to Palestine established laws for managing areas for his service and interest, and today we find that the Palestinian National Authority uses old laws or relies on them. Nowadays there are two different laws, one applied to the areas of the Israeli Civil Administration and other municipal laws in force in Palestinian cities [4].

Actually, the traditional Arab cities were not subject to prior plans in their manner, rather the formal and spatial formation of the urban fabric of this city was the result of man's understanding with the cultural and natural environment, based on experiences and practices, which gave the urban fabric its peculiarity and characteristics [5].

3 THE IMPACT OF THE ISRAELI POLICY ON URBANIZATION IN THE WEST BANK

From the beginning of the occupation, Israel has obstructed the natural growth of urban areas in Palestine trying to disconnect areas and residential communities.

Israel distributed its settlements on the outskirts of Palestinian cities and villages and established camps under the pretext of training and protection for settlers, and surrounding areas around the settlements as natural reserves for future expansion and future geographical contact. Israeli planners prepared local structural plans for Palestinians cities and villages to serve their own interests to occupy more Palestinian lands. In 1981, a 183 structural plan that doesn't even achieve the bare minimum of human needs for Palestinians was approved, even though it definitely was refused by Palestinians themselves [6]. Several years later, the Israeli Planning Central Department of the Military Administration prepared partial structural plans which aimed to narrow and limit Palestinian construction and urbanization for all Palestinians villages in the West Bank, which were approved at the beginning of 1994, after the signing of the Oslo Agreement in 1993. Unfortunately, a large number of these plans, which also promoted the increase of the settlements, are still valid now due to the absence of alternative or new structural plans [7]. These partial plans were prepared by Israeli planners using aerial photos and were limited to residential uses in streets that don't fulfill the needs of these villages, such as the width of the street (16 meters) inside the village, as well as not giving chances of future expansion for the existing residential buildings.

Whereas Israel is still implementing its racist plans by paving roads through Palestinian lands to serve its settlements, as well as demolishing nearby Palestinian houses and working to prevent expansion of Palestinian communities, as these communities surround the roads, their purpose was to stifle them and prevent them from expanding and even prevent landowners from approaching or entering their own lands. Under the rule of the ex-president of the United States Donald Trump, despite the fact that the signed agreements prevent settlement or any activity supporting it, there are still Israeli mechanisms to pave new roads to fulfil its intended plans at the expense of he lands of the Palestinians who are incapable of doing anything [8].

4 THE OCCUPATION AND ITS IMPACT ON THE OLD CITY OF HEBRON

Since the beginning of the occupation Israeli paved a road in the eastern side of the old city, destroying the main entrance to the Ibrahimi Mosque and many neighboring archaeological buildings in order to connect their settlement of Kiryat Arba, east of the city to the Ibrahimi Mosque, which is of special importance to Jewish, and to communicate with the Tel Rumeida area, a settlement outpost in a strategic location (Fig. 2) [9]. This occupation led to many residents leaving their houses and concentrating in old areas such as Haret al-Sheikh neighborhood and Bab al-Zawiya. Other residents built dwellings inside the spaces, squares, and orchards within the urban fabric of the old city center because these families did not have a place to shelter, so they occupied spaces that were formerly used as places for entertainment or green areas, such as Al-Sadaqah Garden (Fig. 3) [9].

So, the city stretched and expanded, but without planning, as many first built houses and then licensed it. Also, several historical buildings were removed and demolished, due to weak existing legislation and laws, to pave the new road or to build commercial buildings These buildings have no connection with the city's past in terms of cultural, architectural and heritage values nor does it even feature the Palestinian identity that it is supposed to protect. From our connection, our identity, our culture and the continuity of our existence, we must work on not eliminating and preserving them and spreading awareness among people of their importance [10].

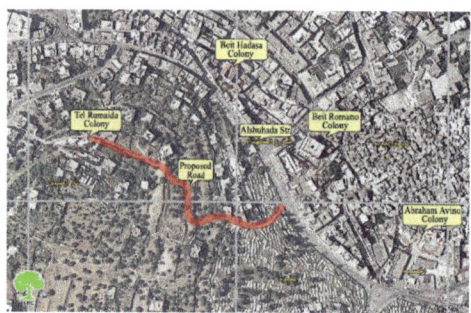

Figure 2: Tel Rumeida, suggested street for the settlement.

Figure 3: Israeli settelments in the old city.

5 THE CURRENT URBAN EXPANSION IN THE CITY OF HEBRON

Nowadays, Hebron expanded in various directions, but very strict restrictions are put by Israel, such as the confiscation of land, preventing construction and even the expansion or restoration of existing buildings. In some areas, construction materials are not allowed inside. Other than working on the displacement of the population, especially in the east and south of the city [10].

In the city center, Israel has built commercial residential units that have nothing to do with the urban fabric and its architectural character, along with the presence of the occupation forces stressing the lives of Palestinians in these areas in order to let them migrate from their houses.

The city lacks stability and political calmness due to these outposts that have been planted in the center, this resulted in moving the historical center to the adjacent northern side in "Bab Al-Zawiya" area, transforming it into the new commercial city center. Urbanization extended densely in this area and in other areas such as "Ain Sarah" (Fig. 4), "Al Salam" Street, "Ras Al-Joura", and others. However, the huge buildings were concentrated closely to "Bab Al-Zawiya" that has become a crowded area with underserved commercial markets. The city already suffers from the lack of street efficiency, which can be limited up to three main axes, starting from the northern entrance to the city, which is the area of "Ras al-Jura", to its southern new center, "Bab Al-Zawiya" (Fig. 5), including: Ain Sarah Street Namira Street, Al Salam Street and secondary streets branching from the previously mentioned streets [11].

Figure 4: Ain Sarah Street "new buildings".

Figure 5: The new commercial district in Bab Al-Zawiya.

The city contains public services like schools, universities, hospitals, small parks, internal and external transport networks, all of the existing services are definitely not enough for the city's population and it's expatriates, so there is a constant need to improve the living conditions of the population, which have become difficult in randomly distributed residential buildings along with the overlapping of residential with commercial and industrial buildings, and for the traffic congestion and parking spots in the middle of the streets, which has led to a complete chaos, due to non-compliance with the developed yet not certified plans [10]. Whoever has the possibility to build, does it randomly an arbitrarily without heading to the municipality's planning institutions to request a license.

Moreover, the abandonment of the old city center and its surroundings and the rapid moving out to other new areas has led to the emergence of a random, irregular and inconsistent structure without the bare minimum of services (Fig. 6). The presence of "Al-Aroub" camp in the north of the city and "Al-Fawwar" camp in its south adversely affect the performance and system of the city. The economic situation was already bad, and due to the emergence of the Coronavirus pandemic, it has become worse and the population has got poorer (Fig. 7).

Figure 6: Israeli settlements, new buildings in the heart of Hebron.

Figure 7: The bad economic situation in the city.

6 PLANNING UNDER THE PALESTINIAN NATIONAL AUTHORITY

The idea of summarizing the challenges and obstacles faced by the Palestinian National Authority with simple plain words is actually difficult due to the existing complex situation, especially that 27 years passed by and what was already agreed on between both Palestinian and Israeli sides is considered to be null, invalid and not even applicable specially from the Israeli side.

6.1 Political challenges

Related to administrative and security divisions and redeployment in various stages, causing geographical contiguity between the Palestinian areas [11].

6.2 Geographical challenges and obstacles

Considered what the Israeli occupation has done on the Palestinian lands, whether by the settlements that are still getting built and expanded, the bypass roads that connect these settlements, and the apartheid separation wall between the Palestinian areas in the West Bank and Jerusalem [9].

6.3 Regulations challenges

The Palestinian National Authority has made various attempts and efforts, especially the planning departments and its institutions, to regulate urban development and land use. The process of planning and organizing for all institutions in all Palestinian cities is still unable and incapable of solving the many problems that can be summarized as follows:

1. The lack of clear policies and laws in urban planning levels.
2. The lack of structural plans, laws and legislations. And even if it is found, it is not followed up and implemented as required.
3. Poor coordination and cooperation between institutions, lack of medical and administrative staff, and lack of experience.
4. Absence of land settlement and classifications, which leads to a lack of clarity about ownership. And weakness and lack of information, data, maps, etc.
5. Lack of funding to make such schemes, and the weakness of the private sector in terms of financing, and even that there is no popular and local participation.
6. Most of the Palestinian cities contain Palestinian refugee camps, as these camps are not subject to the laws and legislation enforcing the cities and villages of the West Bank. Which constitutes the phenomenon of random spread in the architecture [12].
7. In 2000, Israel reoccupied major cities and villages in the West Bank, which negatively affected municipalities and urban planning institutions.

7 THE IMPACTS OF DIVIDING HEBRON CITY

Before the beginning of 1997, Israel controlled all of the Hebron area. On January 17th, The Hebron Protocol was signed, concerning partial redeployment of Israeli military forces from the city. Under this agreement, Hebron was divided into two areas: H1 and H2. Control of H1 (80%) shifted to the Palestinian Authority and H2 (20%) remained under Israeli military control. The H2 area is inhabited by Palestinians and 500 settlers living in four downtown settlements inside the Old City. This area is under Israeli security control, with the exception of some civil powers of the Palestinian National Authority.

As Palestinians, we can be certain that the impact of this agreement on the city was disastrous in all areas, especially the economic aspect. It has negatively affected the urban planning and economic sector of the city, as well as the emotional impact of dividing the city into two parts, cutting off their social contact, in addition to the fact that there are no services for Palestinians as a result of the control of the Israeli army that protects the settlers (Fig. 8).

The main streets around the Area H2 were closed, such as: Al-Shuhada Street, which is the oldest street in the city and passes from its western side, and this closure actually divided the city into two parts (Fig. 9). Othman bin Affan Street, which is located on the eastern side, was also closed [9]. This street is considered as the main link between the north-east of the city and the south and is currently used only by settlers' vehicles between the Kiryat Arba settlement in the east of the city and its historical center, especially the Ibrahimi Mosque, as Palestinians are not allowed to use this street.

The results of that agreement were very bad and negative, as the residents of the Old City deserted their homes for settler attacks on them, poor economic conditions, lack of job opportunities and neglect of the area. This led to the concentration of criminality there, as a result of the lack of Palestinian authorities, in addition to the Israeli complicity with these people and even protect them, which made the area more isolated than it was [12].

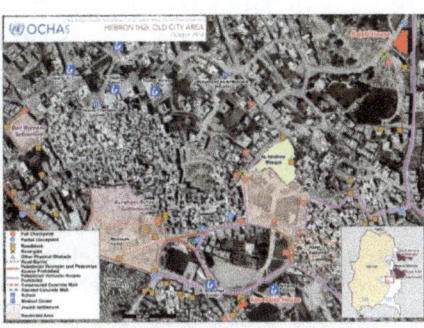

Figure 8:   Settlers and soldiers in Shuhada Street.

Figure 9:   Hebron H2 old city area.

Nowadays, the citizens are demanding the PNA to cancel that agreement especially after seeing its negative results and effects on the economic aspects, which made the Palestinians unable to meet the needs and requirements of families and to provide a living for their children.

8 PLANNING STRATEGIES FOR HEBRON GOVERNORATE

Following the analysis of the different areas in the Hebron Governorate, standards, concepts and theories must be developed to help understanding the use of lands for the governorate in general and for the city in particular. In order to improve the physical, dynamic, social and economic environment. And making these studies feasible with the correct use of the multiple spaces of land, whether it is internal, external, agricultural, industrial, commercial, or residential lands.

It is also necessary to work on creating and improving the infrastructure, improving the internal transportation network, avoiding the depletion of resources, and developing programs and strategies for the continuity of work on the concepts of sustainability, side by side with the competent authorities, the government and local people as well. Land use must be re-studied and its planning and systems must be re-programmed as well according to the new sites of the Palestinian lands [11].

9 FACTORS AFFECTING PLANNING IN HEBRON CITY

1. The presence of settlements and outposts around the city and in its center.
2. Weak city administration due to the Israeli occupation.
3. The lack of application of structural plans and wrong distribution of land use.
4. Weak and old laws and legislations.
5. The political division of the city according to the 1997 Hebron Agreement, known as the Paris agreement.
6. Weak human and material resources and lack of attracting talents.
7. Israeli closures and checkpoints in the city.
8. Narrowing the city by bypassing highways and limiting its natural expansion.
9. Lack of awareness and lack of educating citizens about the value of the historical and cultural heritage.
10. Poverty and bad economic situation.

10 THE PROMINENT PROBLEM IN THE CITY OF HEBRON: STREET VENDORS

Nowadays, due to unemployment and economic crisis the problem of street vendors in the city represents a cultural, local, economic and social disaster. The street vendor is a mobile or fixed point of sale by groups of toiling classes of the people who sell some products of all kinds to make money. Hand drawn vehicles are used in densely populated areas, while for distant neighborhoods they use a cart pulled by the wheel.

Usually, vital sites are chosen or at intersections in the city center, where there are at least 300 street vendors, most of them selling vegetables, fruits, clothes, juices, ice cream, etc.

More than 3 years ago, the Hebron municipality allocated a site for vendors: a covered market located in the city center, near Bab Al-Zawiya with public toilets, a guard room and opened a special street for it that it cut from the playground of the Muhammadiyah School despite the presence of two wide entrances and another on foot. Unfortunately, the market did not work as planned; Vendors have again spread out along the main streets near this market.

These sellers are illegal and so they do not have a license, apart from creating other problems which can be summarized as follows:

1. Obstructing traffic for vehicles and pedestrians;
2. Environmental problems as these vendors leave residues of waste and others;
3. The problem of noise and inconvenience promoting their goods;
4. Social problems caused by sellers' quarrels with each other;
5. Violating the sidewalks and sometimes the street itself, without caring about bystanders or drivers;
6. Indifference to the responsible authorities, instructions and legal texts, refusal to pay violations, and indifference to confiscation of goods and others; and
7. Different prices between the licensed shop and the non-tax-paying vendors.

10.1 Challenges and obstacles

Local authorities find it difficult to solve problems with rover owners due to the difficulty of the mentality of these people who feel to be above the law and have nothing to lose. Many of these salesmen are part of strong and influential family, which is an obstacle for the competent authorities .The large number of street vendors led the municipality to restrict them to one place or even find a suitable place to be accepted by the sellers.

Because of cheap prices, crowds of people arrive to buy, especially on Saturday, Sunday and Tuesday or holidays, where the demand for clothes, food commodities and children's toys increases.

Consequently, there are problems of traffic congestion, especially during the month of Ramadan in old Wadi al-Tuffah Street, the new Wadi al-Tuffah and the Bab al-Zawiya area as well as noise and suffering.

The Coronavirus pandemic came more than a year ago, and it has made life even more difficult than it was, so quick and economic solutions have become urgent because institutions in West Bank in general, especially municipalities, even though are aware of the poverty problem of these street vendors, they turn a blind eye most of the time [1].

10.2 Hebron municipality initiatives to solve the problem of these vendors

10.2.1 First: Arab Bank land and ESSO gas station

The Hebron municipality sought to solve the problem in the city, as it chose a plot of land at the gates of the northern old town near the Arab Bank and carried out a survey for street vendors and set up a program on the basis of creating a market that serves all sellers of various commodities and products, for example, vegetable carts come on Friday and Saturday, clothes carts take on Sunday, Monday, etc. (Figs 10 and 11).

Figure 10:   The fruit and vegetables   Figure 11: Peddlers in the street of Hebron.
             market.

The proposed land was planned and divided into a number of subdivisions, then counted and given the numbers of the specified locations, and a nominal fee was imposed for use in exchange for municipal services, Unfortunately, the project failed due to the refusal of the vehicle owners to move to the place for the following reasons:

1. After the proposed plot of land from the vital commercial center of the city.
2. Transferring the transportation line to feed the villages east and north of Hebron from the proposed location to the new car complex, which negatively affected the proposed area due to the lack of access to and abandonment of people to that area.
3. The car owners completely depend on their stalls to support their families daily, while the municipality wastes only two days of the work week.

The municipality, based on the vehicle owner's request, searched for a closer place that matches the arguments of the vehicle owners' demands, and found land near the city's commercial center.

10.2.2 Second: The land of Dais

The municipality re-classified the statistical classification of the owners of vehicles who relied only on it to support their families as a main source of income and found that the real percentage of vehicles was only 60% of the total existing vehicles, since some of them owned and rented more than one vehicle.

The municipality studied the project for the plot of land that was more capable than the owners of the vehicles, but the project stumbled and was not implemented, since the proposed land is private land and not government in a very vital location where no agreement was reached between the two parties (the owner and the municipality).

We find that the carriage problem has stumbled, a solution to this day due to the lack of available lands and the lack of compliance by the sellers with the law. Therefore, another area must be studied, even if it is far away, and institutions should be placed next to it that help in the activity of movement in it, as Jordan has done the popular market in the Ras Al-Ain area in central Amman near bus stations the main market is considered successful.

11 CONCLUSIONS AND RECOMMENDATIONS

Urban planning is the formation of a physical environment in line with the requirements, needs and behaviors of its users, therefore we find that the urban spaces that prevent the requirements of the users lead to negative behaviors that may be environmental, behavioral, economic, social or cultural. Therefore, this will affect by following other methods such as abandoning the area or making another arrangement or modification and addition to this urban area to match their requirements and behaviors, or individuals acquiring new behaviors to adapt to those spaces, which leads to deformation and loss of the physical environment.

The city's regulations, antiquity and modernity created imbalance and homogeneity in the urban space, whether in terms of form or function. Therefore, the research reached the following convictions and results:

1) The absence or non-adoption and approval of the structural plan in the city of Hebron, whose importance lies in controlling the quality and efficiency of the visual image of the urban landscape, controlling land use and building heights, and working on providing services, will lead to more negatives and complications that affect Urbanism, the environment, people's behavior, and more suffering and poverty.
2) The need to accelerate the review of the unapproved structural plan and put it to the specialists, satisfy the needs of the city and the citizens, and issue a plan that meets the population needs and the comprehensive development of the city.
3) The necessity of drafting new laws to work on the city in accordance with the law and urban and building requirements.
4) Resorting to international and human rights institutions to inform them of the racism that Israel does towards the Palestinian people and their land.
5) The necessity to allocate funds to make plans and to implement development projects, especially infrastructure and road networks, and to support local authorities and institutions and develop their cadres to enable them to work and implement these plans.
6) Confronting, by all means and capabilities, the Israelis who keep sabotaging the city of Hebron by means of settlements, bypass roads, and outposts, where the political situation plays a major role in the urban structure, its expansion and spread.

REFERENCES

[1] Chamber of Commerce and Industry of Hebron. 2017 (Arabic).
[2] Palestinian Central Bureau of Statistics. p. 29, 2010 (Arab).
[3] https://www.progettodreyfus.com/gli-accordi-di-oslo-il-testo-integrale/.
[4] Al-Kanani Camel, K.B., *Arab Islamic Town Planning*, p. 6, 2006.
[5] A sequence of Palestinian history since the First World War. https://www.bbc.com/arabic/40739743. Accessed on: 10 Apr. 2021.
[6] Applied Research Institute Jerusalem-Arij-pp.7-11-13, 2009.
[7] Isleimieh Mahmoud, A.M., Israeli colonies impact on physical expansion of Palestinian Settlement Hebron Governorate. Master thesis degree, unpublished test, pp. 69–74, 2006 (Arabic).

[8]   Palestine Ministry of Local Government. https://geomolg.ps/L5/index.html? viewer=A3.V1. Accessed on: 15 Apr. 2021. Hebron (H2) – Old City Area (October 2013).

[9]   Al-Tafkaji Khalil, *Israeli Settlements in the West Bank*, Palestinian Geographical Center, Jerusalem, Palestine, 1994 (Arabic).

[10] Abu Serrieh, A.H., Urban design. Unpublished test, Palestine Polytechnic University, p. 53, 2019.

[11] Hebron-Architecture-Urban Planning-ar.wikipedia.org. Accessed on: 10 Apr. 2021.

[12] Vv.Aa., *Hebron Ancient Charm of the City and Historical Architecture*, the first edition, Hebron Rehabilitation Committee: Palestine, 2008 (Arabic).

EVOLUTION OF THE TRADITIONAL TURKISH HOUSE

TAMARA KELLY
Abu Dhabi University, UAE

ABSTRACT

Cities or its urban fabric arose over many centuries but during the evolution period, architecture lacked its identity and deform premises erected. Architecture progresses according to many factors and among those influences are intangibles as political drive, inspirations from other civilizations, religious guidance. Whereas physical sways such as climate or availability of materials and skilled labours are behind other great architecture formation. Despite all those influences cities nowadays are similar and buildings are identical regardless of their location on the map. Turkish houses changed during various eras to meet the needs of different generations yet maintained robust characteristics and responsiveness. Hayat house is the traditional residential premise and a significant element in Turkish architecture, it is the core of this article. This paper is an attempt to help students and researchers to learn lessons from our ancestors who managed to advance the components of their house and yet created iconic innovative architecture. Furthermore the paper will examine the impact of diverse forces in shaping the distinct character of Turkish houses. The methodology of this paper is based on a literature review of various references and a theoretical analysis of several case studies in Turkey.

Keywords: Hayat house, Sofa house, Ottoman architecture, vernacular architecture.

1 INTRODUCTION

Houses in Anatolia diverse in materials and layout according to various entities including their location in the mainland, climate, and cultural influences. For instance, a flat roof is a choice for hot regions in Eastern and south Anatolia, sloped roof optimal in northern, western Anatolia and Balkans or black sea plateau. In southern Anatolia, main building materials are stone or mud-brick walls reinforced by wooden beam over a stone foundation. Whereas in northern plateau stone are dominant and houses vary in sizes from small to great palaces. Even with the differences in materials, Ottoman houses share a similar core layout and aspects in many areas. Anatolian Turkmans erected the basic element of a Turkish house in eastern Anatolia, the house layout suits the agricultural environments of Turkmans but in later stages, the house components developed to meet other cultural and religious considerations. Despite the similarity in houses layouts and components between Turkish houses and neighbouring countries, the ottomans created a distinct typology of houses on no occasion that exists anywhere. For instance, Hayat house appears related to many houses in Mesopotamia and Syria but innovated into a unique ottoman style. The following section discusses the fundamental features of Turkish houses and illustrates the advancement of its elements to reach various configurations. Furthermore, many case studies are examined in central and southern districts showing to what extent the character of Hayat house was altered to meet diverse considerations and requirements.

2 HAYAT OR SOFA HOUSE

2.1 First phase, lack of privacy

Hayat house represents the core and foundation of various Turkish houses styles, the word Hayat is Turkish means Life, and Sofa is a word refers to a common central area veranda or an open gallery. The Hayat house establishes by Turcoman and most Turkmen were nomads migrated from Central Asia or Western China. Turkmen settled in the Anatolian basin in around the 11th century and turned into Islam they lived on harvest, breeding and weaves. Hayat house main components are the Ewan (Evan) leading into two-room on its sides and open gallery fronting the Ewan but in later stages, the gallery developed into several modes (Fig. 1). The ground floor raised partially to create a half basement that can be used to suit the rural life such as stable, warehouse or service area. Halil Agha house in Mudanya (Bursa) 1640 reflects Hayat house with a wooden colonnade on the two-tier galleries and the front facade comprises several openings while the side are blinds without any views. The notion of moving between spaces in Hayat house through the exposed or open area is normal for Turcoman since they originally used to live in tenets overlooking their belonging. This approach still exists in houses located in Balkans and many Anatolian regions. As the nomads settled and no longer migrates between lands they applied their custom into new houses where the tents turned to rooms and as a result, the Hayat house initiated. The outline of this house was developed by Turcoman and it is appropriate for their way of living by farming and cattle herders. The open galleries of the house provide a great connection with the surrounding site and overlooking or engagement with farming activities. And this is crucial for ranchers, therefore moving through open spaces between the house components is reasonable for the Turk tribe the occupants of Hayat house.

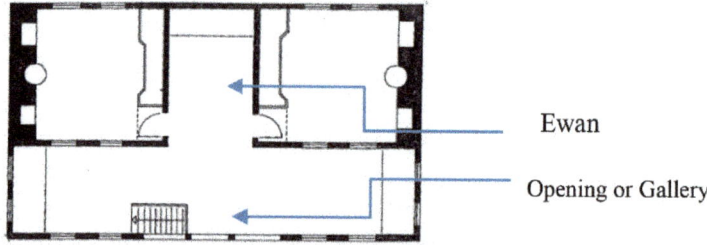

Figure 1: Halil Agah house in Bursa, typical Hayat layout.

But the layout of Hayat house resulted in lacking privacy of women consequently, Hayat house in Islamic norms is unfitting with the privacy concern and the open gallery was undesirable.

2.2 Stage two, multiple occupations and segregation between male and female sections

In Islam, the house components must conserve the privacy of women and it is crucial to provide an introverted lifestyle away from the overlooking of outside world, this led to the transformation of the open gallery into an inner courtyard garden creating a semi-closed or semi-open arrangements. Furthermore, new notion elaborated and segregated sections created or allocated for a male and female, this was common in palaces and the houses of the wealthier owner. Likewise, the house of the middle-class family consists of spaces with multi occupations during the day and night. Karel in his book imagine the Turkish house state that the finest room in the house in terms of best lighting, finest motif or furniture and location is dedicated to male entertainment with guests during the day while in the night the room dedicated into the family.

The idea of two separated sections dedicated to males and females commonly found during the 18th and 19th centuries as Muslim, Turkish believe that houses are for family and belong to women in particular since men normally out during the day time. Women in the agricultural community engaged heavily with outdoor activities whereas in urban environments spending most of the time at home hence privacy is a key factor in urban residential premises. Consequently, many measures articulate this notion including high walls, concealed opening, lattice or stain glazed windows, introverted plan, an ultimate separation between male and female sections. The guest room and other spaces on the first floor are projected over the street allowing the occupants to overlook any events outside without being seen. Murat house in Bursa is a sample of minor changes to Hayat house to address privacy, therefore all rooms are open towards the gallery *Hayat*, and side walls extend to the roof reflects the continuity of blind opening on side blank walls a feature of 18th and 19th centuries (Fig. 2).

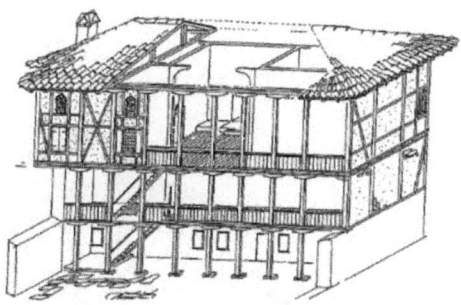

Figure 2: Murat house, blind side without opening.

2.3 U shape Hayat house and the crucified layout

The shape of Hayat house progressed further over the years and this can be witnessed in Beyoglu house situated in Manisa (Kula village). The house is a typical model of Hayat all the rooms open into the gallery (Fig. 3). The first floor projected over the lower floor and two rooms or only the gallery extend further with wooden colonnades.

The additive expanded areas were a great new element creating three-dimensional effect and characterized Hayat house. A similar approach was in Cakiraga konab house in Izmir (Birgi village), the house was built at the end of 18th century but substantially rooted in the antiquity reflecting cultural relations with early nomadic. The plan of Cakiraga Konab house is U Shape extended towards the court but yet influenced by the original shape of Hayat layout which consists of two-room flanked around the Evan (Fig. 4). In Cakiraga house two Evan inserted at the end of the gallery and leading into new featured additive rooms forming *U outline* and facing the garden. The house is extending further inside the court by projected arear erected from the gallery. This outspread space creates newly shaded areas overlooking the garden for socializing, family gathering and provided interaction platform with the external but private scenery. These new components including: projected zone, multiple Evans with the rooms attached into the gallery regarded as a great feature of Turkish houses.

The open Hayat house was transformed in Cakirage case study (Fig. 5) into a semi-open or semi-closed house and this notion will be evolved in later centuries. "*The Cakirage Konab in Birgir represents a stage in the development of Hayat ev at which the rural character of the house remained unimpaired and Hayat displayed its most highly developed characteristics*" (*Ottoman Architecture*, p. 478). Nonetheless, soon and in 18th century, the great influence of European culture is commenced on Turkish housed, this resulted in the transformation Hayat house plan from leaner synchronizing into crucifix shape.

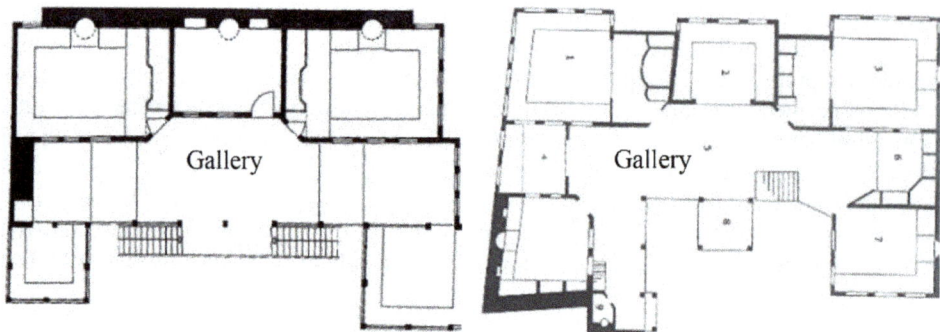

Figure 3: Beyoglu house, u shape.   Figure 4: Cakiraga Konab.

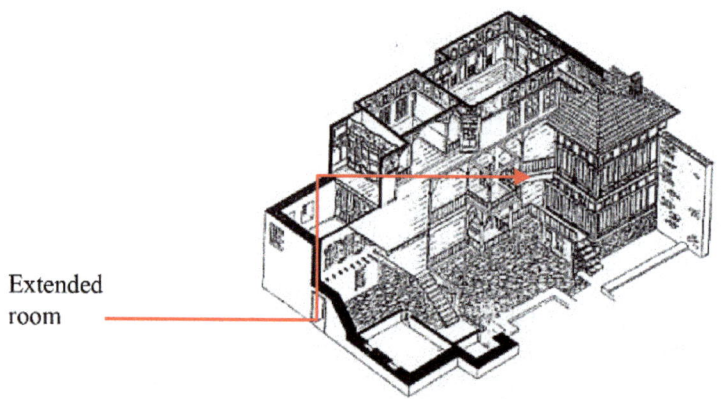

Figure 5: U shape Hayat house, semi open.

This plan based on centralizations or twin access is common in dense urban fabric where the sizes of garden or courtyard minimized or even diminished. According to Dougan, the influence of western architecture was evident in Sipahi Omar house in Izmir (Fig. 6). He believed that the house looks like a church with the crucified opening middle part and four-cornered rooms with great symmetry and axial arrangement *"A second reason was the European influences on the house design. Features such as an axial arrangements, symmetry and central halls plan"* (*Ottoman Architecture*, p. 483).

In fact the house is a transformation of Hayat house but as mirrored version around horizontal axes without the opening gallery, which is usually exposed into the surrounding field, however, the four Evans are altered and converted into enclosed spaces. The cornered rooms are projected further and the four Evans appears as crucified spaces inside a square.

In densely populated cities in Central Anatolia or Balkan regions and Istanbul, the Early mentioned centralized house plan was altered and converted into a central sofa extend along the entire length of the house with clustered rooms on each side of the sofa. Furthermore enclosed shared areas are located at the end of the rectangle sofa but with windows allowing connection with the external environment. Although In Istanbul and after world war two massive destructions effected the city centre and vast areas of Istanbul were damaged but this

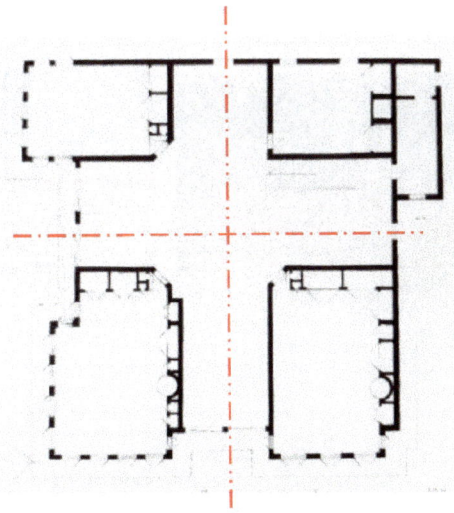

Figure 6: Sipahi house, crucified plan.

house according to many Turkish authors is widely spread in Istanbul and early stated districts. At the end of 19th century new luxurious waterfront houses emerged in Istanbul, it consists of Hayat house layout but clustered around multiple horizontal or vertical axis. Koceoglu palace built-in 18th represents such an approach where the components of Hayat house mirrored around vertical Axis furthermore new elements were added such as the columns in the middle of the main sofa, segregation between male and female is obvious and minor corridors connect these two configurations.

In Koceoglu palace, the distant lower cluster is the female section consisting of multiple private or socializing rooms with the baths (hammam). And follows a centralized arrangement whereas the top units are allocated for male nearer to the entrance welcoming the male visitor (Fig. 7). The plan of Koceoglu palace developed further in Yasinci palace erected at the end of the 18th century.

In Yasinci palace, the edges of the staircases on the sides and main wide Sofa corridors were softened and curved. The palace consist of female section (no male is allowed) and the male section however the arrangement of the sections are mirrored along vertical axis (Fig. 8). The female harmmlek part is clearly distinguished in places from the male salamlik section but in the lower rank of society, this segregation is less obvious nevertheless, the separation approach still exist in the finest living room but the space is alternated between the two genders during the day and night.

Saffet Pasha palace built in Istanbul during the 19th and reflected the notion of seclusion of male and female but the two parts are opposite with the middle service area on the ground floor (Fig. 9). The place follows a cruciform plan, the central outline typology of the palace is a development of Hayat house but paralleled on horizontal Axis. The residence is raised and the entry is through many steps showing the influence of European architecture *"in this typology, the basic nucleus of the Hayat ev with central sofa is placed on two sides of the axis and thus converted into a cruciform plan central space, sometimes elliptical, or very rarely circular form"* (*Ottoman Architecture*, p. 486). The layouts of earlier case studies, mostly used in larger residential premises of the wealthier sector of society and merchants who commence commercial trading from home, hence privacy was key factor.

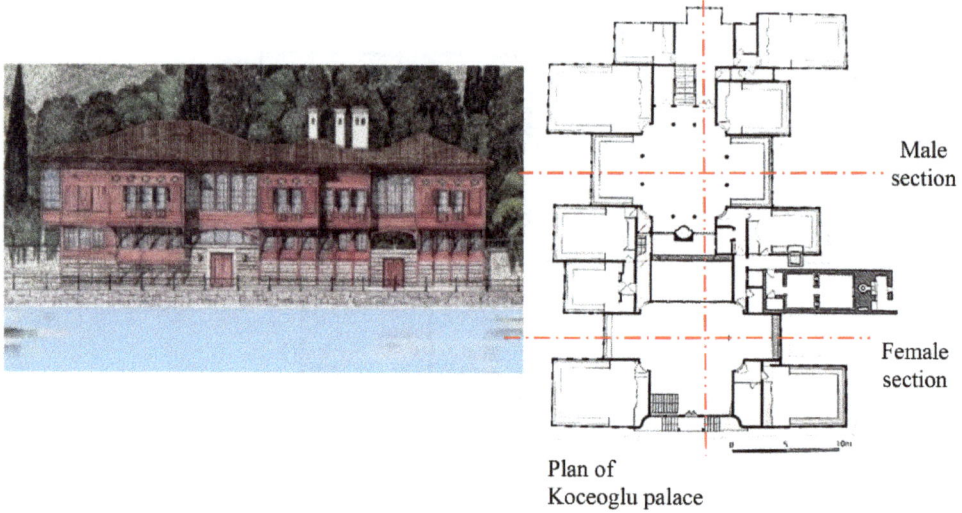

Male
section

Female
section

Plan of
Koceoglu palace

Figure 7:   Left is photo Koceoglu palace showing waterside façade reflecting diverse finishing materials.

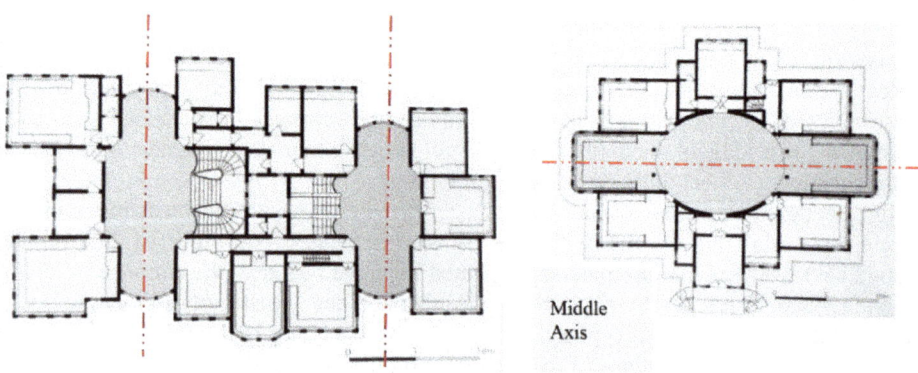

Middle
Axis

Figure 8: Yasinci palace.

Figure 9: Male opposite female sections, middle sofa Saffet pasha palace.

The Examination of earlier case studies bring to light that diverse arrangements evolved from the original units of Hayat house, the basic components were altered, rearranged or enclosed reflecting religious, social and cultural influences. The numerous units might be dissimilar, inversed and partially reused or arranged around horizontal or vertical Axis Figs 10 and 11. The sophisticated mixture of units in Hayat house plan is reflected in the structural system, and this was evident in the elevations of Koceoglu palace, or Saffet palaces, as a stone is the foundation materials, the roof is sloped and walls filled by mudbrick between timber frames.

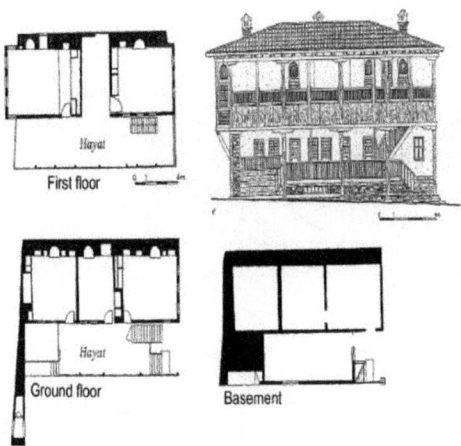

Figure 10: On the left, typical hayat house proposals.

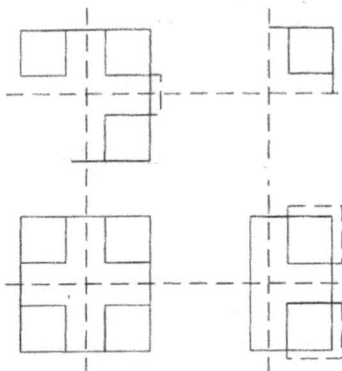

Figure 11: On the right, samples of Hayat house diverse arrangements evolved from the original units of Hayat house, reflecting the influence of many aspects.

3 THE IMPACT OF NEIGHBOURING SETTING OVER THE HOUSES IN SOUTHERN REGIONS OF TURKEY, DIYARBAKIR CITY

Turkey surrounded by civilized nations in particular southern provinces, Mesopotamia and Syrian culture cast its shadow over the Turkish architecture. The influence is evident over the house in bordered Turkish regions, for instance, houses in Diyarbakir adopted courtyard layout which is a feature of Syrian traditional houses. Moreover stone and timber are the main materials in Syria and likewise in Diyarbakir and Mardin city. Turkish author Dogan in his publication Turkish Hayat house claimed that Turkish Hayat House is a unique invention of Turcoman but in reality, its components exited before in various neighbouring regions and in particular Mesopotamians. For instance, the Mesopotamian open porticos *Tarma house* consist of the main features of Hayat house such as; narrower opened colonnade and single Evan with flanked rooms (Fig. 12). In fact, Mesopotamian culture is former than the Turkoman epoch hence the Hayat houses are interrelated with nearby Ancient civilizations and its components inspired by Mesopotamian houses.

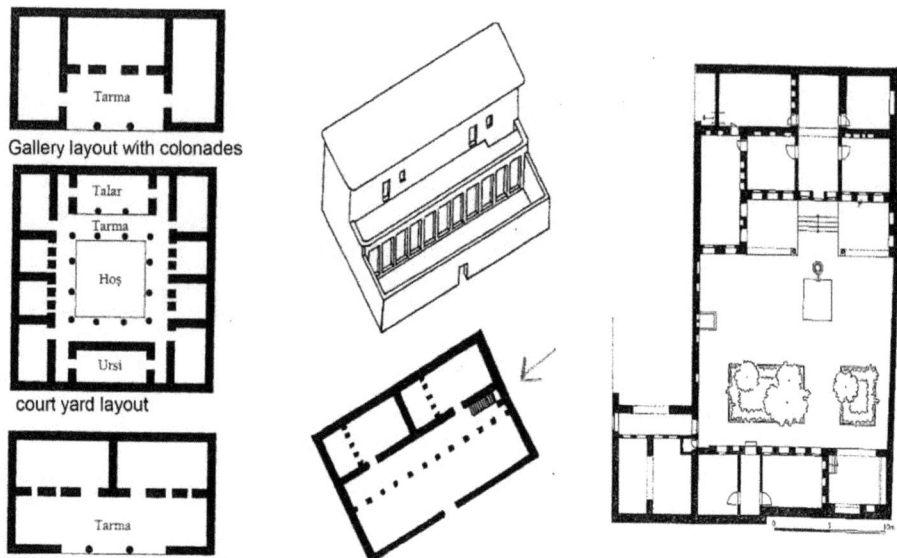

Figure 12: Mesopotamian houses.

Figure 13: Syrian or Lebanon house.

Figure 14: Ocakhouse in Diyarbakir.

Moreover, houses in north Lebanon and Syria accommodates similar arrangement but lacked Ewan (Fig. 13).

However, Dogan underlined minor transformations in Hayat and in particular in the first floor, the gallery is much wider and used as living space instead of a connector or passageway and in later phases, other elements or extensions were added at the ends of the gallery as family gathering areas. In his publication the Ottoman architecture, Dogan believes that the garden courtyard commonly used in Iraq and Syria is replaced by the open gallery were daily activates occur and the portico overlooking the garden. Turkish rural settlements commonly located on sloped terrain and courtyard assemble is challenging hence Hayat house is suitable in villages and agriculture town. Whereas the courtyard and flat roof are common in urban architecture and in hot climate regions located south Turkey.

Diyarbakir is a substantial southern city situated on the banks of the Tigris River closer to Syria and Iraq, the layout of Diyarbakir's house influenced by the neighbouring culture of early mentioned countries. But the Turk added great aspect to the houses, monumentally is key element of the houses in Diyarbakir region reflecting the political and commercial role of the city as informal capital of Kurdish citizens. Houses of this city are prior to Hayat layout but contributed to the initiation of Haya house, however, the layout is still based on the basic design of Hayat plan with a slight alteration of Ewan location. The hot climate resulted in narrow deepening Ewan with a colonnade portico and consequently, the courtyard accommodates fountain to enhance air quality. Privacy and climate are crucial elements in shaping the identity of the house in this city hence windows in all the floors are only inwards looking and the court is enclosed by high walls. Two types of Evan emerge in some houses for winter and summer seasons furthermore a semi-underground basement exit and used during the hot season. The high ceiling is a feature of houses in Diyarbakir and underscored the monumental appearance contradicting Hayat house. Fig. 14 shows Ekrem house in Diyarbakir it is a u-shape Hayat house overlooking the garden with other components located beneath the courtyard. Entry to the house is through the opened courtyard and similarly moving between the spaces is within the open-air a feature of Hayat house which is lingered in Diyarbakir residential premises.

3.1 Mardin and Urfa houses

Mardin city is located in southern Anatolia and close to the Syrian border, the city is named Cradle of Cultures since it was the centre and a crossroads between different civilizations. Hence the influence of Syrian and Mesopotamian cultures is evident through the choice of materials and house layout. Stone and courtyard are the main features of Syrian houses and likewise, the houses in Mardine adopt inward courtyard arrangement and constructed of stone.

But Houses in Mardin are massive, monumental be like castles and characterized by levelled terraces. The house consists same components of Hayat and Diyarbakir's house inducing covered Evans in the basement and other floors, with numerous terraces overlooking the featured garden (Fig. 15). Mardin houses situated on steep land hence the house is layered following the distinct feature of the terrestrial. The city skyline shows the characteristic houses of Mardin comprising of tiers of courtyards rising above various layers of terraces in great response to the inclined site which resulted in multiple entrances to the house through different floors.

The house of Mardin appears as the foundation of Hayat house with marginal differences for instant the basement in Mardin houses may comprise featured Evan overlooking the water element.

 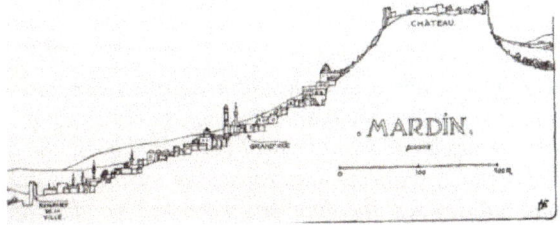

Figure 15:   On the left view of Mardin city, houses are massive and monumental. The right is the sections, across urban fabric of Mardin City.

The gallery narrower or replaced by wider terraces Furthermore the court is new elements and all the opening are towards the court (Fig. 16).

3.2 Urfa

Urfa is a city of ancient traditions, and great commercial centre during various eras, many civilizations inhabited the city including Roman, Byzantine and Islam moreover, the city is located on the trading paths with southern nations. The houses in this region are distinct and the architectural details were crucial factors in shaping the character of residential premises however courtyard and Evans remain the primary elements of the houses *"The Urfa houses*

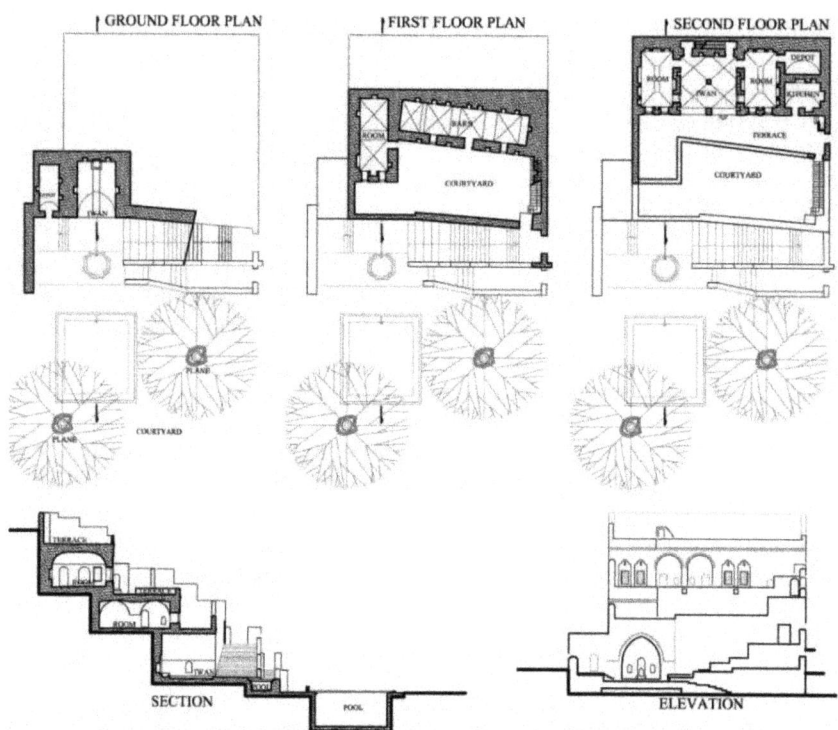

Figure 16:   At the top plans of Bahriltor premise. The bottom is the sections showing the distribution of the floors over multiple layers.

display the richest examples of decoration in south-east Anatolia. The heavy consoles with multiple profiles, the cornices composed of rows of small arches, the ornamental tracery in the unglazed head windows of various shapes, the aviaries over the windows, the gezenek balustrades, the windows arches, the medallions adoring the blank walls, the iron railings and balustrades, the door knockers, the woodcarving on the doors in the interior and the wooden shutters are all components of a rich ornamental vocabulary." (*Ottoman Architecture*, p. 492). The ground floor is above half basement stables with a featured staircase leading into the upper floor and dominating the courtyard. Summer and Winter Evans are additional elements and located in ground level with kitchen open into the courtyard where cooking and food preparation activities take place. In fact, the residue of Hayat house is obvious through Evans and first floor galleries although some house lacks the porticos, yet courtyard is the main connector between the spaces and movements through the open-air likewise Hayat house. The external envelope is concealed and the houses are inwards looking since the climate is hot and dry. Many activities take place in the flat roof during the day including drying the clothes, fruits or herbs whereas socializing or sleeping arises in the evening. Alike the court or Evans serve many uses and living space in summer or winter, but predominantly are employed for food preparation or cooking, seating, playing areas for kids, sleeping, family gathering and wedding or celebrating platform, and circulation between spaces, vegetation, washing and drying clothes, a place for animal and farming, circulation zone, a welcoming entrance and a place for carpet cleaning or weaving (Fig. 17). Hence the size of the courtyard is bigger accommodating family needs, social and economic necessities. In wealthy residential premises in Urfa two segregated sections for males and females are common and a barrier as a wall may exist between the haremlik and Salamalik.

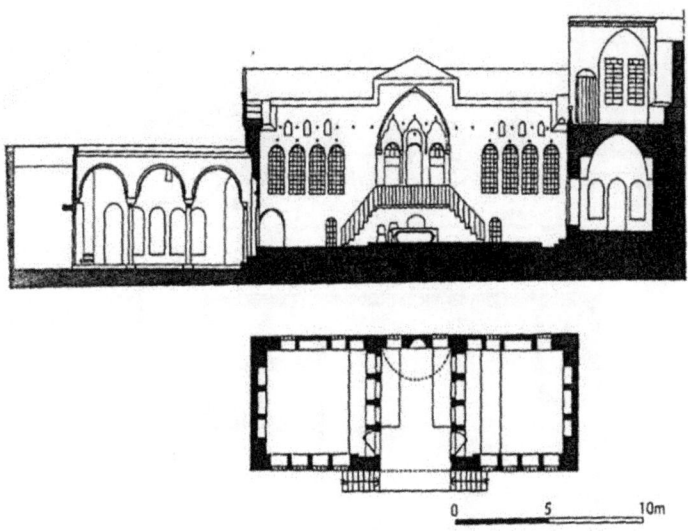

Figure 17: Izgordu house in Urfa.

3.3 Harran houses

Evan House is dominated in southeast Anatolia and existed in urban and rural regions but in Harran town, a distinct domed house found consisting of a complex room with various uses. Harran is a plane town located Southeast Urfa, between the two rivers of Euphrates and Tigris. In ancient languages Assyrian and Hittite, the name Harran meant the land between different routes. Harran was established as a merchant centre since its location along the trading path and between the Mediterranean basin and the plains of Tigris River. Harran erected since the third millennium BCE when the area was the cultural and religious centre of settled societies. In the last five hundred years, the city was a temporary settlement of nomad cultures. Therefore and according to many Turkish authors, the existing houses were rebuilt several times during the last centuries. The rudiments of the house are; the courtyard, domed room, and the flexible layout allowing more rooms to be added if needed (Fig. 18). The occupants of the houses are part of the same tribe and live together hence Harran houses are adjacent or nearby resulting in an organic path like the forms of the houses. Inhabitants farm the land and live on stock-breeding, therefore, they live close to the production sources. Consequently, the house consists of many spaces for their animals and storing crops. Each room covered by a square dome and may lead to the adjunct room or the court. Some of the rooms are opened to the courtyard with a small door and in winter, the kitchen is also used as a living room. The hole at the peak of the dome for lighting and as a chimney, while the shape allows for excellent air circulation. The earnings of the building owners, family size and needs govern the number of rooms. As the family members increased, new rooms are added to existing ones or to the court. These domed roofs with opening and according to many historians represent a notion or a practice exited before thousands of years in Northern

Mesopotamia. Many authors recommend studying the foundation of Harran houses through the Mesopotamian civilizations where excavations showed similar square dome roofing exited during 7th century BC. The influence of Mesopotamian architecture is evident in Harran houses through the domed roof, courtyard, and choice of mud as main construction materials.

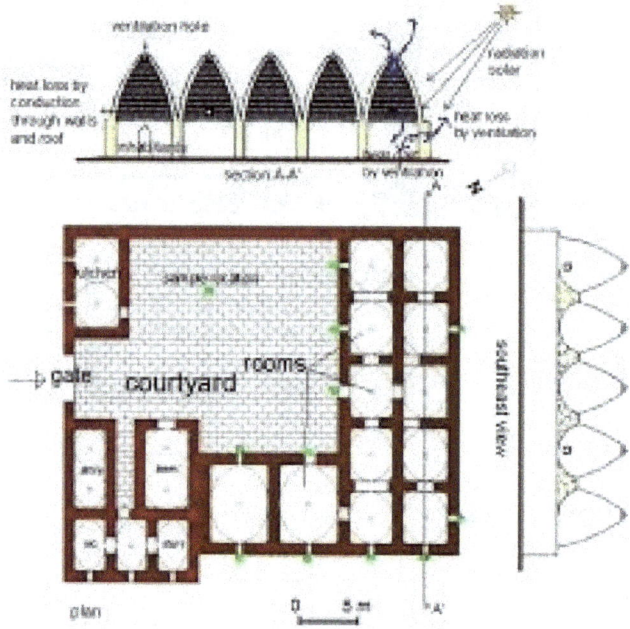

Figure 18: Haran houses.

4 CONCLUSION

Houses are key factors of Turkish culture and as important as monumental architecture. Residential premises, in particular, were rooted in agricultural societies and Hayat house erected emphasizing the great connection with surrounding environments. The culture of the Turkmans and according to many Turkish authors can be seen through the great connections with surrounding environments. And the open gallery overlooking the farms in Hayat house represents this conception. In fact the influence of tribe dominating internal details, for instance, the roof of the guest room in many palaces and residential premises accommodates small size dome extended in the middle emitting the tent shape. Moreover, the seating layout is over projection similar to the way adopted by the clan and resulted in creating a divan room known as a male gust room or large assembly room. The head of the tribe normally meets up with the community in his large size tents where law cases heard and the general topic discussed regarding needs and future of the kinfolks. The equivalent approach followed in wealthier residential premises or palaces and Divan space created as a reception room where the Sultan meet with administrators and visitors. The open connecter or gallery with middle Ewan and two rooms on the opposite side underscoring great connection with various activities occurred in the surrounding environment. Furthermore, the open Hayat with Evans regarded as athletic features of Turkish houses and a symbol of historical and cultural constituents. During the Islamic reign, the need for privacy surfaces and overshadowed the layout of Hayat house. Privacy considered by various means and the notion development further resulted in deep segregation between male and female sections and in particular in wealthier palaces. In rural areas summer and winter Evan was enclosed and new elements arose including the courtyard with garden or water features. In the 18th and 19th centuries, the sophisticated composition of Hayat cells developed in Istanbul palaces attunes with Islamic considerations yet following western culture. The character of Hayat house reformed under the new social and cultural needs but its components are still the core of various configurations. Nevertheless, and over centuries, Turkish learnt from ancient civilization and managed to address variable needs and advance the elements or notion of Hayat house creating a distinct Turkish style houses.

REFERENCES

[1]   Kuban, D., *The Turkish Hayat House*, Eren, 1995.
[2]   Bertram, C., *Imagining the Turkish House: Collective Visions of Home*, University of Texas Press, 2013.
[3]   Kuban, D., *Ottoman Architecture*, Antique Collectors Club Ltd. Publisher, 2010
[4]   Aabidin, M.Z., *The Architecture of Ottoman Mosques*, Dar Qabess House Publisher, 2005.
[5]   Fazio, M., Moffett, M. & Lawrence Wodehouse, *A World History of Architecture*, Laurence King Publisher, 2008.
[6]   Nehru, J., *Glimpses of World History*, Asian Publishing House, 1934.
[7]   Freely, J., *A History of Ottoman Architecture*, WIT Press Publisher, 2011.
[8]   Necipoglu, G., *The Age of Sinan: Architectural Culture in the Ottoman Empire*, Reaction Book Ltd., 2007.
[9]   Kuran, A., *The Mosque in Early Ottoman Architecture*, University of Chicago Press, 1968.
[10]   Goodwin, G., *A History of Ottoman Architecture*, Thames & Hudson Publisher, 2003.

WOODWORKING TRADE AT THE TURN OF THE 19TH AND 20TH CENTURIES AND ITS PROTECTION

KLARA KROFTOVA[1] & LUKAS HEJNY[2]
[1]Faculty of Civil Engineering, Czech Technical University in Prague, Czech Republic
[2]National Technical Museum, Czech Republic

ABSTRACT
Part of modern monument care is not only the protection of the buildings themselves, but also their components and details created by specific craftsmanship. We can see the importance and influence of traditional crafts both on historical monuments themselves and in the collections of museum institutions. Knowledge and application of these traditional ways of processing building details will make it possible to achieve quality restoration results and also to preserve traditional crafts and sets of their individual skills and habits, usually passed down from generation to generation. The craft tradition is not only hidden in the manual way of processing, but also in the overall approach to the craft and its products. Among traditional historical crafts, woodworking trade occupies an important position, whose products – fillings of openings, shutters, wall and wainscoting, floors and others – are closely connected with the construction of the house and significantly affect both its overall architectural expression and its useful properties, e.g. the well-being of the indoor environment. The works of woodworking trade belong to the building components that wore out quickly and were subject to functional and stylistic change. These factors are the main reason why the number of authentic preserved elements is rapidly declining. In many cases, it is precisely these components of buildings that are being replaced by available industrially produced modern elements, which, however, damage the expression and monumental value of historic buildings. When restoring historic buildings, emphasis should be placed on preserving the historic original and restoring it by traditional historical methods so as to preserve the *genius loci* of the historic building (and the whole environment) and the craft tradition.
Keywords: woodworking trade, 19th century, cultural heritage, protection.

1 INTRODUCTION
Woodworking trade is one of the basic crafts that has always been and is an integral part of construction production. Fine woodworking joinery products form part of the construction and include not only doors and door frames, windows and shutters, shop fronts, wainscoting, but also stairs, balustrades, elevator or telephone cabins and more. Joiners (Latin *cistator*, resp. formerly *mensifices, mensatores*) have been associated since the Middle Ages in professional guilds ("brotherhoods of craftsmen"), which, among other things, recorded the numbers of joiners and carpenters, supervised and controlled the work performed and recorded the list of products. For example in 1419 there is only one carpenter registered in the city of Prague (Bohemia), 100 years later, in 1526, 11 carpenters and joiners are mentioned [1]. Within this profession it is then possible to record a number of narrow craftsmen specializations, such as table makers, bench makers, board makers and others (Fig. 1).

The craft of a fine woodworking joiner was very often supplemented with the craft of a carpenter who prepared a rough basic construction (e.g. carpenter's door frame, carpenter's floor) and the construction carpenter supplemented them with a final layer (i.e. carpentry lining, floor friezes). In some cases, these two close trades complemented each other, in others it depended on the purpose and requirements (operational, aesthetic, etc.) for the design of the element or structure – for example, stairs that do not require fine joiner's work were performed by a carpenter and vice versa.

WIT Transactions on The Built Environment, Vol 203, © 2021 WIT Press
www.witpress.com, ISSN 1743-3509 (on-line)
doi:10.2495/STR210071

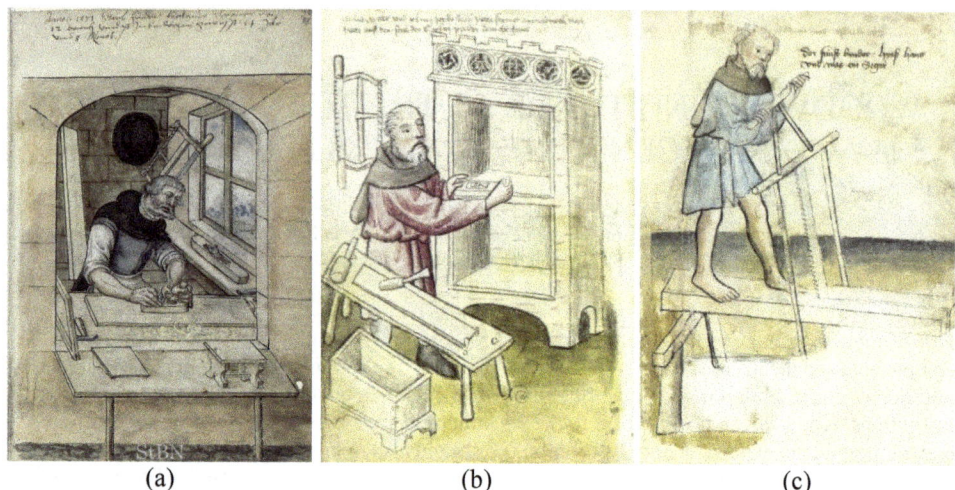

(a)   (b)   (c)

Figure 1:   Carpenters brothers, resp. joiners and carpenters in historical illustrations in the Nuremberg Book of Crafts by Mendelschen and Landauer's Twelve Brothers. The illustrations show the practice of the craft, with characteristic production processes, typical tools, workshop technique, materials and products. (a) Brother Fridrich Punkauer, 1571 [2]; (b) Brother Petr, 1444 [3]; and (c) Brother Hanspilar, 1425 [4].

The basic source of information about the craft of works of woodworking trade are the works themselves. Historical iconography and period literature also play an important role, which is mainly represented by construction manuals and textbooks for university students.

2 WOODWORKING TRADE

The main raw material for the production of the range of construction joinery was wood, but since the second half of the 19th century in construction we have encountered a large number of new materials and structures, the introduction of which was given primarily by economic advantage. For example in the field of construction of urban tenement houses, the new innovations only manifested themselves when they became more financially advantageous than the existing solution, or when it was not possible to achieve the desired goal in any other way. The ever-tightening fire and building regulations were also a significant impulse (comprehensive fire protection regulations date back to the 17th century, the first major positive legal acts are the laws of 1873 and 1876, which issue the Fire Police Order and the Fire Order for the Kingdom of Bohemia, Moravia and Silesia). The industrialization of production and mentioned innovations have led to a reduction in the share of joinery products in construction and, as a result, to simplifications and a certain decline in the joinery trade in the 20th century (Fig. 2).These changes were also closely related to the development of architecture and means of expression.

The choice of a tree species was primarily influenced by the requirements for the properties of wooden structures and individual elements. A secondary role was played by the composition of forest stands at a specific moment, and thus also by the availability of the necessary dimensions and the amount of selected material. "*For joiners and carpenters, wood is always sought after, mature, dense, straight, healthy and, as far as possible, without*

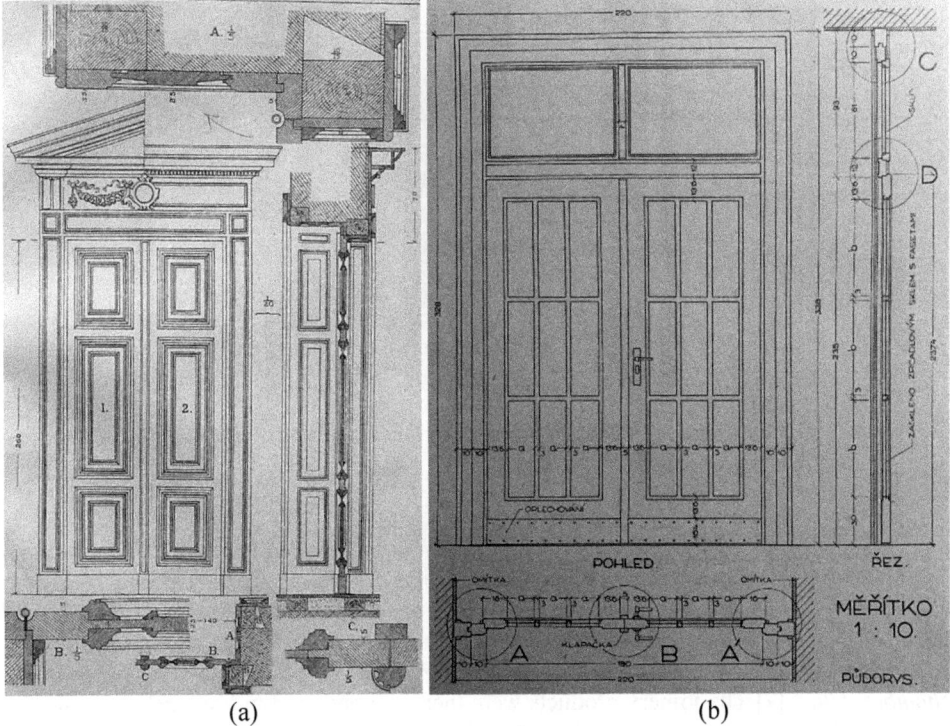

(a)   (b)

Figure 2:   The shift from ornate design to simplicity on the design of a double leaf entrance
              door. (a) Design of a door with a rich profile for a city apartment building in the
              construction manual from the end of the 19th century [5]; and (b) Simplified
              design of a door to an apartment building from a textbook for industrial schools
              from the 1940s [6].

knots…" [7]. Among the types of wood used in construction joinery was mainly mature core
wood of pine, spruce, larch and oak, which shows very good mechanical properties and
higher durability of structures made of it (Fig. 3). Pine, larch and oak have a more durable
and colorful core wood compared to spruce wood, whose wood is light, soft and easy to work
with. Pine wood is heavier than spruce wood, shows smaller dimensional changes and, thanks
to its higher resin content, is characterized by a longer durability. Larch wood, like pine,
contains more resin, is well machined, but tends to twist when not sufficiently dried. Due to
its properties, heavy, strong, flexible and biotic wood-destroying agents, oak wood is suitable
for use in constructions with an increased demand for their mechanical properties and
especially durability in an environment with increased humidity [8].

Vitruvius already states in his Ten Books on Architecture [9] that wood should be felled
in the winter, when it has the least sap. The rules for the drying process then applied to the
individual woody plants: e.g. *"The deciduous trees are either whole or helically cut from the
bark to make the wood dry faster and the membrane to harden. the moisture did not evaporate
quickly, otherwise the trunk would rupture."* [7].

In order to avoid dimensional changes in already finished products, it was necessary for
the wood for joinery structures to be dried to 8–12% moisture content before its further

Figure 3:   Types of wood used in parts of traditional building joinery. From left spruce, pine, larch, oak [8].

processing. *"The wood to be used for joinery work must be dried properly beforehand; air dried wood – i.e. slowly – always has better properties than wood dried and roasted in chambers. However, it needs to air dry: 3–4 years of soft wood, up to 5 years of hard wood, 8–10 years of thicker boards from the time of felling; in addition, it is advisable and highly recommended to dry the wood before processing on a bunk bed or worse, or even in a chamber, kiln."* [7] The joinery products were therefore prepared in a workshop from dried planed lumber and were subsequently installed in their place on the construction site.

The prepared parts made of dried and processed wood were joined by structural joints so as to form a solid unit. Joints, which varied according to the type of joint, the type of construction and others, were made as dry or secured with bone glue. If necessary, the connection was reinforced with pins, wedges, tapes, screws or nails. One of the basic joints used was a ledge joint, which, for example, connected the boards of the door leaf. Among others, it is possible to mention mainly the mortise and tenon joint, open-slot mortise joint, bridle joint and dovetail (Fig. 4). Compared to the carpenter, the joiner used a relatively limited number of joints, which he adjusted according to the specific purpose of use.

The demands on the necessary space of the manual joiner's workshop were small, its center was the so-called workbench, on which the entire production process took place. The basic equipment also included the so-called shaving horse work bench – a bench for securing the product when working with a timber shave and an adjustable woodworking bar clamp for supporting the processed element. Each joiner's workshop then contained various types of tools – saws, planes, chisels, gimlets, drill bits, joiner's clamps and more (Fig. 5). With the overall industrialization of the company during the 19th century and with the advent of machining, a number of changes entered the joiner's workshop. In manual joinery production, the need for space necessary for the placement of machines and handling space around them was eliminated. The integration of the machines into production was reflected in the change of the spatial scheme of the workshop and the requirements for the floor area increased to ensure its efficient operation. During the 19th century, woodworking machines for cutting material gradually appeared in carpentry shops, mainly band saws (banding machines), spindle moulder (leveling) and thicknesser machines (planers and broaching machines) and others (Fig. 6).

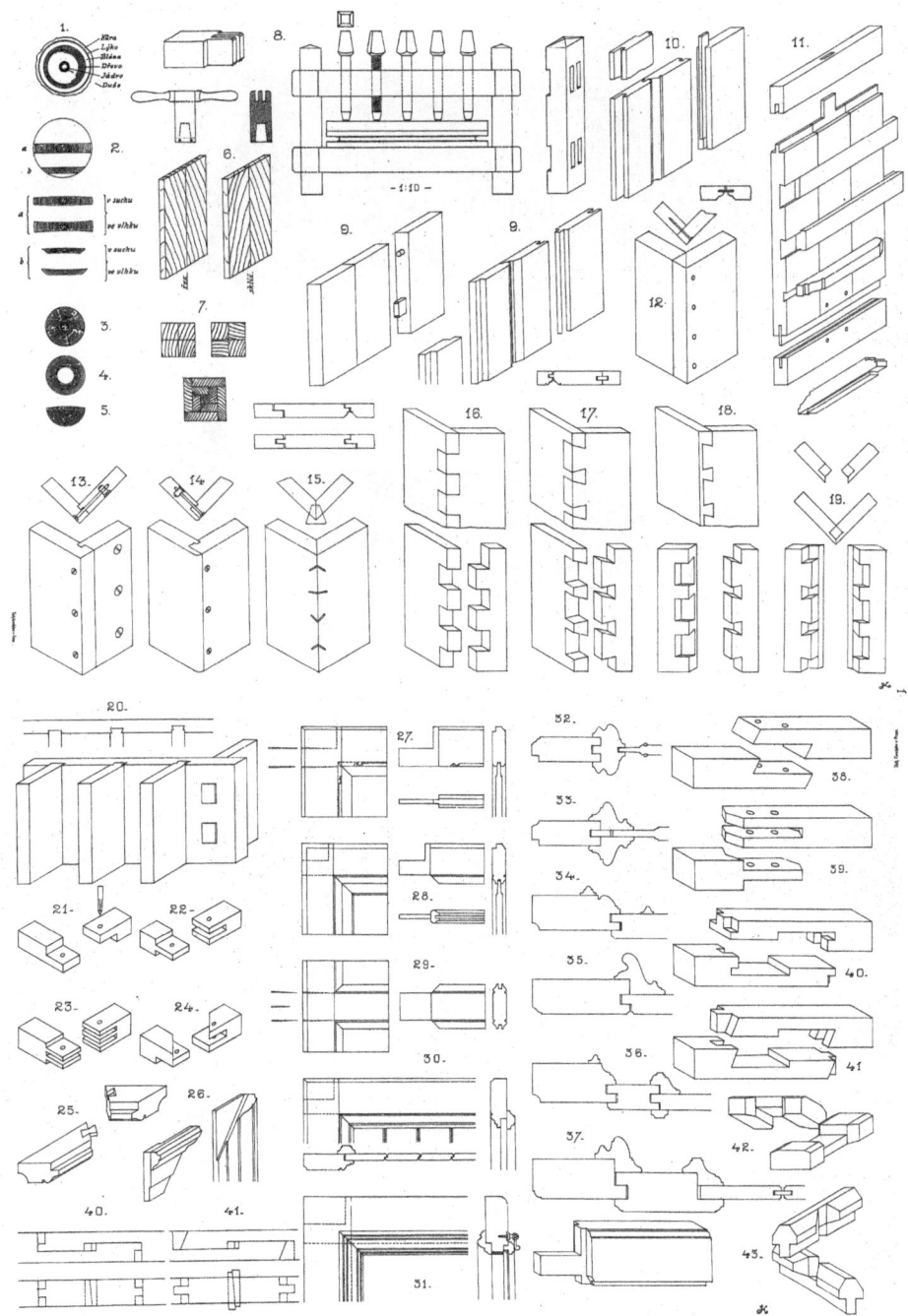

Figure 4:   Details of the design of the most frequently used joinery joints, such as ledge joint, mortise joint, tenon joint, bridle joint and dovetail, given in the worksheets of the professor and government commissioner of industrial schools, Ing. Jan Kubeš [7].

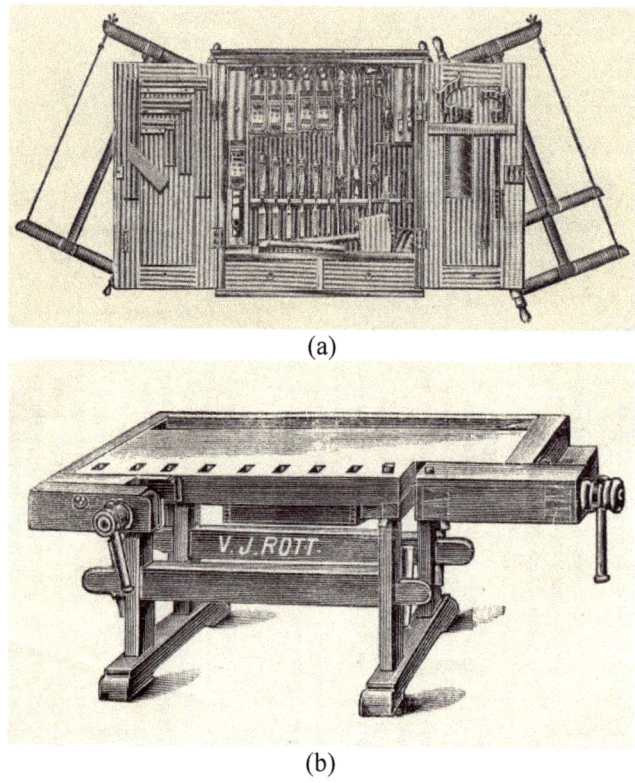

Figure 5: Characteristic equipment of a carpentry workshop with manual machining of material. (a) Basic tools for manual woodworking and their practical storage [10] and (b) Workbench – the necessary equipment of every manual carpentry workshop [11].

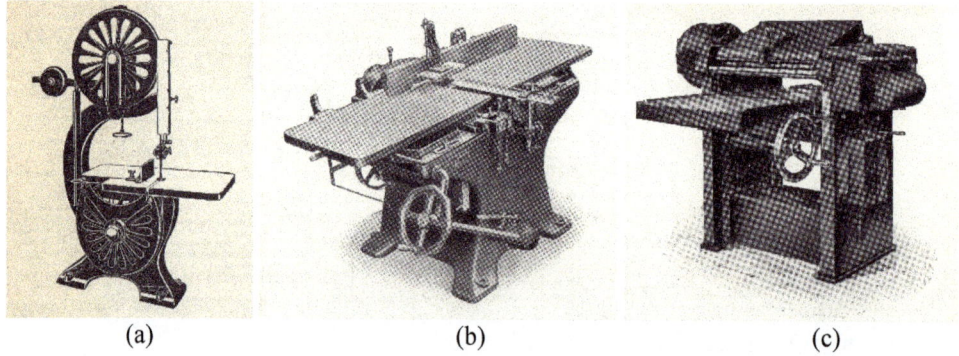

Figure 6: Examples of carpentry workshop first machinery for cutting and planing material from the beginning of the 20th century. (a) Band saw [12]; (b) Spindle moulder; and (c) Thicknesser [13].

3 PRODUCTS OF CONSTRUCTION JOINERY

As mentioned above, joinery products form an integral part of buildings and play a significant role in their architectural and artistic expression. Simultaneously with the elements of construction joinery, we are regularly in direct contact – we take the door handle on a daily basis, open windows daily, pass shop fronts daily, etc. In addition to the aesthetic function, the utility and economic functions also play an important role here. Not only for these reasons, emphasis has been placed in history on both the durability and quality of construction joinery products as well.

On individual products – doors, windows, shop fronts, wainscoting, etc., we can observe their design, surface treatment and the development of architecture and artistic perception, given that all elements of construction joinery reflected the contemporary aesthetic opinion (Fig. 7). In general, in everyday production it is possible to notice a change from decorative to elegant and simple design, not only in artistic expression, but also in the detail of profiling, surface treatments and accessories.

Especially, the production of the 19th and early 20th century occupies an important position due to the fact that a large number of buildings of urban and village architecture come from this period. Unlike most older monuments, however, these buildings have largely preserved fillings of openings and other joinery components.

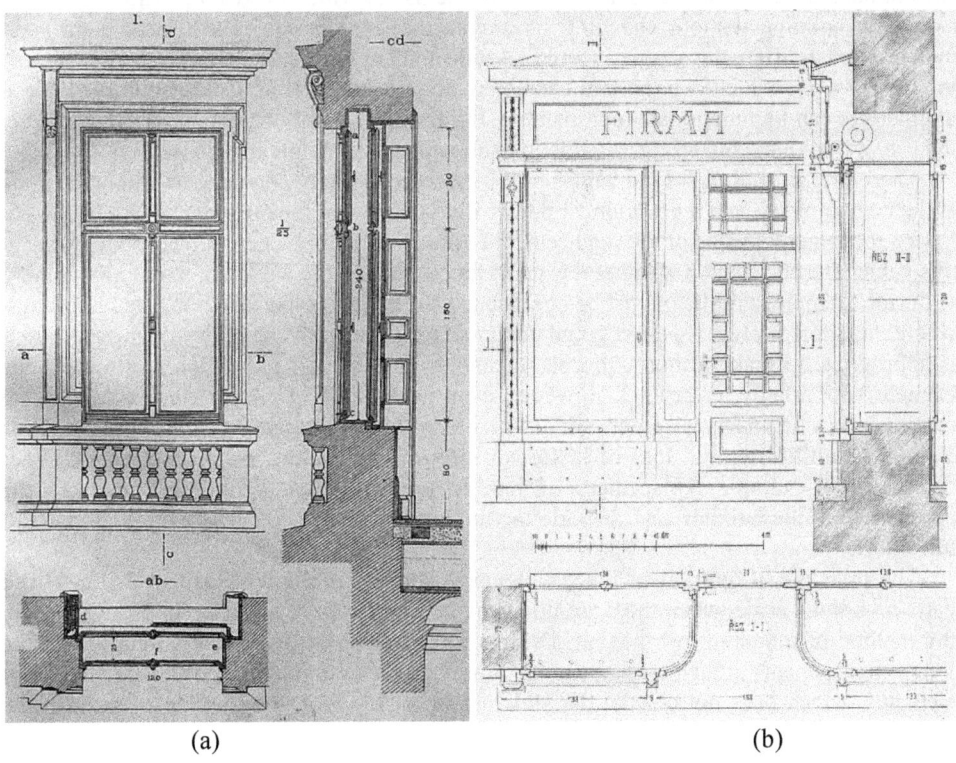

(a)   (b)

Figure 7: Design of joiner's products. (a) Design of a double window in a town apartment building from the end of the 19th century [5]; and (b) Design of a shop window for a town apartment building given in the construction manual from the beginning of the 20th century [14].

4 PROTECTION AND RESTORATION OF CONSTRUCTION JOINERY WORKS

An important cultural and historical value is the artistic contemporary expression, while each period had its artistic ideals, which it tried to achieve. The created artistic quality retains the ability to address contemporary society, even if the current expression is different. Historic architecture offers details (e.g. brass window handles), elements (e.g. door panels or window frames), facades and, last but not least, the overall environment, the preservation of which makes sense for our society. Likewise, care for the preservation of traditional crafts, knowledge of the production and processing of traditional materials should be part of the cultural heritage.

The rescue of valuable works of construction joinery is one of the most difficult tasks with which we can meet in the field of monument care. At present, unfortunately, quality joinery products are still being removed, the qualities of which – artistic, architectural, craft and especially monumental – can hardly be achieved by replacing them with new products. In the context of monument care, it is necessary to emphasize the value of the historical original, which tells about the time of its origin and is the bearer of the mentioned cultural and historical values. Only from the original it is possible to draw information about the aesthetic feeling, creative abilities and historical technical skills. Closely related to this requirement is traditional crafts and materials, the use of which makes it possible to restore a historic building without changing its architectural expression. The use of a different material and a different treatment would be reflected not only in the appearance but also in the way of aging which is important characteristic of each material. Contemporary material alternatives can be functionally comparable, but from the point of view of monument care, the use of traditional material has value in itself and is in principle irreplaceable.

Therefore, an analysis of the values of the preserved joinery components should be done at the beginning of each renovation, and only if it is clear that the preserved elements do not have a monumental value or the requirement to preserve them would not be proportionate to this value. The decision on whether it is necessary to replace the work of construction joinery with, for example, a replica, must be approached sensitively, especially for listed buildings, where the effort to preserve the original components is one of the priorities. In these cases, it is appropriate to make a thorough documentation of the finding situation and replace the element with a copy, or replica. However, even in the case of unprotected buildings, the refurbishment of joinery elements should be preferred to replacement, with which there is always simplification and loss of historical information. At the same time, an improper decision, for example replacement of joinery products, can significantly damage the appearance of the building and degrade the monument, respectively the historical value of the object.

The principle of renewal of joinery elements consists in the replacement of long-lived parts on a small scale, when there are no changes in structure or material, in profiling, or in the method of mounting or opening. During this process, the technical parameters of the joinery components of the building can be improved, for example, by additional insertion of elements, which does not change the structure or profile (e.g. interruption of the thermal bridge, or reduction of air permeability of windows).

5 CONCLUSIONS

The works of construction joinery represent parts of buildings with an unforgettable cultural value, which is very important for the quality of our environment. The priority interest of society and monument care should be the protection of preserved historical elements and the extension of their lifespan.

When restoring the works of construction joinery, it is important to respect the requirements for preserving the principle of authenticity and monumental values (value of age, artistic value, historical value). The preferred approaches include the refurbishment of joinery elements, in which a substantial part of the original material is preserved, and thus also the authenticity and monumental values.

ACKNOWLEDGEMENT

This article was written as part of the NAKI DG18P02OVV038 research project "Traditional City Building Engineering and Crafts at the Turn of 19th and 20th Centuries" (2018–2022, MK0/DG).

REFERENCES

[1]   Winter, Z., Dějiny řemesel a obchodu v Čechách v XIV. a v XV. století, Praha, p. 449, 1906.

[2]   Amb. 279.2 Folio 49 recto (Landauer I) [online], Die Hausbücher der Nürnberger Zwölfbrüderstiftungen (nuernberger-hausbuecher.de). Accessed on: 13 May 2021.

[3]   Amb. 317.2 Folio 66 verso (Mendel I) [online], Die Hausbücher der Nürnberger Zwölfbrüderstiftungen (nuernberger-hausbuecher.de). Accessed on: 13 May 2021.

[4]   Amb. 317.2 Folio 1 recto (Mendel I) [online], Die Hausbücher der Nürnberger Zwölfbrüderstiftungen (nuernberger-hausbuecher.de). Accessed on: 13 May 2021.

[5]   Pacold, J., *Konstrukce pozemního stavitelství. Práce tesařské, pokrývačské a truhlářské, podlahy, stropy a schody dřevěné*, part I, Praha, p. 71, 1900.

[6]   Ondřej, S., *Stavba domu v praksi*, Praha, pp. 321, 1932.

[7]   Kubeš, J., *Truhlářství stavební i nábytkové*, Plzeň, pp. 2–21, 1895.

[8]   Hejny, L., et al., *Stavební truhlářství a parketářství*, Praha, pp. 59–63, 2020.

[9]   Vitruvius, M.P., *Ten Books on Architecture*, Translation D. Rowland, Cambridge, 1999.

[10]   Technický obzor pro moderní zpracování dřeva, 1. ročník, Klášterec nad Ohří, 1929.

[11]   Axamit, J., *Domácí dílna, její zařízení a práce v ní. II. díl, Zpracování dřeva*, Praha, 1928.

[12]   Technický obzor pro moderní zpracování dřeva, 1. ročník, Klášterec nad Ohří, 1929.

[13]   Kouřil, J. et al., *Odborná nauka truhlářská pro odborné pokračovací školy*, part 1, Praha 1934.

[14]   Kohout, J. & Tobek, A., *Konstruktivní stavitelství, Tesařství a stavební truhlářství*, Jaroměř, p. 245, 1915.

SECTION 2
MAINTENANCE AND
CONSERVATION ISSUES

TOWARDS AN AUTONOMOUS MANAGEMENT MAINTENANCE MODEL APPLIED TO A HERITAGE BUILDING: THE CASE OF HERNANDO COLÓN COLLEGE, UNIVERSIDAD DE SEVILLA, SPAIN

MIGUEL LEÓN-MUÑOZ, DAVID BIENVENIDO-HUERTAS & CARLOS RUBIO-BELLIDO
Department of Building Construction II, Universidad de Sevilla, Spain

ABSTRACT

Maintenance in buildings is crucial to assure a proper use during their lifespan. However, both unqualified managers and a lack of commitment to develop a maintenance management plan is a challenge to establish Total Productive Maintenance (TPM) and European Foundation for Quality Management (EFQM) models. Moreover, if a building comprises heritage special requirements, specific actions must be considered. In this context, this research proposes a maintenance methodology by unqualified managers to facilitate decision-making for qualified agents. For this purpose, process management provides the bases for an autonomous management maintenance model characterisation and its application to historic buildings. Hernando Colón College (CMHC by its abbreviation in Spanish) of the University of Seville is a superb example of a heritage building with maintenance data and track record in process management approach, which supposes a starting point to apply the proposed methodology. The model has been implemented in the CMHC since 2015. The results showed that the maintenance optimisation reduced the number of corrective maintenance actions by 13% in comparison with the preventive actions. The results also indicated that the total maintenance actions were reduced by 37%. This study demonstrates, by collecting data based on quantitative measures, that it is possible to apply an autonomous maintenance management model in a historic building in use.
Keywords: building maintenance, management, TPM, human factor, lifespan.

1 INTRODUCTION

Building elements are degenerated during buildings' lifespan because of the pass of time, external agents, sporadic accidents, and buildings' use, so they should be appropriately repaired. This degradation, generally of lesser importance, does not impede the correct operation of the element, so users are not aware of the degradation from the first moment. Occasionally, this fact results in more serious damages, even irreversible, thus taking more expensive actions. However, this cost increase could be avoided by previously detecting it and taking the appropriate measures. Equally, owners consider some of these damages as inevitable, just repairing the elements that affect building's habitability.

It is therefore essential to take a set of actions that always ensure the good operation, predicting the possible risks that could affect building's security and protecting environment from possible aggressions. Building maintenance management has not been deeply studied as it is thought that, after buildings are constructed, their durability is guaranteed forever.

In Spain the law [1] forces buildings' owners and users to correct and to maintain buildings appropriately; however, there is no legislation that regulates and indicates the professional who should manage this aspect or the requirements that should be fulfilled.

From the industrial point of view it is said that "business suffers because maintenance is not enough considered" [2], so from the building point of view it could be said that "the lifespan of our buildings is reduced because their maintenance is not enough considered".

WIT Transactions on The Built Environment, Vol 203, © 2021 WIT Press
www.witpress.com, ISSN 1743-3509 (on-line)
doi:10.2495/STR210081

If the maintenance of our in-use buildings with certain heritage value is managed, time of lifespan is gained and, to a lesser extent, the capacity, the use and the improvement of rooms and installations are increased, thus guaranteeing a comfortable environment, and facilitating safer spaces.

The aim is not just the cost reduction by adjusting the maintenance effectiveness, but achieving a total productive maintenance (TPM), which is already implemented in the industrial sector by easily making all managers, owners, and users party to heritage buildings. All this is based on management strategies and processes to be studied in relation to the various uses, future strategies, and construction possibilities.

This research work presents an intuitive tool easy to use. This tool determines various states of maintenance management and develops control processes and indicators to obtain values to establish various strategies to be adopted. A model that allows unqualified workers, users, and owners to manage the maintenance of heritage buildings simply and effectively by developing the habit and the intention to carry out an optimal maintenance.

2 AUTONOMOUS MAINTENANCE MANAGEMENT OF A BUILDING

Maintenance is a building support process considered as a cost. This is the reason why maintenance is minimised, or the maintenance function is outsourced without assessing its impact on the continuous improvement of the building's use and lifespan.

The current market implies to reconsider the current maintenance system in the historic building sector. Maintenance efficiency and competitiveness are achieved by managing building uses appropriately, as well as the maintenance management of the installations, spaces and elements to meet the quality, performance and comfort goals expected [3].

In the industrial sector, pioneer in maintenance management, this development has led to the implementation of the TPM, which emerged in Japan in the core of the Japan Institute of Plant Maintenance [2].

Today, given the importance of continuous improvement management programmes, such as the Lean methodology or the EFQM model that are applied in many companies and entities, it is crucial to consolidate an effective maintenance management of heritage buildings used as assets. This leads to using the TPM as an improvement programme in relation to the maintenance management of any building.

The TPM is the result of the evolution of maintenance management systems: from systems that have been standards for many years to more complex, but highly effective, systems [4]. Traditional maintenance systems have not been removed but included with a new approach.

Based on the philosophy of the TPM applied to buildings, users (workers or others) are the responsible for the rooms, elements, and installations of the building, particularly to detect problems or anomalies before use difficulties are generated. These difficulties could alter the service of the building, as well as its cleaning and appropriate use.

The building sector suggests the application of the TPM as a basis to develop an autonomous maintenance management model of both the various heritage buildings and their assets. An effective and integrated maintenance management is suggested, including the previous types of maintenance, as Fig. 1 shows.

2.1 Autonomous maintenance of a heritage building

The importance of the TPM lies on the incorporation of autonomous maintenance by workers themselves. In our adaptation, autonomous building maintenance is carried out by users themselves, playing their respective roles, as far as possible. Thus, all employees are actively

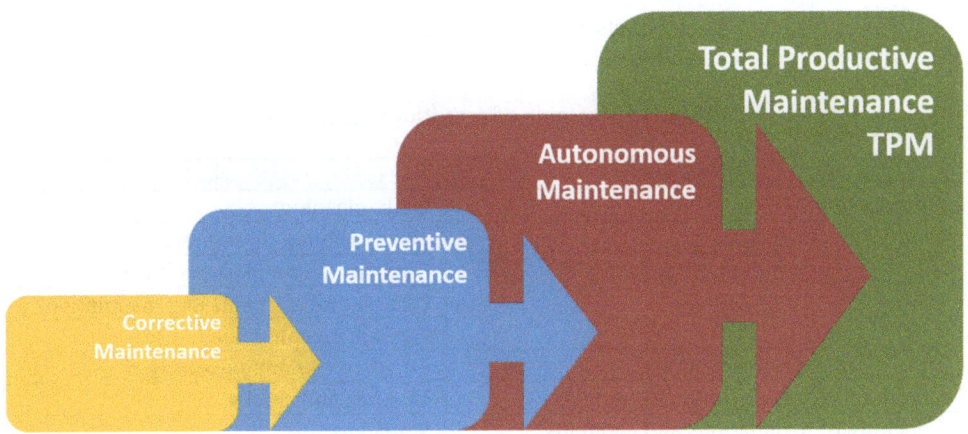

Figure 1: Evolution of maintenance management.

involved, from senior officials to maintenance labourers, going through users. In addition, the goals proposed by the manager of the building are fulfilled, and an own culture that stimulates the teamwork and raises the staff's spirits is developed.

Therefore, the philosophy of the TPM is applied [2]. It adapts the concept of continuous improvement from the maintenance point of view to the total building management, thus leading to the Autonomous Building Maintenance, which will be a new maintenance concept in the building sector. By including the Autonomous Maintenance as an integral and fundamental part of the TPM, the total balance of the maintenance tasks continuously and jointly managed, with the building being used, will be achieved.

Thus, the autonomous maintenance applied to a building is the maintenance in which any worker, even sometimes users, takes on maintenance management tasks based on noticing the need for or the ineffectiveness of maintenance. There is therefore a basic task: noticing the maintenance deviations that could be generated in a building.

For this purpose, the procedures required to contribute to the effectiveness of autonomous management that leads to an autonomous maintenance of heritage buildings are established in each building.

Talking about use in a TPM context is based on finding defects, anomalies, and needs for maintenance that guarantees the use of the building. The sequential description of this philosophy is shown in Fig. 2, which shows that a good use, i.e., a committed work/use, facilitates inspection. Thanks to that inspection, anomalies and precautions that guarantee the use of the area could be found, thus reducing intervention times.

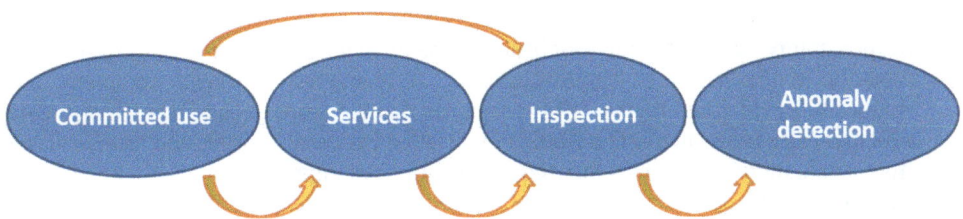

Figure 2: Evolution towards anomaly detection.

An example of an Autonomous Maintenance action is shown in Table 1, which includes the activities conducted in a possible stage.

Table 1: Example of activities in each stage.

Committed use	Services
Appropriate use	Daily cleaning
Immediate notification	Deep cleaning throughout year
	Cleaning of the inaccessible zones
	Cleaning of elements and installations
	Generic cleaning due to use
Inspections	Anomaly detection
Search of visible and invisible defects	Fissures and cracks
Checking of installations and elements	Scratches
Identification of obstacles hindering cleaning	Hits
Checking of measure devices	Broken pieces
Checking of tools	Leaks or escapes
	Humidities
	Weaknesses making tasks difficult
	Inaccessibilities

2.2 Goals of the TPM in the building sector

The major goals of the TPM applied to a building should be as follows:

- To maximise use effectiveness and guarantee of the building or parts of it and to remove all losses.
- To develop a sense of property in the employees and users of a building though a training and involvement programme.
- To promote a continuous improvement through the activities of small groups, including owners, users, managers, and those responsible for maintenance.

Buildings should have their own definition and view of the TPM, but most should maintain common elements and issues. There are seven key elements in any TPM programme applied to building management (Fig. 3) [5]:

1. Physical resource strategy. The TPM is commonly used to support or to facilitate the principles of quality and Just-in-time (JIT) management. This includes having structures in a cellular configuration and removing redundant steps or tasks. They are also a common part of the plan to include radical improvements in the management methods of a building asset.
2. Responsibility and autonomy. Self-management. The TPM puts the improvement capacity in the hands of employees/users. It guarantees workers' autonomy and responsibility. In addition, the TPM recognises that the employees of an area can teach a lot and learn from others. The whole organisation becomes stronger and receives ideas from small, motivated groups focused on a continuous improvement.
3. Resource planning and programming. The assistance demand for the maintenance department will be significantly increased during the introduction of the TPM, particularly while workers are being trained to increase their awareness and knowledge

about the building. Afterwards, they will want to quickly correct the causes of any pathology as they discover them.

4. Systems and procedures. A systematic maintenance management should be as effective as possible to reduce the risk of failures. First, the nature of the failure of a specific case should be understood. Then, the solution could be selected based on time, use factors, condition, or any other tactic.

5. Measurements. Measuring the progress of a continuous improvement is also useful for the TPM, including the number of small active groups and their individual and collective progress. From the beginning, disseminating beyond each affected area the good news about the progress will motivate all the members involved in the management of both the maintenance and the building.

6. Continuous improvement teams. An example is RADAR, a concept established by the ISO 9001 standard on Quality Management Systems. RADAR corresponds to the abbreviation that includes five elements: Results, Approach, Deployment, Assessment, and Review.

7. Processes. Processes should be reviewed and rectified in the new climate of responsibility, flexibility, and response capacity. They should be analysed, clearly understood, and redesigned to support the goals of the TPM. Each step should add value and reduce any waste of cost, time, service, quality, or other resources.

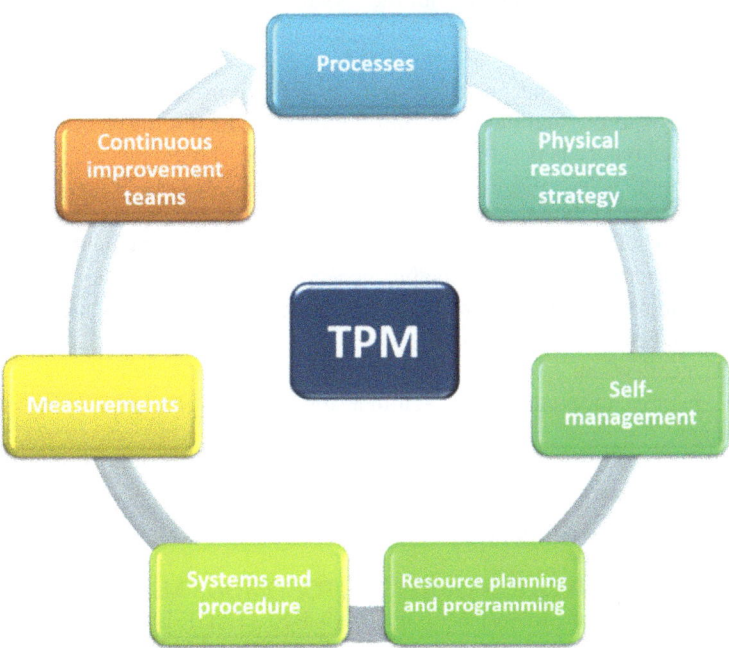

Figure 3: TPM wheel.

2.3 Implementation of the TPM in a heritage building

The meaning of the TPM and what can be achieved with it is different in each application. Likewise, the implementation plan should be specific according to each building.

An implementation plan of the TPM model should consist of four stages that must be carried out by all the people involved in the maintenance management process.

Awareness, education, and training. These are not just essential improvement activities of the TPM but a key pillar that supports the others. Directors, maintenance staff, team leaders, and workers should be involved in the learning process.

Education and training should support the following elements:

- Decentralisation of decisions and authority delegation to employees and users. This will imply to act in an autonomous way, with knowledge and confidence, and as members of a team they know where and when ask for help.
- Maintenance prevention or reduction of the maintenance intervention without sacrificing reliability. This is achieved through both standard operating procedures and systematic analyses of the indicators established.
- Workers' polyvalence (numerous skills) to maximise flexibility, efficiency, and satisfaction in the work of the people involved.
- Result measurements to assess the costs/benefits of the TPM. The successes of the programme are disseminated, and the obstacles are removed to improve the productivity of resources.

Apart from understanding the underlying theory of the TPM, some practical knowledge is required before making changes in the system. Carrying out a pilot maintenance management project will provide ideas and experiences, and confidence will be inspired in the implementation team.

Conducting a previous and posterior study in detail will be very useful for a test. An effective method is to photograph or to film the area to identify defects and damages. The creation of a visual record is part of the process composed of eight steps in a pilot project:

- Education (bases). Internal seminars about the elements and goals of the TPM, and how it is related to both management models and continuous improvement programmes.
- Survey. To determine the more suitable areas for a pilot programme.
- To select the pilot area based on both the successful probability and the potential to improve maintenance management.
- Initial data collection.
- Specific maintenance education. Seminar for the staff responsible for maintenance management.
- Photo tour.
- Training. To establish a training period on maintenance management for the staff.
- Beginning. To select a formal beginning date and to allocate responsibilities.

3 IMPLEMENTATION OF THE AUTONOMOUS MAINTENANCE MANAGEMENT MODEL IN A HERITAGE BUILDING

The results are shown in the steps followed to develop the Autonomous Maintenance Management Model in a heritage building. This proposal is the result of the methodology developed. This methodology is based on three pillars:

- Adaptation of the TPM model to the building sector, including autonomous maintenance.
- Exhaustive knowledge of the building, its uses, and possible variables.
- Order of criteria, indicators, measurements, control ways with easy and accessible tools, and result assessment.

Thus, the autonomous maintenance management model in heritage buildings is obtained. This model aims to be as effective as possible by using the existing resources. The results are presented in the autonomous maintenance management process of a building.

3.1 Description of the case study

The autonomous maintenance management model was implemented in the building of the Hernando Colón College (Fig. 4) of the University of Seville due to both the ease to access to maintenance data and the development in process management according to the EFQM excellence model. It is a specific case study, within the heritage buildings of the University of Seville, in operation for more than 70 years.

Figure 4: Front view of the Hernando Colón College.

Since the inauguration in 1948, it provides residence to the students and researchers. The building was designed by the architect José Gómez Millán, located in the Reina Mercedes Campus, in a square projected for the Universal Exhibition of 1929, which was called Plaza de los Conquistadores, nearby the Basque Pavilion or the Cordoba Pavilion. This location allowed to provide the Hernando Colón outdoor facilities integrated with the university campus (Fig. 5).

The 3,500 square meters building was designed in a T-shape, with four floors and basement. The last two floors occupy the front part of the building, with a rectangular shape. The ground floor (Fig. 6) and basement are destined to common rooms and services. The first, second (Fig. 7) and third floors allocate 128 rooms, common toilets and showers. The distribution outdoors comprises green areas, swimming pool and sports courts, as well as car parks.

Figure 5: Aerial view of the Hernando Colón College.

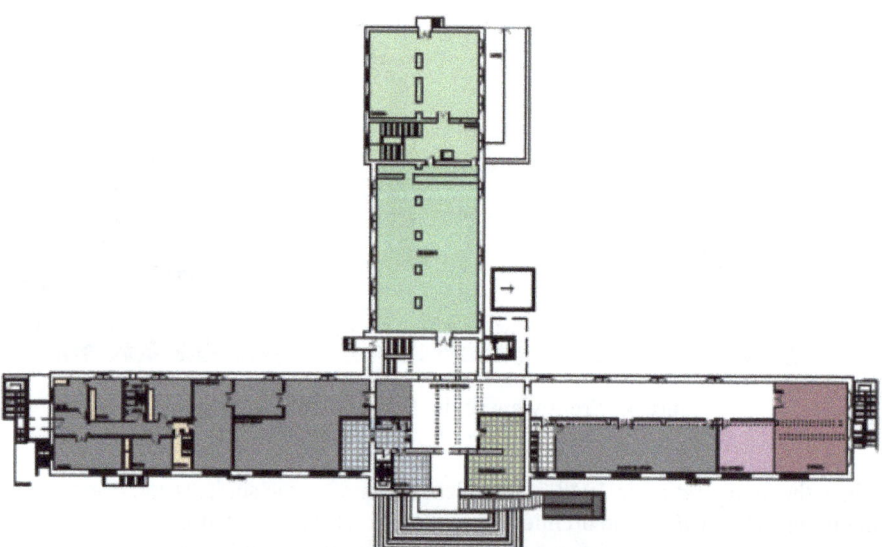

Figure 6: Ground floor plan of the Hernando Colón College.

The structure of the building is made up of load-bearing brick walls and a central reinforced concrete portico on which two-directional slabs rest. The foundation is formed by continuous footings. Two types of roof can be distinguished, a gable roof made of ceramic tiles, which covers the third floor. And a flat roof, which covers the rear area up to the first floor. The building's partitions are mainly made of clad brickwork, with the exception of the third floor, which is made of plasterboard partitions.

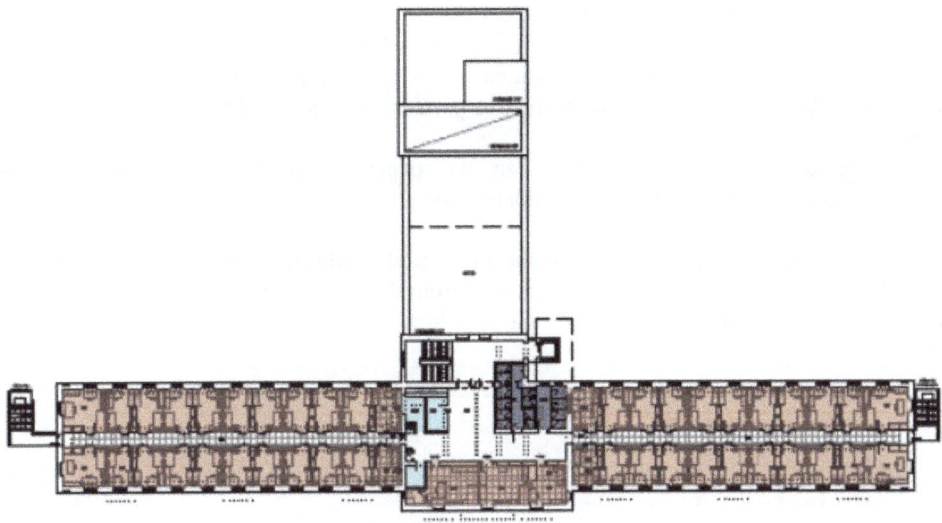

Figure 7: Second floor plan of the Hernando Colón College.

3.2 Implementation of the process

The model was implemented as an opportunity to improve the building and was included in the proposals established in the strategic plan in 2015. The starting point was the expertise of the staff, together with the historical data obtained until 2014.

The autonomous maintenance management model defines maintenance criteria, actions, and control indicators, with all of them being both basic to start managing maintenance in a useful way and applicable to the resources available in the centre.

This maintenance management model is susceptible to be modified, thus generating new actions, criteria, and indicators, or modifying those existing. Today the version 02 is applied.

In this version 02, the model is called as Autonomous Maintenance Management Process.

The sequence of the works developed in the experimental stage was based on the sources established by the EFQM Model, perfectly implemented in the centre, and assimilated by all those responsible for its management:

- Performance of the SWOT analysis.
- Performance of the Improvement Opportunity as part of the 2015 strategic plan.
- Writing of the Maintenance Management Model, version 00-2015.
- Implementation of the Maintenance Management Model.
- Analysis of maintenance indicators 2015.
- Review of the Maintenance Management Model, version 00-15.
- Writing of the Maintenance Management Model, version 01-2016.
- Analysis of maintenance indicators 2016.
- Review of the Maintenance Management Model, version 01-16.
- Writing of the Autonomous Maintenance Management Process of the TPM, version 02-2017.

Based on the data obtained from the experimental phase (from 2015 to today), in comparison with data from the analytical phase (from 2012 to 2014), maintenance is optimised according to the criteria included in the process that supports the goals of this research work, as Fig. 8 shows. The corrective parts are reduced, and the preventive parts are increased.

Having ended the year 2016, when the experimental phase was finished, the implementation of the model was consolidated, so the model was reviewed by those involved in their implementation.

Like in the previous period based on the analysis of results, there is an improvement (Figs 8 and 9) in comparison with the reference data from the previous years: more information is demanded to establish more references and to achieve more goals.

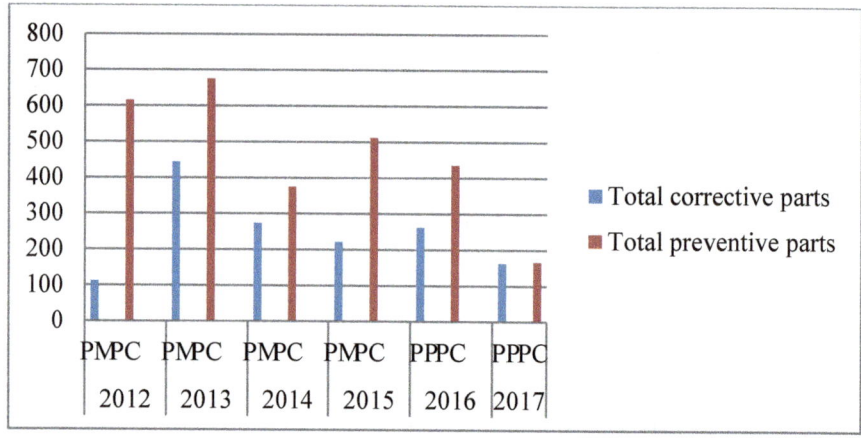

Figure 8: Maintenance tasks carried out in the period 2012–2017 (CMHC).

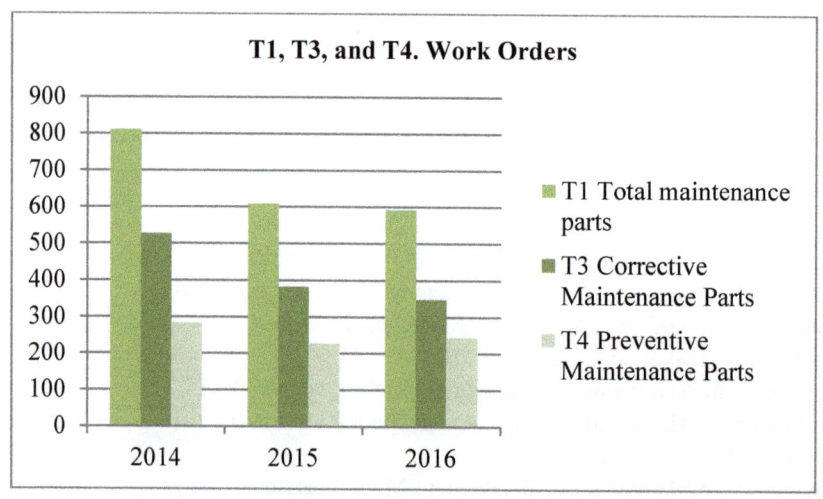

Figure 9: Example of the data obtained in the experimental stage 2 (CMHC).

3.3 Model assessment

The reviews carried out in the implementation of the model and in its maintenance as operating process determined the following aspects to be progressively implemented:

a) Consolidation of the design of the structure and classification of areas, assets that should be maintained, and uses of each area.
b) Programming of the hours of human resource available according to the working day.
c) Implementation of maintenance criteria according to the structure of the model:

- Strategy of the building. Structure of the building.
- Application of self-management.
- Training.
- Planning. Legal preventive maintenance according to frequencies and schedules allocated to the human resource. Gantt diagram.
- Optimal measurements.
- Review at the end of the period established.

d) Increase of the selection of key maintenance management indicators.
e) Use and control of the Excel application to develop a maintenance management database.
f) Control of the annual maintenance. Adding programmed preventive maintenance actions to legal preventive maintenance actions.
g) Control of non-programmed maintenance according to the classification established in the study of the model.

With all this and after the review conducted by the agents involved, the improvement group from the Hernando Colón College assessed the implementation of the model, determining the autonomous maintenance management model implemented in the centre after the experience in the two experimental stages. Moreover, an experimented technical document was developed, which fulfilled all the aspects established in the goals of this work.

The autonomous maintenance management process in a heritage building as a possible, realistic tool directly used in built and in-use buildings, using the human resources available.

By way of summary, the autonomous maintenance management process of the Hernando Colón College is today part of the management operating processes. This process is based on the strategies of the centre and is susceptible of the implementations demanded because of the continuous search of excellence in management.

4 CONCLUSIONS

The need for building maintenance management is supported by many research works.

Current concern on sustainability in architectural constructions, apart from the attention paid to energy efficiency, should be included in a minimum maintenance management framework that provides the building with an appropriate life cycle durability, thus reducing the consumption of resources from an ineffective management and increasing at the same time users' satisfaction.

According to our goals, the proposals of this research work are reasonable and in favour of an effective and autonomous maintenance management, thus contributing to the continuous improvement of the management processes implemented.

These results give response to the generation of an autonomous maintenance management model in heritage buildings that is based on use, the human resources available, and building constructive characteristics. A model mainly based on the human factor, assimilating the criteria established in the Total Productive Management (TPM) and establishing a continuous improvement process as a global part of the building management.

The model is completed by applying directly the methodology developed, thus leading to the Autonomous Maintenance Management Process of the Hernando Colón College. The goal is to test its implementation in a building in use by both assessing its global effectiveness and proving a possible methodology that can be directly used in heritage and in-use buildings. Furthermore, the human resources available are used, and indicators to establish strategies for an optimal maintenance management are considered, thus impacting the use of the various buildings as less as possible.

REFERENCES

[1] Ley 38/1999, de 5 de noviembre, de Ordenación de la Edificación, 1999.
[2] Dixon Campbell, J., *Orgenización y Liderazgo del Mantenimiento*, Edición en., Toronto, 2001.
[3] Martin Ezama, L., *Gestión del Mantenimiento en Edificación*, COAATIE AL.: Albacete, 2015.
[4] Cuatrecasas, L. & Torrel, F., *TPM en un entorno Lean Management*, 2010.
[5] Gonzáléz Fernández, F.J., *Teoría y Práctica del Mantenimiento Industrial Avanzado*, 4ª Edición, Madrid, 2011.

OPTIMIZATION OF HYDRATED LIME PUTTIES AND LIME MORTARS USING NOPAL PECTIN FOR CONSERVATION OF CULTURAL HERITAGE

ANGÉLICA PÉREZ RAMOS[1], JOSE LUZ GONZÁLEZ CHÁVEZ[2], LUIS FERNANDO GUERRERO BACA[3],
MIGUEL ÁNGEL SÁNCHEZ ESPINOSA[1] & ANAÍ CHIKEN SORIANO[2]
[1]Iberomaerican University Puebla, Mexico
[2]National Autonomus University of Mexico (UNAM), Mexico
[3]Metropolitan Autonomus University of Xochimilco (UAM-X), Mexico

ABSTRACT

Hydrated lime as a putty has had a very important role as a binder for mortars and plasters in which hundreds of years of archeological, artistic, and historical heritage have been preserved. There is evidence that the Maya improved the properties of their mortars by adding gum from different Mayan tree barks during lime hydration process. This interdisciplinary investigation made it possible to demonstrate that nopal mucilage can optimize slaked lime properties during lime hydration process, starting from the idea that the cactus produces both, gums and mucilage. During the chemical experimentation, it was possible to isolate the main component of the optimal physical and chemical interaction between lime and nopal mucilage. The main objectives of this research were to rediscover the value of an ancestral production technique, to position hydrated lime putty as an attractive product to builders that work with historic structures by optimizing its properties and lime mortars' behavior, and to reduce hydrated lime putties production time, specifically by speeding up the aging process from months to hours. Rheological and mechanical experiments succeeded in proving that the pectin of the nopal has a direct influence on the slaking lime process, improving lime putties and mortars' properties. Lime putties and lime mortar specimens were tested in XRD (X-Ray Diffraction) and SEM (Scanning Electron Microscope 7800) where crystallographic changes related to the optimization of the properties of lime mortars and lime putties were verified. The results established a new paradigm for the slaking lime process as the main production phase for lime putties and their mortars. Nopal pectin improves lime putties and lime mortars' plasticity, mechanical strength and proclivity to carbonation, while it also produces a reduction in capillarity absorption and mortar cracks.
Keywords: lime mortars, slaking lime, lime putties, nopal pectin, rheological, mechanical, lime properties, hydrated lime, mortar properties, Mayan.

1 INTRODUCTION

It is well known that hydrated lime is one of the basic materials in the archaeological and architectural legacy of the whole world [1].

Since ancient times, hydrated lime could be obtained through different procedures that depended mainly on culture. One of the most common procedures consisted on hydrating quicklime using a great amount of water (greater than necessary) in order to obtain hydrated lime putty. The process begins when the limestone ($CaCO_3$) is calcined at 1,100°C to generate quicklime (CaO) that, when cooled, is added to water. This causes an exothermic reaction (up to 100°C) followed by a constant stirring phase that after about an hour, results in hydrated lime putty ($Ca(OH)_2$) [1], [2].

When hydrated lime putty is obtained, it must remain under a mirror of water to protect it from carbonation. This phase is known as the aging period of hydrated lime putty and the purpose is to optimize its properties. The longer the time it remains in this phase, the better the properties it develops [2]–[5].

There is a large number of investigations that have already demonstrated the chemical and physical evolution of calcium hydroxide upon aging of lime putty and its influence on lime properties [3], [4]. Basically, an aging process provides better plasticity for lime putties and mortars. In later investigations it was also proved that not only putties, but mortars as well, acquire a better plasticity and strengthen behavior [4], [5].

In our country, the procedure of hydrating quicklime (CaO) at construction sites currently takes place in large rectangular brick containers where, in addition, the aging of hydrated lime putties (Ca(OH)$_2$) is carried out. Unfortunately, the traditional process of quicklime hydration has been neglected due to the long aging period of lime putties, which takes at least 2 months as indicated by the authorities. This long aging process can have a negative impact on a project's budget; therefore, many builders would rather use other binders – such as cement – that are incompatible with traditional construction systems. This doesn't only affect the results of restoration work on historic buildings, but it also contributes to accelerating the loss of an ancestral production technique and of the material itself. As part of this loss, the lime slaking technique is sometimes negatively altered and lime putties hydrated at the construction site lack adequate rheological and mechanical properties. This results in constant empirical experiments with natural additives in order to improve lime and mortar properties for conservation purposes [1], [6], [7].

2 THE ANCIENT MAYAN TECHNIQUE

Mayans were known experts in lime production for construction purposes. In recent investigations, extensive areas of limestone quarries have been located, which might show how essential of a material lime was for the city-states of the Mayan civilization. According to archeological findings, the use of lime as a stone and as a binder allowed them to build famous arches and temple crests, while they also developed lime mortars with extraordinary properties. The use of lime as a binder, implied a specific form of production that included the gum from the bark of some trees to obtain hydrated lime putties that optimized their mortars' properties [7].

As it has been exposed, the phase of quicklime hydration is essential to obtain a lime putty with good binding properties. In the past, the Mayans from the northern region used a slightly different procedure to obtain lime putties. This procedure did not only consist on adding water to the quicklime (CaO), but also adding the bark of some trees into the water so that their gums exuded and thus serve as a hydration agent.

Gum from tree barks improved plasticity, cohesion, strength and porosity in mortars that were prepared using the resulting hydrated lime putty as a binder. This improvement was explained as an increase in the solubility of quicklime (CaO), caused by monosaccharides as main components of gums from the bark of the trees [1], [7], [8].

3 THE OBJECTIVE OF THIS INVESTIGATION

Based on the Mayan technique for lime hydration, it was proposed to use nopal mucilage in the process, considering its similarities in chemical composition – such as polysaccharides – with the gums of the barks of the Mayan trees.

Besides being affordable for the entire population and growing practically throughout the entire Mexican territory, nopal mucilage has been used as an additive during the formulation of mortars for restoration work as it has been demonstrated that it optimizes certain properties of the mortars in which it is used. Since nopal is mainly composed by polysaccharides, it was expected to have an influence at the early stage of lime putties and mortars' production – the quicklime hydration phase – as a hydrating agent [6], [9].

In line with the above, the main purpose if this research was to demonstrate that nopal mucilage can work as a hydration agent for quicklime (CaO) to improve the properties of slaked lime putties (Ca(OH)$_2$) [6], [9], [10].

During the first experimental phase in the chemistry laboratory, it was discovered that, in addition to polysaccharides being contained in nopal mucilage, a specific substance, galacturonic acid, was the one responsible for enabling the successful interaction between nopal mucilage and quicklime (CaO). This substance was isolated to test it as a hydrating agent for quicklime (CaO) in parallel with nopal mucilage and water. The results were focused on the use of nopal pectin as a hydration agent compared to water [11].

3.1 Material and methods

Galacturonic acid was isolated from cactus pectin. A solution with a specific concentration was prepared with this substance to enable the hydration of quicklime (CaO) and to obtain hydrated lime putty (Ca(OH)$_2$) in nopal pectin [6], [11].

The lime hydration process was carried out using 1:2 proportion by weight of the ratio between quicklime and the pectin solution. At the same time, quicklime (CaO) was hydrated in water as a control solution to compare the behavior of both products. The hydration process of the quicklime (CaO) in each solution was carried out with constant stirring for 1 hour. Once the putties were obtained, they were protected with a mirror of water to avoid carbonation, and finally experimentation with each of them was carried out [6].

Lime putties were analyzed in different phases, from their recent production (one hour old) through all the different periods of their aging process [6]. The lime pastes were analyzed at 28 days and then at 400 days of aging. The analysis sought to characterize the crystallographic behavior, for which a 7800 SEM was used. The physical phases were analyzed in an XRD Bruker D8 and the Abrams cone was used to determine the plasticity of the lime putties as shown in Figs 1–3 [6].

Figure 1:   (a) Lime putty taken from its container during aging phase; (b) and (c) Determination of plasticity using Abrams cone.

Figure 2: (a) and (b) XRD and lab material for testing lime putties.

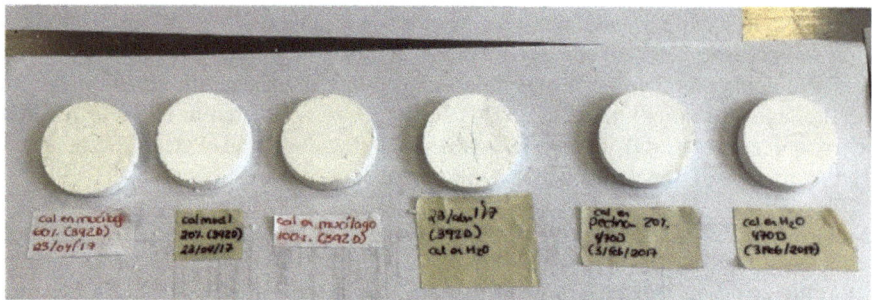

Figure 3:   Different lime putty samples to be analyzed in the XRD at different periods of aging.

Different mortar samples were prepared with each type of lime putty produced in the two selected aging periods. The analysis of the mortars had the objective of characterizing their mechanical and rheological behavior based mainly on the ASTM-C-109 and NMX-C-082-1974 standard. The mortar specimens were also analyzed to determine their propensity for carbonation using phenolphthalein and the calculation of their capillary absorption capacity was based on the NORMAL 11/85 standard [6].

The mortars were formulated using a ratio of 1:3 by weight, with regard to the lime and arid ratio. This ratio was determined after several previous experiments with different ratios. The procedure for preparing the mortars and their specimens (cubic specimens about 125 cm^3), was based on ASTM C270 – 02, ASTM D752000 and ASTM C-109 standards mainly (Fig. 4(a)–4(g)) [6].

Figure 4:   (a) Lime putty taken from its container during aging phase; (b) and (c) Lime mortar preparation process; (d) and (e) Cubic specimens for experimentation; (f) Specimens to test the adhesion capacity of mortars; and (g) Specimens prepared to measure cracking control.

The mortar specimens were subjected to different analyses that would measure their behavior at 45, 100 and 200 days of aging. The capacity of resistance to compression and adhesion of the mortars were evaluated using a manual digital press type machine model E657-2, their plasticity was measured by using the Abrams cone, the capillary absorption capacity was measured with the help of an Ohaus Precision Balance in accordance with the procedure established in the NORMAL 11/85 standard, and finally the mortars' availability to carbonation was determined by using the phenolphthalein method [6].

4 RESULTS AND DISCUSSION

The results of the different tests carried out on the putties and the mortars that were manufactured with them made it possible to conclude that the pectin extracted from the nopal mucilage improves both the putties' rheological properties and the early carbonation phases. The diffractograms showed calcite phases in lime putty samples that had been hydrated in nopal pectin for only 28 days, which suggests the immediate start of the carbonation process. In contrast, in the putties that were hydrated in water, only portlandite phases could be detected, that is, without initiating the carbonation process. Both putties were exposed to CO_2 for only 15 minutes while preparing to be examined on the XRD (Figs 5 and 6).

Changes in the crystalline morphology could be identified in the SEM 7800 from the first day of their manufacture in the lime putties that were hydrated with nopal pectin. The crystalline morphology tended towards flat hexagonal patterns that usually appear when the lime putties have undergone an aging process of 12 to 14 months, while in the putties that where hydrated with water, the crystalline morphology preserved the prismatic hexagonal patterns that are characteristic of a paste in the early stages of aging [6].

The observations in the SEM, showed that the crystallographic transformation during the aging process from prismatic hexagonal to flat hexagonal patterns is more evident in the putties that were hydrated in nopal pectin (Figs 7–9).

Regarding the mechanical behavior of mortars, those prepared with lime putty hydrated in nopal pectin showed an increase of 67% in resistance to compression and better plasticity than mortars prepared with lime putty hydrated in water (Fig. 10). The results presented here were obtained from specimens tested at 45, 100 and 200 days of aging [6].

Compared to the mortars that were made using lime putties hydrated in water, mortars prepared with lime putties hydrated in nopal pectin for 28 and 183 days showed a 50% reduction in their capillary absorption capacity. The evaporation of the water in the samples took place 35 minutes faster and the carbonation process was carried out 85% faster, even in the areas with less exposure to CO_2 (Fig. 11(a)–11(d)). Moreover, the cracks in the mortars used as coatings showed a reduction of 50%.

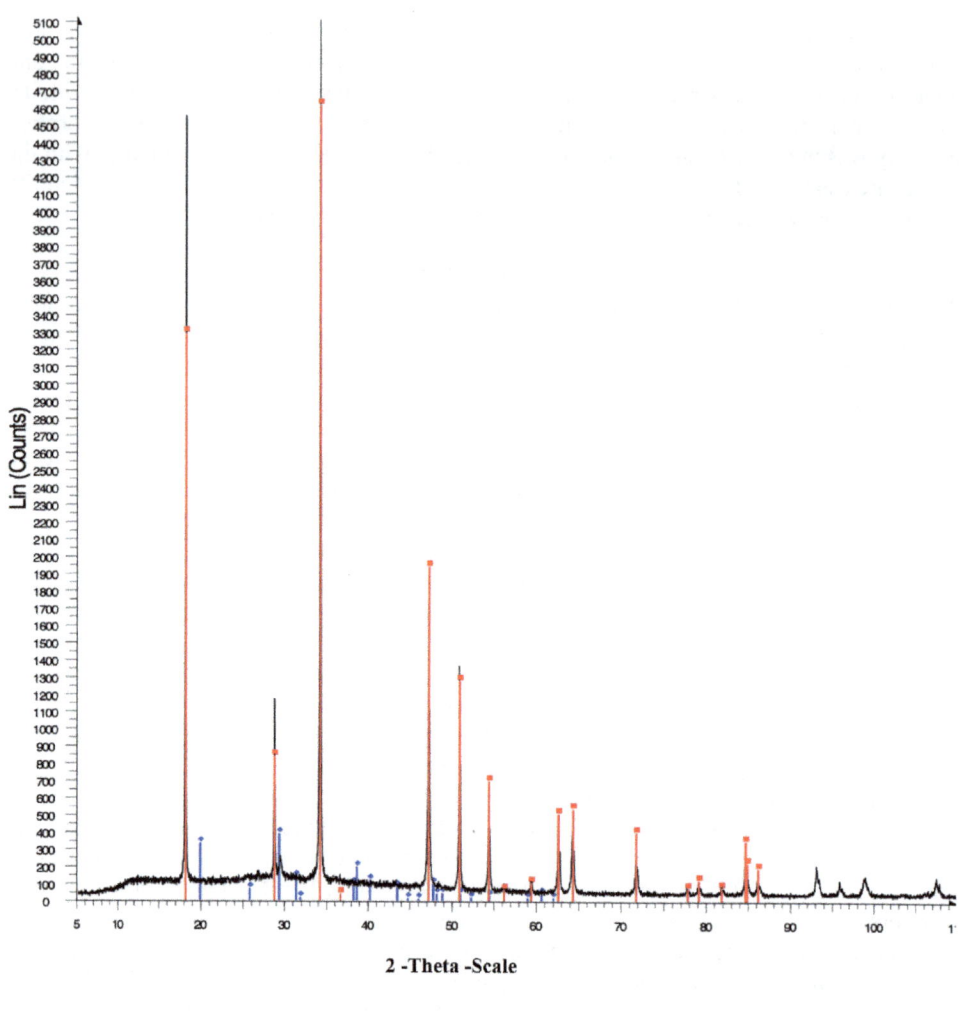

Cal-H2O22-02-18-File: Cal-H2O-22-05-18.raw-Type: Locked Coupled- Start: 5.000°-
End:109.999° - Step: 0.021° - Step time: 38.4 s- Temp.:25°
Operations: Import

01-084-1263 © - Calcium Hydroxide – Ca(OH)2 – Y: 90.34 % - d x by: 1. – WL: 1.5406 –
Hexagonal – a3.59180 – b3.59180 – c4.90630 – alpha 90.000-

00-024-0223 (*) – Hydrophilite [NR] – CaCl2 – Y: 7.57 % - d x by: 1. – WL: 1.5406 –
Orthorhombic – a 6.26100 – b6.42900 – c4.16700 – alpha 90.000 - bet

Figure 5:   Diffractogram corresponding to lime putty hydrated in water of 28 days of aging
where portlandite phases were identified.

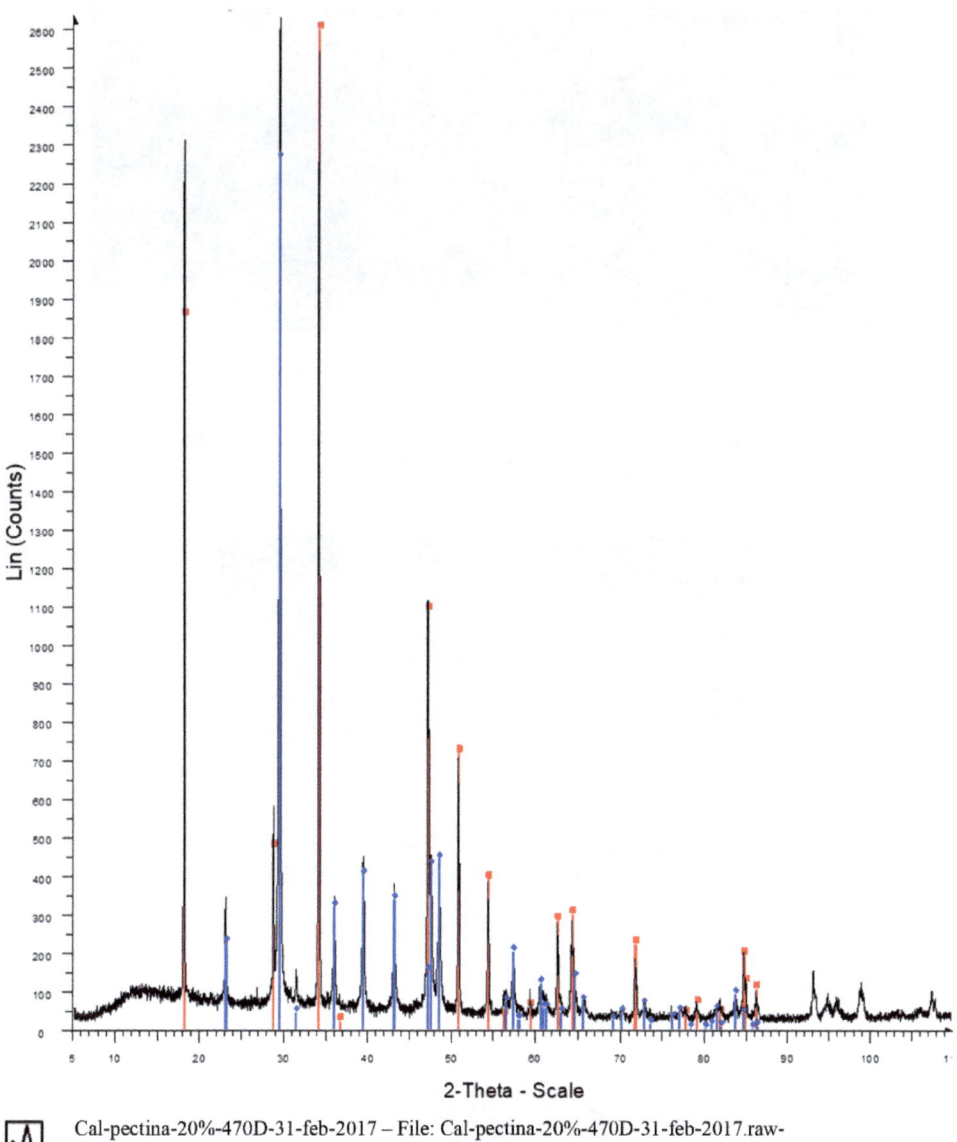

Cal-pectina-20%-470D-31-feb-2017 – File: Cal-pectina-20%-470D-31-feb-2017.raw-
Type:Locked Coupled – Start:5.000°-End: 109.999°-Step:0.021 Operations: import

01-084-1263 ©- Calcium Hydroxide- Ca(OH)2-Y:98.80%-dxby: 1-WL:1.5406-Hexagonal-a
3.59180-b 3.59180-c4.90630 alpha 90.000-

01-072-1937 ©- Calcite CaCO3-Y: 86.06% - dxby 1. – WL:1.5406- Rhombo. H. axes -a
4.99400-b-4.99400 – c17.80100 – Alpha 90.000 -beta 9

Figure 6: Diffractogram corresponding to lime putty hydrated in nopal pectin of 28 days
of aging where calcite phases were identified.

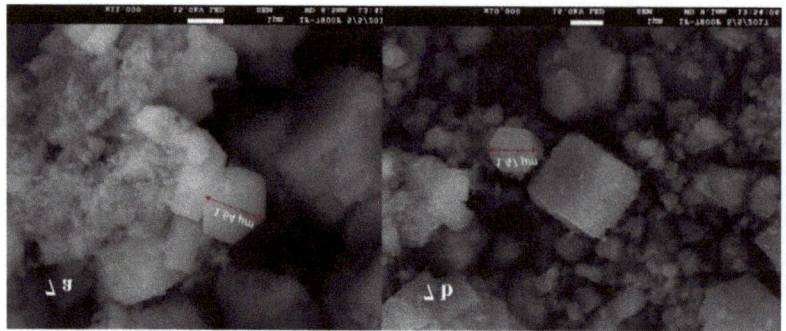

Figure 7:   (a) SEM image corresponding to lime putty slaked in nopal pectin at 28 days; and (b) SEM image corresponding to lime putty hydrated in water at 28 days. In both images, crystallographic structures of prismatic hexagonal morphology micrometric size can be identified.

Figure 8:   (a) SEM image corresponding to lime putty slaked in nopal pectin at 28 days; and (b) SEM image corresponding to lime putty hydrated in water at 28 days. In both images, micrometric sized crystallographic structures of prismatic hexagonal morphology can be identified.

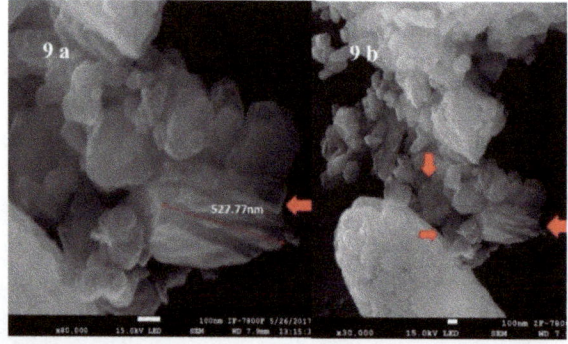

Figure 9:   (a) and (b) Both SEM images show lime putty hydrated in water for 110 days. Nanosized hexagonal plate patterns can be identified in the two pictures.

Figure 10:   (a) Mortar specimen in the machine to test its compressive strength and (b) and (c) Compression tested mortar specimens.

Figure 11:   Mortar specimens tested for phenolphthalein to determine their capacity for carbonation. (a) and (b). In images (a) and (b) the magenta of phenolphthalein in the internal faces of the specimens indicate a lack of carbonate; (c) The faded magenta shows partial carbonation of the internal faces of these mortar specimens; and (d) The magenta color is almost imperceptible, which reflects a total carbonation of the internal face of the mortar specimen that used slaked lime putty in nopal pectin.

5 CONCLUSION

Nopal pectin as a hydrating agent in concentrated solutions significantly improves plasticity and makes the lime putties more reactive to carbonation when calcite phases occur at very early stages, as shown in the corresponding diffractograms. The SEM images demonstrated that there is a true evolution in the crystalline morphology from prismatic hexagonal to flat hexagonal patterns and a much more significant reduction of micrometric to nanometric dimensions in lime putties that have been hydrated in nopal pectin, compared to the ones that were hydrated in water. The foregoing allows us to conclude that the aging period that enables the optimization of the lime putties has been reduced by 92% using nopal pectin in solution to hydrate quicklime [6].

Although the properties of hydrated lime putty depend on all the phases of the production process, including the selection of the limestone and its calcination, it was concluded that the phase of quicklime hydration is not only susceptible to control, but also the most outstanding in terms of optimizing the period of aging and significantly improving the properties of the lime putty and its mortars [6].

Compared to mortars that were prepared with water-hydrated lime putty, the mortars that were prepared with lime putty hydrated in cactus pectin showed a notable increase in their adhesion and resistance to compression, greater plasticity, faster evaporation of water during capillary absorption, greater capacity for carbonation – even in areas with less exposure to CO_2, and a significant decrease in cracks in coating mortars during the carbonation process [6].

The ancient practice of slaking quicklime inherited by the Mayans can now be reproduced with a cactus that is affordable in economic and territorial terms in practically the entire country. Nopal mucilage improves the properties of slaked lime putties when it is used as a hydration agent, but it achieves greater effectiveness when its pectin is isolated and used in solutions at a specific concentration to hydrate quicklime. The reduction of the aging period of the lime slaked in nopal pectin by 92% compared to that hydrated in water could be the most attractive of all the properties that were successfully optimized, especially for builders who are involved in the restoration of historic buildings, as this significantly reduces costs and allows revaluing and rescuing a technique that is about to disappear – not without ensuring the compatibility of the construction systems that originally contain hydrated lime putty. Above all else, however, this practice rescues valuable knowledge and a constructive memory that deserves be preserved [6].

ACKNOWLEDGEMENT
To Universidad Iberoamericana Puebla, Puebla, Mexico, to which I am attached for the support of the mobility program for full time professors.

REFERENCES
[1]   Barba, L. & Villaseñor, I., *La cal, historia, propiedades y usos*, Instituto de Investigaciones Antropológicas, UNAM: Mexico, pp. 60–102, 2013.
[2]   Rodríguez, C., Ruiz, A., Elert, K. & Hansen, E., Crystallization and colloidal stabilization of $Ca(OH)_2$ in the presence of nopal juice (Opuntia Ficus Indica), implications in architectural heritage conservation. *Langmuir, American Chemical Society*, **33**, pp. 10936–10950, 2017. DOI: 10.1021/acs.langmuir.7b02423.
[3]   Rodríguez, C., Cazalla, O., Sebastian, E., De la Torre, M. & Cultrone, G., Aging of lime putty effects on traditional lime mortar carbonation. *Journal of the American Ceramic, Society*, **83**(5), pp. 1070–1076, 2000.
DOI: 10.1111/j.1151-2916.2000.tb01332.x.
[4]   Rodríguez, C., Calcium hydroxide cristal evolution upon aging of lime putty. *Journal of the American Ceramic, Society*, **81**(11), pp. 3032–3034, Nov. 1998.
DOI: 10.1111/j.1151-2916.1998.tb02735.x.
[5]   Rodríguez, C., Ruiz, E. & Hansen, E., *Nanostructure and Irreversible Colloidal Behavior of Ca(OH)₂, Implications in Cultural Heritage*, Departamento de Mineralogía y Petrología Universidad de Granada and The Getty Conservation Institute: Granada, España, 2005, pp. 10948–10957. http://www.ugr.es/~grupo179/pdf/Rodriguez%20Navarro%2005.pdf.

[6]   Pérez, A., El Mucílago de Opuntia como Agente de Optimización de las Propiedades de Morteros y Pastas de Cal Apagada en la Restauración del Patrimonio Edificado. PhD thesis, Universidad Nacional Autónoma de México, Ciudad de México, México, 2019. 132.248.9.195/ptd2019/agosto/0794434/Index.html.

[7]   Bedolla, J.A. et al., *Aditivos Orgánicos en Morteros de Cal Apagada en la Edificación Histórica*, Ciencia Nicolaita: México, pp. 51, 153–166, 2009. http://www.cic.umich.mx/documento/ciencia_nicolaita/2009/51/CN51-153.pdf.

[8]   Carrascosa, B. & Lorenzo, F., *Estudios Previos en Morteros Tradicionales de Cal para la Evaluación de su Comportamiento Hídrico y la Idoneidad de ser Empleados en Clima Tropical*, Instituto Universitario de Restauración del Patrimonio de la Universitat Politécnica de Valencia: España, 2002, pp. 55–61. https://riunet.upv.es/bitstream/handle/10251/33046/2012_6-7_55-62.pdf?sequence=1.

[9]   Pérez, A., Nopal mucilage as hydration agent of quicklime; extraction methods. *Grupo Español de Conservación*, International Institute of Conservation of Historic and Artistic Works: Madrid, España, pp. 189–195, 2017. https://ge-iic.com/ojs/index.php/revista/article/view/475.

[10]   Vargas, L. et al., Adhesivo de nopal en Pinturas a la Cal. *Revista Salud Pública y Nutrición Edición Especial*, (4), pp. 165–174, 2012. DOI: 10.29105/respyn.

[11]   Chiken, A., Influencia del Mucílago de Nopal Sobre las Propiedades Fisicoquímicas de la Cal y su Aplicación para Materiales de Restauración, Facultad de Química. Master's degree thesis, Universidad Nacional Autónoma de México, UNAM, Ciudad de México, México, 2017. 132.248.9.195/ptd2017/mayo/0759816/Index.html.

SECTION 3
MONITORING AND
DAMAGE DETECTION

MONITORING AND STUDY PROCESS
OF GOTHIC BUILDINGS

SERGIO COLL-PLA, GUILLEM MATEU, AGUSTÍ COSTA JOVER & JAUME ROSET-CALZADA
Unitat Pre-departamental d'Arquitectura de la Universitat Rovira i Virgili, Spain

ABSTRACT

The energetic understanding of medieval buildings requires a joint study, because in this building typology, the structure acts as a climatic protection as is known. From another point of view, medieval buildings tend to be grouped in a pineapple, conditioning the climatic and structural understanding. Therefore, the knowledge and study of the old urban models is crucial for their understanding. The urban model and its surroundings determine the climatic demands of the interior spaces. In the same way the construction characteristics and materials. The article aims to develop an open system of hydrothermal measurements that allows hydrothermal comprehension of spaces. The result will be a sensor system with interpretable data that will allow us to understand the hydrothermal system of a medieval building located in its context. The interpretation of these data will allow making a comparison between different construction solutions that allow improving the hydrothermal conditions of the interior according to the climate.

Keywords: *open source, hydrothermal sensors, temperature, inertia.*

1 INTRODUCTION

Understanding the energy performance of buildings is essential to adapt to climatic conditions, which is why current regulations respond to these conditions. It should be noted that the regulations are prepared for new buildings or major renovations, especially since the entry into force of the Spanish Technical Building Code. The difficulty of adapting buildings to current energy regulations often undermines the architectural principles of the building. Therefore, the Technical Building Code does not prescribe specific standards for intervention in existing buildings and allows these standards to be waived in heritage buildings. European regulations also exempt heritage buildings from compliance with reducing consumption. Therefore, medieval city centres lose their character in favour of their habitability.

Speaking about city centres, Europe has many medieval city centres. The main feature of medieval urbanism is the occupation of open spaces, especially public spaces. As far as the division of land into plots is concerned, these acts by means of the fragmentation or gathering of neighbouring plots. Specifically, in terms of urban planning, this is the phenomenon of the invasion of private space over public space, of the appropriation of private spaces which have been destroyed in favour of the location of new roads, or more rarely of cleared land. The medieval urbanism has its decline with the book *De Re edificatoria* (1485) by Leon Batista Alberti. With it, during the Renaissance, the notions of proportion, regularity, symmetry, and perspective gave birth to what in the following centuries would become known as urban art. The purpose is explicitly aesthetic: visual pleasure (the search for beauty) is privileged over necessity and comfort. These ideas spread throughout Europe at the end of the 16th century, finding their end as a dominant tradition in the search for utility imposed by the industrial revolution in the 19th century.

There is currently a considerable amount of medieval building stock, while Spanish building regulations encourage energy efficiency in new and existing buildings. This means that construction solutions accepted by the regulations for new buildings are applied to pre-industrial buildings without questioning the energy benefits they bring. It should be borne in mind that these buildings have historical, aesthetic and functional value [1]. Therefore, parameter monitoring using technological components is on the increase [2]. The main objective is to generate a system of free access for the study of the climate of old buildings, which will allow us to study the advantages and disadvantages of urban planning and historical buildings in order to optimise their advantages and minimise their disadvantages. This urban planning will condition the architectural typology, as well as the construction details.

2 OBJECT OF STUDY

The object of study (Fig. 1) is a building integrated into a medieval urbanism, in this case Gothic, in the town of La Llacuna. This town is mentioned in documents from 987 as belonging to the county of Barcelona. In 1079, it was owned by the Cervelló family, who became the castle's mistress. In 1347, the barony of La Llacuna was formed, a noble title that lasted until the end of the old regime [3].

Figure 1:   Medieval urbanism of La Llacuna.

The building under study is a building with a gothic elements that has remained unchanged over the last 60 years (Fig. 2), so that it has not suffered any kind of intervention with current materials that provide thermal modifications. The main thermal feature is the conservation of original woodwork and insulation systems.

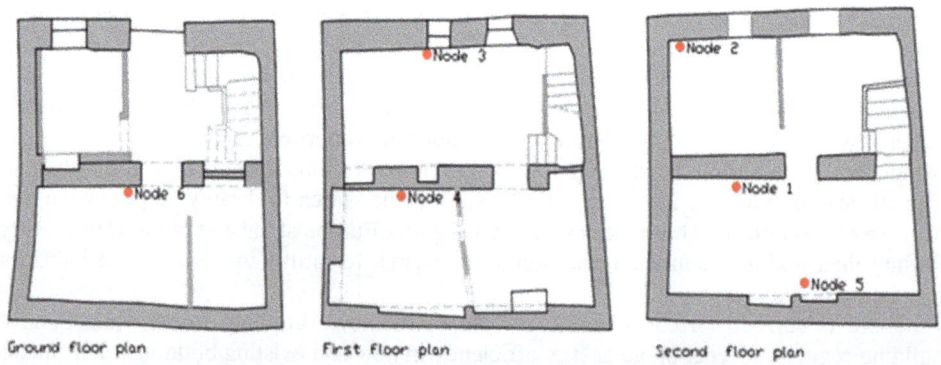

Figure 2: Plants of the building under study and location of the sensors.

3 STATE OF THE ART

A complete study of the building requires an analysis of the history and functions, energy, climate change audit and an evaluation of the compatibility and reversibility of each action [4]. The study will focus on the type of system to be used, as well as the types of sensors to be chosen. Building modelling can be carried out using commercial sensors and systems, the main problem being that they are not modifiable [5]. Therefore, the use of freely available hardware [6] is chosen for the remote collection of climate data [7]. The sensors to be chosen are essential. Francisco Javier Mesas-Carrascosa used temperature and relative humidity sensors in the Mosque-Cathedral of Córdoba [8]. Martin Consuegra studied energy consumption (diesel, electricity) as well as comfort parameters such as interior versus exterior temperature in the main building of the Eduardo Torroja Institute [9]. A complete study refers to the study carried out by Mahdjoubi, where the following sensors are applied: air temperature, relative humidity, CO_2 level, sensor of volatile organic components, light sensor and study of energy consumption [1].

4 DESIGN OF THE SYSTEM

The study method will take into consideration the objective to be achieved and will require the creation of a system for taking climate data in real time by giving it an internet connection. Another requirement is that the resulting system must be easily expandable and adaptable to any monitoring needs that may be necessary to prevent damage to historic buildings, which is why the Open-Source system is used [10]–[14]. The need to design hardware must also be taken into consideration. This point requires the choice of components, sensors, and microcontrollers to be used.

4.1 Selected sensors

The temperature and humidity sensors to be used will be: AM2302 [15], also known as DHT22. This sensor communicates with the microcontroller via a OneWire1 communication bus. According to the manufacturer's specifications, this sensor allows us to measure humidity from 0% to 99.9% with an absolute error of 2% and a resolution of 0.1. As for the temperature, it has a range between −40°C and 80°C with an absolute error of 0.5°C and a resolution of 0.1; AM2320 [16]: This sensor is practically the same as the previous one with the difference that it has an absolute error of 3% humidity and that, besides being able to communicate in OneWire protocol, it can be connected to an I2C bus.

The tilt sensors to be used will be the chosen sensor is the ADXL345 [17]. It is a three-axis accelerometer that allows us to obtain two inclination angles. It gives us the readings in digital format through an I2C or SPI bus, eliminating the need to add an ADC2. Furthermore, it allows us to reach a resolution of 3.9 mg^3 thanks to which we will be able to detect changes in the inclination of less than 1°C. We will use a module called GY-291. This module has the sensor soldered onto a board, as well as additional passive components and allows us to make the connections through pins that are much more convenient to use on prototype boards. With trigonometry, we can find the angles that we want to calculate from the reading of the clarification in each axis. Specifically, the formulas we have used are as follows:

Monitoring the opening of a crack once its appearance has been detected can be very important, since small variations undetectable to the naked eye or even with basic measurement tools, can be useful in detecting movements of the building before irreparable damage occurs. It is interesting to be able to reference this measurement with the ambient temperature, as it will allow us to see how this affects expansion by seeing the differences throughout the day or between seasons. To measure linear displacement, we have decided to do so with an LVDT4. It is an absolute linear displacement transducer that has a very high precision and a very low hysteresis. The LVDT we have chosen is the SM1 from Solartron Metrology; we have chosen this model because of its lower price. It has a lower travel than others (1 mm), but we do not need more, as if the crack exceeds this opening the situation would be critical anyway.

The measurement of the surface temperature is interesting to check the thermal inertia of the walls. Our intention is to measure it on the interior and exterior surface of the façade and see the variation throughout the day and between seasons. For this measurement we have chosen to use thermocouples because of their low cost and the ease with which they can be mounted on any surface with the appropriate adhesive. Its low mass makes its thermal inertia negligible and its temperature is the same as the surface on which it is in contact. Thermocouples consist of the union of two conductors made of different metals. This union produces a voltage difference that varies according to the temperature at each of the ends. When using them, it must be considered that another temperature sensor is needed to compensate for the reference temperature. Thermocouples have a very large measurement range, up to thousands of degrees Celsius, and along this range they do not have a linear behaviour.

By isolating, we obtain the equation of the temperature as a function of the stress, the temperature of the cold junction, and the Seebeck coefficient. We will use type K thermocouples, which have a coefficient of 41 V/°C.

To connect the thermocouples in the microcontroller, we have done so with an ADS1118. This chip is an ADC of the same family as the ADS1115 [18] that we use for LVDT measurement. The difference is that the ADS1118 includes an internal temperature sensor that can be used to compensate for the cold junction temperature of the thermocouples. If we conjure it up to obtain the maximum possible resolution, which is 7.8125 V, we will have a resolution of approximately 0.2°C in the temperature measurement.

4.2 Hardware design

When monitoring the environment health of a building, we often find ourselves in the situation where we must take measurements in multiple locations, in some cases we may be close (measuring temperature at different points on the surface of a wall), other times we may be distant (ambient temperature in different rooms or floors). It would not be surprising to develop sensors where those that are the farthest apart are tens of meters away. That is why it is not viable to have a single microcontroller making the readings of all the sensors.

On the one hand, the distances are too great for the communication buses used by some sensors to work without problems (due to voltage drops and very high capacitances because of very long cables).

On the other hand, it is likely that, as we are dealing with historical buildings, for reasons of aesthetics or conservation, we will not be allowed to install many cables in sight or make holes in the walls to pass them through, and we will be asked to keep the number of visible system elements to a minimum. It is therefore clear that the solution is to use communication without cables. The option chosen is radiofrequency, specifically Wi-Fi technology, because the wide range of hardware available makes it an economical option and the amount of documentation available makes it an easy option to implement. Our network will have two nodes, one of them will act as a bridge between the local network and the Internet.

We have sensors distributed in six nodes (Fig. 3). This allows us to distribute them as we see fit throughout the building where they are easy to install: Node 1 consists of a NodeMCu plate with two different environmental temperature and humidity sensors, one of which is an

AM2302 and the other an AM2320. Each is connected to a OneWire bus; Node 2 apart from the temperature and humidity sensor AM2302 and AM2320, this node also includes the accelerometer for measuring the inclination of a wall. Like the previous one, it uses a NodeMCU board; Node 3 is like the other two previous nodes: NodeMCU board and temperature and humidity sensors AM2302 and AM2320. It also measures the two surface temperatures using two thermocouples; Nodes 4, 5, 6 are identical, only using an AM2320 sensor to measure temperature and humidity. As they only use one sensor, we have chosen to use an ESP-01 module, which is cheaper than a development board.

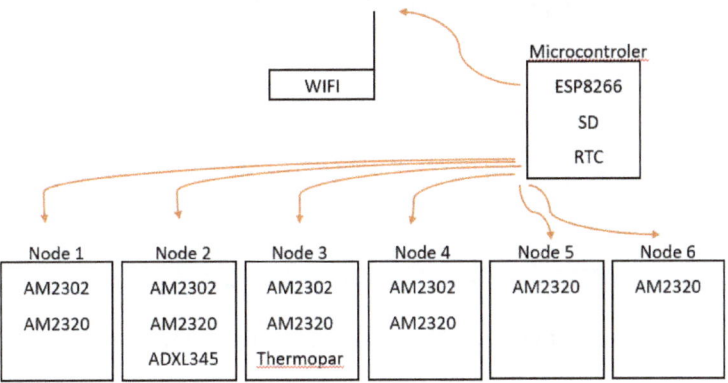

Figure 3: Scheme of monitoring system.

4.3 Design of the network structure

The ESP8266 is a SoC[6] (Fig. 3) that integrates a 32-bit microcontroller and components required for 2.4 GHz Wi-Fi connectivity: as well as other peripherals common to most microcontrollers such as UART, I²C, PWM and ADC.

The use of these chips has been widespread since its creation thanks to the Maker community, which saw potential in it and immediately integrated it into the Arduino environment making it compatible with the hundreds of existing libraries, making its use in hobbyist projects skyrocket. It is because of this large amount of information, documentation, libraries, and examples; as well as his low cost and previous experience, which we have chosen him for this project.

The system is complemented with a SD memory card and a Real-time Clock.

5 RESULT

The results obtained from August 2nd to February 15th (Figs 4 and 5) show the evolution of the knots over two days, being shown in node 1: the graph of humidity (range of 51.1–49.6), temperatures (range of 26.7–20.2); in node 2 the graph of humidity (range of 74.2–49.2), temperatures (range of 27–7.3) and inclination (range of 175.59/172.82–175.17/172.27); in node 3 the humidity graph (range of 59.5–64.1), indoor temperature (range of 8.6–25) and outdoor temperature (range of 33.47–6.45); in node 4 the humidity graph (range of 66.5–66.1) and temperature (range of 24.3–8.7); in node 5 the humidity (range 85.2–75.5) and temperature (range 7–9.9) graphs and in node 6 the humidity (range 64.8–70.1) and temperature (range 7.6–22.4) graphs.

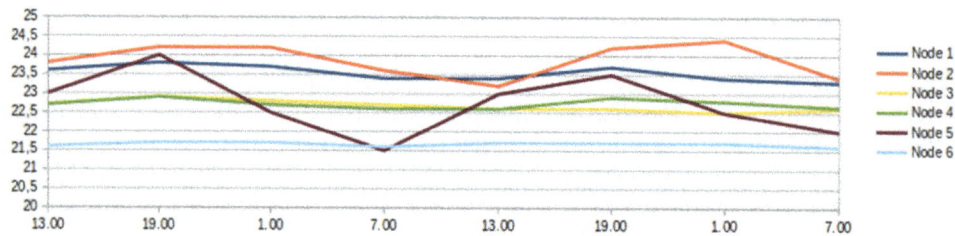

Figure 4: Temperature data of all nodes.

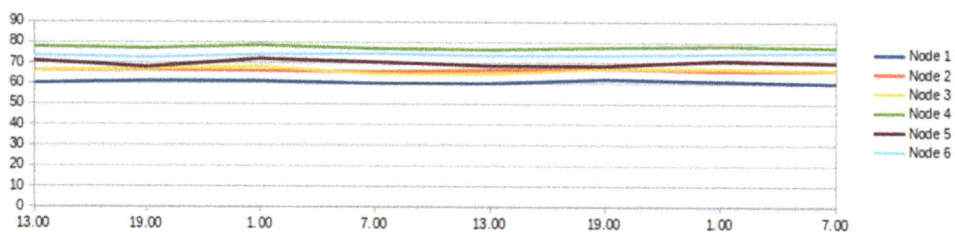

Figure 5: Humidity data of all nodes.

6 DISCUSSION

Regarding the system of chosen nodes: If we study the humidity graphics, we see that the nodes with a type of sensor AM2320 have a different reading than those with the sensor AM2302. They do retain the daily oscillation, but not the oscillation range; the expansion and inclination sensors require a complete structural understanding of the building and its environment in this case.

The study of the temperature on August 15 indicates that the inertia of the building is high since the temperature of the ground floor is 23.2°; nodes 3 and 4 located on the first floor have a stable temperature throughout the day of 25.3 and 25.7; nodes 1 and 2 already have the oscillations caused by the outside temperature, however it should be explained that node 1 (26.3–26.7) is in a room with a ventilated air chamber, avoiding direct radiation; and node 2 (27.2–28.2) is located in a room with a compact floor without insulation. Node 5, installed in the ventilated air chamber, indicates that this is useful to avoid direct impact from the sun.

The same happens in the study of the temperature on January 15 with the following data: node 6 has a temperature of (7.7), nodes 3 and 4 located on the first floor have a stable temperature throughout the day of 7.3 and 8.4; and node 2 (7.2–8). Node 1 has the same temperature and behaviour as the outside temperature with a range of (6.45–16.11).

If we take a reading from the thermal inertias, we can see that the study of 15 August (Fig. 6), where the outside temperature is generally higher than the inside temperature, node 1 (located in a room with a ventilated air chamber on the roof) is better insulated than node 2, located in a room with a compact floor on the roof. This same study to January 15 is observed that in the daily study, the sensor 1 even maintaining the oscillation of the inertia, the temperature is near to the outside, on the other hand the node two, also maintains the oscillation, but the temperature is lower.

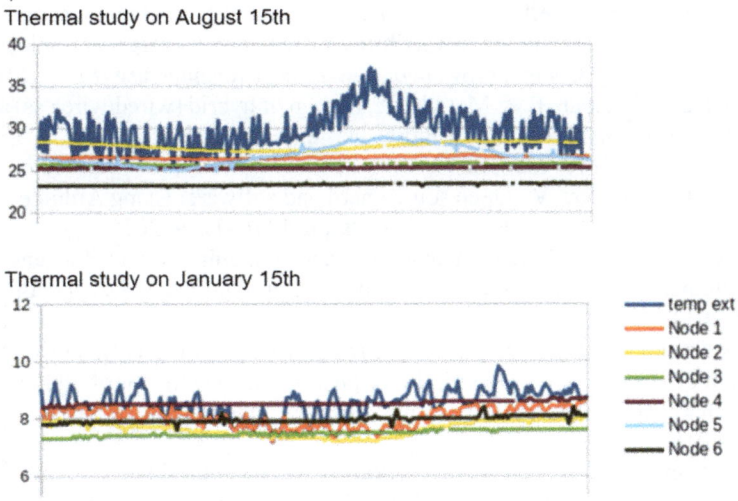

Figure 6: Study of temperature of all nodes.

In summer, nodes 3 and 4 follow the same temperature range, favoured by the low incidence of the sun on the façade and node 6 is the coolest room, the furthest from the façade and the roof. On the other hand, in winter, node 3 (closest to the façade) is the lowest temperature, and node 6 (furthest from the façade and the roof) is the highest temperature.

7 CONCLUSION
The system designed for free access is useful for heritage building modelling. One of the main concerns the usefulness of the ventilated roof. In summer, it helps to keep temperatures low and in winter it helps to maintain the temperature interior of the building. A good example of this type of roof is the Catalan-style roof or the ventilated tile roof. So, this is an element to be preserved in the building. The thermal inertia is more favourable in summer than in winter, as studies show that point 6 has a range of difference of 3°–15° with the outside climate and in winter point 6 has a range of difference of 1°–2° with the outside.
The future works, it will be necessary to propose a system that always provides information on the surface temperature of each room, as well as the hygrometric oscillations of the subsurface in the case of humidity on the ground floor.

ACKNOWLEDGEMENTS
Collaboration, writing and technical assistance are grateful to Melanie Bas, Ferran Modinos, Hajar Oubaddi, Maria Margalef, Pol Fernandez, members of the PATRIARC-CAIT research group.

REFERENCES
[1]   Mahdjoubi, L. et al., A guide for monitoring the effects of climate change on heritage building materials and elements, 2017.
[2]   Llie, M., Smith, R., Reed, J. & Inglis, R., Southwest Scottish crannogs: Using in situ studies to assess preservation in wetland archaeological context. *Journal of Archaeological Science*, **35**, pp. 1886–1900, 2008.
[3]   AAVV, Enciclopedia Catalana. Accessed on: Aug. 2020.

[4]   Lucchi, E., Energy efficiency in historic building: A tool for analysing the compatibility, integration and reversibility of renewable energy technologies. *World Renewable Energy Congress. Sweden. Low-Energy Architecture (LEA)*, 2011.

[5]   Diego, F.J.C., Esteban, B. & Merello, P., Design of hybrid (wired/wireless) acquisition data system for monitoring of cultural heritage physical parameters in smart cities. *Sensors*, **15**, pp. 7246–7266, 2015.

[6]   Faugel, H. & Bobkov, V., Open source hard and software: Using Arduino to keep old hardware running. *Fusion Engineering*, **88**, pp. 1276–1279, 2015.

[7]   Leccese, F. et al., A new acquisition and imaging system for environmental measurements: An experience on the Italian cultural heritage. *Sensors*, **14**, pp. 9290–9312, 2014.

[8]   Mesas-Carrascosa, F.J., Verdú Santano, D., Meroño de Larriva, E., Ortiz Cordero, R., Hidalgo Fernangez, R.E. & Garcia-Ferrer, A., Monitoring heritage building with open-source hardware sensors: A case of the Mosque-Cathedral of Córdoba. *Sensors*, **16**, p. 1620, 2016.

[9]   Martin-Consuegra, F., Oteiza, L., Alonso, C., Cuerdo-Vilches, T. & Frutos, B., Análisis y propuesta de mejoras para la eficiencia energética del edificio principal del Instituto c.c. Eduardo Torroja-CSIC. *Informes de la Construcción*, **66**, pp. 536, eo43, 2014.

[10]   Chang, P.C., Flatau, A. & Liu, S.C., Review paper: Health monitoring of civil infrastructure. *Structural Health Monitoring*, **2**(3), pp. 257–267, 2003.

[11]   Basto, C., Pelà, L. & Chacón, R., Open-source digital technologies for low cost monitoring of historical constructions. *Journal of Cultural Heritage*, **25**, pp. 31–40, 2017.

[12]   AdCiv, Open collaborative design. http://www.adciv.org/Open_collaborative_design. Accessed on: Aug. 2018.

[13]   St. Laurent, A.M., *Understanding Open Source and Free Software Licensing*, O'Reilly Media, 2004.

[14]   Gibb, A., *Building Open Source Hardware: DIY Manufacturing for Hackers and Makers*, Addison-Wesley, pp. 253–277, 2015. Accessed on: Aug. 2018.

[15]   Aosong Electronics, AM2302 datasheet. https://akizukidenshi.com/download/ds/aosong/AM2302.pdf. Accessed on: Aug. 2018.

[16]   Aosong Electronics, AM2320 datasheet. https://akizukidenshi.com/download/ds/aosong/AM2320.pdf. Accessed on: Aug. 2018.

[17]   Analog Devices, ADXL345 datasheet. http://www.analog.com/media/en/technical-documentation/datasheets/ADXL345.pdf. Accessed on: Aug. 2018.

[18]   Texas Instruments, ADS1115 datasheet. http://www.ti.com/lit/ds/symlink/ads1115.pdf. Accessed on: Aug. 2018.

INDOOR CORROSIVITY IN KLEMENTINUM BAROQUE LIBRARY HALL, PRAGUE

KATERINA KREISLOVA[1], PAVLINA FIALOVA[1] & TEREZA BOHACKOVA[2]
[1]SVUOM Ltd., Czech Republic
[2]VSCHT, Czech Republic

ABSTRACT
Air pollution in museums, libraries, churches, and other indoor environments of culture heritage objects was studied since 1980s. For classification of corrosivity of such type of indoor environments the ISO 11844 series was developed. The indoor corrosivity for basic metallic materials had been studied for Baroque library hall of Klementinum areal, Prague, using exposure of metallic coupons and measuring the climatic and pollution parameters. The Baroque library was first opened in 1722 year as a part of the Jesuit university based in Klementinum. It houses over 20,000 books, from the beginning of the 17th century until recent times. As the building is located in the centre of Prague, the outdoor environment is polluted mainly by nitrogen oxides from traffic source, but in 1990s of last century the sulphur dioxide pollution was also on high level. The Baroque hall is cladded by wooden panels which maybe a source of vapour organic acids corrosive for some metals and other materials as fresco paintings. The indoor gaseous air pollution was measured for 2 years with passive samplers – SO_2, NO_x, vapour acetic acid. The new method for acetic acid measurement was verified. Metallic coupons of zinc, coper, silver and lead were exposed in this hall for one year and their mass increases and corrosion mass losses were evaluated. For metallic coupons the colour change is one of measured parameters, too. Repeated one-year measurement data are presented and indoor corrosivity according to ISO 11844 was determined.
Keywords: indoor corrosivity, air pollution measurement, organic acid vapours, metallic coupons corrosion loss.

1 INTRODUCTION

The large complex of Klementinum was founded by the Jesuits after their arrival in Bohemia in 1556 year. Initially, members of the order lived in a former Dominican monastery, but in 1653 year began expanding their premises. The reconstruction lasted over 170 years, so there is a variety of architectural styles in Klementinum. With more than 2 hectares it is also one of the largest building complexes in Europe and a part of Prague UNESCO cultural heritage.

The Baroque library was first opened in 1722 year as a part of the Jesuit university based in Klementinum (Fig. 1). The interior of the Baroque library has remained intact since the 18th century. The library currently contains a number of important and unique works, that have global significance. It houses over 20,000 volumes of mostly foreign theological literature, coming into Klementinum from the beginning of the 17th century until recent times. Books with white painted spines and red marks have been in the library since the time of the Jesuits.

As the building is located in the centre of Prague, the outdoor environment is polluted mainly by traffic source, but in 1990s of last century the sulphur dioxide pollution was also on high level. Both these air pollutions affected the indoor environment. Presented article gives the new results about the characterisation of indoor atmospheric corrosivity in Baroque library hall. The climatic parameters and air pollution were measured for long period and show the effect of changing outdoor air pollution onto their indoor level.

The Baroque hall is cladded by wooden panels and equipped by wooden bookcases which maybe a source of vapour organic acids corrosive for some metals used in historic books as decoration and painting pigments and mainly the paper itself. The specific pollution by organic vapour acids (formic, acetic) was measured together with metallic coupons exposure to determinate indoor corrosivity. Repeated one-year measurement data are presented and indoor corrosivity according to ISO 11844 was determined.

Figure 1: Klementinum Baroque library hall.

2 METHODOLOGY OF INDOOR CLASSIFICATION

Air pollution in museums, libraries, churches, and other indoor environments of culture heritage objects was studied since 1980s [1], [2]. One significant source was outdoor air pollution, mainly by SO_2 and NO_x, which concentration decreased according to distance of outdoor source and type of building. Contemporary outdoor air pollution situation in Europe has not significant effect on indoor environments. For classification of indoor environments corrosivity of storage rooms, museums and churches the ISO 11844 *Corrosion of metals and alloys – Classification of low corrosivity of indoor atmospheres* series was developed. The evaluation of low-corrosivity indoor atmospheres can be accomplished by direct determination of corrosion attack of selected metals (see ISO 11844-2) or by measurement of environmental parameters (see ISO 11844-3) which may cause corrosion on metals and alloys. This classification and mainly the methods for determination and/or estimation of indoor corrosivity had been applied in many cultural objects [3]–[7].

The determination of corrosivity of indoor atmospheres is based on measurements of corrosion attack on triplicate standard specimens of five reference metals (carbon steel, zinc, copper, silver, lead) after their exposure for one year. In 2020 the revision of this ISO 11844 series had been finished, where lead was included as standard specimen with high sensitivity to vapour organic acids. Rectangular coupons in the form of flat sheets shall be exposed vertically.

Environmental characteristics are informative and allow assessment of specific corrosion effects with regard to individual metals. Corrosion for many of the metals is significantly influenced by the synergistic effects of different pollutants. Due to the permanent exchange between indoor and outdoor air caused by infiltration and ventilation processes, it may be important to supplement indoor air measurements with a simultaneous measurement of the outdoor air. For SO_2 and NO_x the passive sampling was used on the monthly basis for long-term period at outdoor environment and in indoor environment (Baroque library hall). The model estimates the indoor concentration of pollutants originating from outdoors was derived for the steady-state indoor/outdoor (I/O) relation as 0.5 for sulphur dioxide and 0.6 to 0.8 for nitrogen oxides.

But indoor environments are sources of specific type of pollution themselves. Especially applied materials for construction of showcases are potential emission source for variety of volatile organic compounds [8], [9]. Formaldehyde, formic and acetic acids belong to the most discussed pollutants in the museum and similar environments. Their corrosion impact increases in the following order: formaldehyde < formic acid < acetic acid. The active sampling and long-term passive sampling were used for acetic acid measurement in Baroque library hall.

3 MEASURED ENVIRONMENTAL AND CORROSION DATA

3.1 Environmental data at Baroque hall

The Baroque library hall is not heated space, but other part of building is heated. Climatic data in library were measured in hours intervals. Temperature was in range 11°C to 26°C and relative humidity in range 37% to 62% (Table 1). The corrosion level can be expected to increase for each 10% rise in RH above 50%, and also for each rate of change of RH greater than 6% per hour. In Table 1 the climatic data are summarised and classified levels are indicated.

Table 1: Summarising of environmental parameters in Baroque hall.

Level	Temperature (°C)	Relative humidity (%)	Concentration of gaseous pollutants, c ($\mu g \cdot m^{-3}$)		
			SO_2	NO_x	H_2S
I	$T < 15$	$RH < 40$	$c < 1$	$c < 1$	$c < 0.3$
II	$15 \leq T < 20$	$40 \leq RH < 50$	$1 \leq c < 5$	$1 \leq c < 5$	$0.3 \leq c < 1.5$
III	$25 \leq T < 30$	$50 \leq RH < 70$	$5 \leq c < 10$	$5 \leq c < 10$	$1.5 \leq c < 3$
IV	$T \geq 30$	$RH \geq 70$	$c \geq 10$	$c \geq 10$	$c \geq 3$

The SO_2 air pollution in the centre of Prague was relatively high till 1990s of the last century and then decreased (Fig. 2(a)). The NO_x air pollution depended on traffic intensity in this locality which decreased with changes in traffic policy and building new roads. Air pollution by SO_2 and NO_x is measured by passive samplers according to ISO 9225 since 2007 in outdoor localities of the Klementinum buildings and in selected indoor localities, including Baroque library hall (Fig. 2(b)). In 2006/07 the SO_2 concentration was ca 9.0 $\mu g \cdot m^{-3}$ at outdoor atmosphere and 3.5 to 4.0 $\mu g \cdot m^{-3}$ at indoor atmosphere, i.e., ca 40%–45% from outdoor pollution level. Contemporary SO_2 at outdoor atmosphere decreased on ca 5 $\mu g \cdot m^{-3}$ and the pollution level in indoor is on the same level so the I/O ration is only 0.8. Contemporary NO_x air pollution at outdoor atmosphere is ca 25 $\mu g \cdot m^{-3}$ and the pollution level in indoor is ca 14 $\mu g \cdot m^{-3}$ so the I/O ration is 0.6. In Table 1 the pollution data are summarised and classified levels are indicated.

The pollution effect of indoor corrosion is specific for each metal and interdependent (combination of contaminants, humidity and temperature effects). The concentration of all

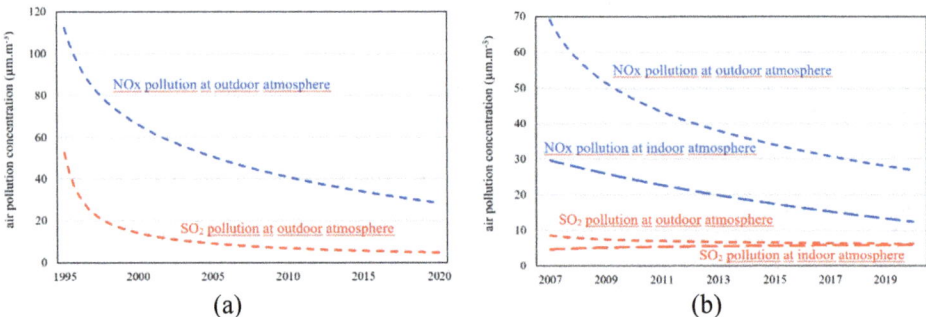

Figure 2: Trends in air pollution in locality. (a) Long-term trend in outdoor air in Klementinum locality; and (b) Comparison between outdoor and indoor pollution in Klementinum.

Figure 3: Passive sampling of acetic acid.

measured corrosive gaseous pollution was very low. Only exception is nitrogen oxide which is not so corrosive as other acidic pollution.

The standardised approach does not include concentration of vapour organic acids which are corrosive to metal as zinc and lead. There are no classification levels for organic vapour acids. Formic and acetic acids' concentrations in Baroque library hall were measured by using active and passive sampling by Gastec Dosimeter Tubes. CH_3COOH concentration was measured as 1,228 $\mu g \cdot m^{-3}$ (0.5 ppm ~ lower limit for detection) and HCHO concentration was 377 $\mu g \cdot m^{-3}$ (0.2 ppm). This method was used for H_2S with under lower detection limit (<0.01 ppm).

For more precise and long-term measurement of acetic acid vapour, the passive samplers were used (Fig. 3). Samplers are stainless steel mesh discs in plastic tubes containing 40 μl of 1 M potassium hydroxide solution and 10% (v/v) ethylene glycol dimethyl ether [10]. After exposure the absorbed acetic acid content is analysed from water extract by ion chromatography method. Based on the obtained mass m, known sampling rates v and sampling time t, the concentration c of analytes in the sampled air in mass per air volume was calculated with equation:

$$c = \frac{m}{v \; x \; t}.$$ (1)

The yearly average of monthly values of vapour acetic acid is ca 280 μg·m^{-3} (0.11 ppm) in Baroque library hall, but it shows very significantly dependence on temperature (Fig. 4). In summer months the average acetic acid concentration was ca 460 μg·m^{-3} (0.20 ppm). During this period the temperature in the Baroque hall was ca 25°C. In other months when average temperature was 15.5°C the acetic acid concentration was 225 μg·m^{-3}. Although the wooden cladding and furniture in Baroque hall is created some centuries ago there is still significant source of vapour organic acids. The differentiation of passive measurement of acetic acid by passive tubes exposed only few days in indoor environment and passive samplers exposed for one month gives different values, but passive samplers are more sensitive method.

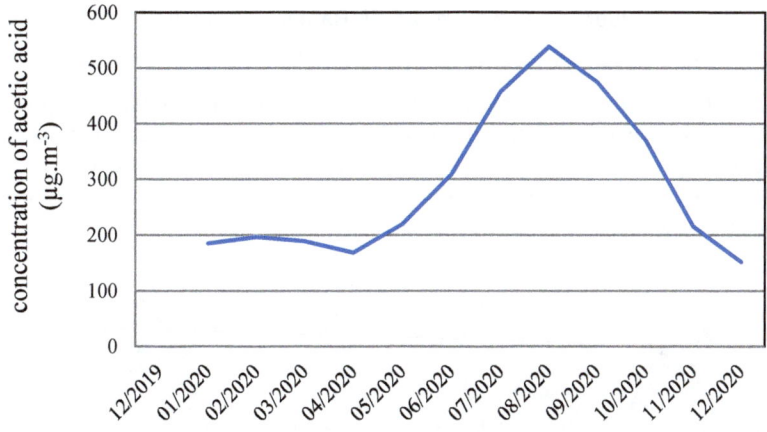

Figure 4: Acetic acid concentration measurement.

3.2 Corrosion data at Baroque hall

The metallic coupons (zinc, copper, silver, lead) with dimension 30 × 80 mm and 50 × 50 mm were prepared from metallic sheets. The weight loss coupons were abraded with silicon carbide paper to 320 grids, cleaned with de-ionised water, and degreased with ethanol prior to exposure. The triplicate coupons of each metal were exposed vertically mounded on small plastic racks in selected indoor environments for 1 year period (Fig. 5). The coupons designed for electrolytic reduction were polished with P1200 emery paper, rinsed with de-ionised water and degreased with ethanol in ultrasound bath for 5 minutes.

The colour (ΔE) and gloss (ΔG) changes were measured in measured by spectrophotometer Specro Guide Gloss S, fy BYK-Gardner, USA to use CIEL*a*b* regular coordinates. Change of colour is expressed as ΔE value calculated according to equation:

$$\Delta E = \sqrt{\Delta a^2 + \Delta b^2 + \Delta L^2},$$ (2)

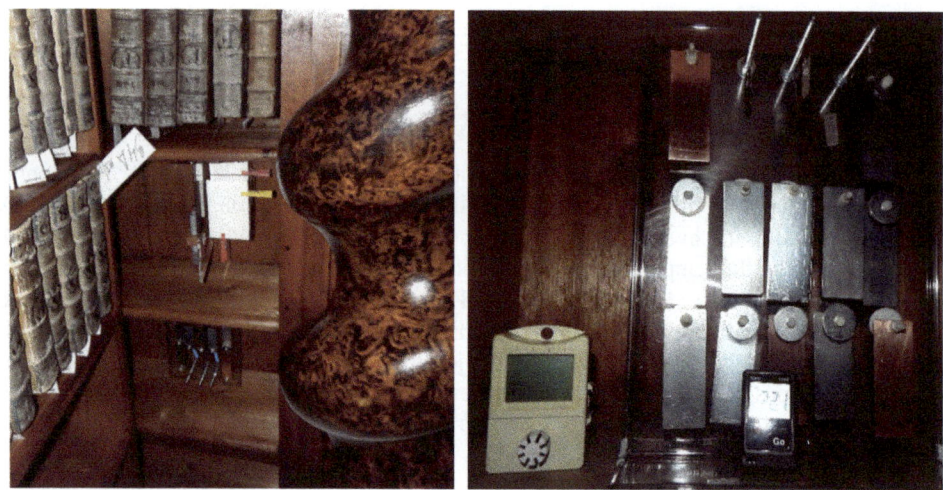

Figure 5: Exposure rack in Baroque library hall.

where a, b, L are coordinates of three-dimensional space for each metallic coupon measured before and after one-year exposure. The appearance of 1 year's exposed coupons is shown in Fig. 6. The most visually evident change of coupons occurred on silver coupons. The colour changes of exposed coupons are given in Table 2. The most intensive changes occurred on zinc and lead on L scale (white/black) and on silver on b scale (yellow/blue).

Figure 6: The appearance of exposed coupons (zinc, copper, silver, lead).

Table 2: Visual changes of exposed coupons.

Material	L	A	b	ΔE	ΔG
Zinc	−13.38	0.08	1.79	13.02	89.17
Copper	−0.87	−0.89	−0.27	1.86	25.23
Silver	−3.00	1.95	6.24	7.72	27.10
Lead	−21.37	−0.49	−1.43	21.43	6.33

The mass gains were evaluated after exposure by weighting and corrosion mass loss were estimated after removal of corrosion products by interval pickling procedure according to ISO 8407 [11] and by electrolytic cathodic reduction (for 50 × 50 mm coupons only) according to ISO 11844 [12]. Values and classification of corrosivity of indoor atmospheres based on rate of mass increase measured and based on corrosion rate measurements by mass loss determination of standard coupons are given in Table 3.

Table 3: Corrosion data of exposed coupons.

Material	Mass increase		Corrosion rate	
	$(mg \cdot m^{-2} \cdot a^{-1})$	Corrosivity category	$(mg \cdot m^{-2} \cdot a^{-1})$	Corrosivity category
30 × 80 mm coupons				
Zinc	111.1	IC2	201.4	IC2
Copper	13.9	IC1	131.9	IC2
Silver	−13.9	−	83.3	IC1
Lead	715.3	−	5,645.8	−
50 × 50 mm coupons				
Zinc	75.0	IC2	61.6	IC2
Copper	−96.7	−	216.9	IC3
Silver	−57.7	−	114.6	IC2
Lead	629.1	−	1,702.3	−

The results are affected by method of estimation – coupons size, corrosion products' removal, etc. Each exposed metal reacts to specific kinds of pollutants which may be present in indoor environments. Copper corrosion is often associated with chlorides, sulphides, and acidic pollutants such as NO_2 and SO_2. Silver reacts with sulphides such as carbonyl sulphide (COS) and hydrogen sulphide (H_2S). Lead reacts with organic carbonyl pollutants and acidic pollutants, but the classified intervals proposed in ISO 11844-1 are lower than estimated at Klementinum Baroque library hall. Zinc is also sensitive to organic acid pollutants and it is metal mostly reactive in high humidity.

Although the corrosion layer forming was evident from colour and mass changes the layers were so thin than any analytical methods (EDX, XRD, etc.) cannot be used for their identification.

The corrosion rate of copper, silver and lead was measured by resistive AirCorr sensors in period 09–11/2020 (period with average acetic acid concentration 300 $\mu g \cdot m^{-3}$) too. Courses of corrosion rates measured by sensor is given in Fig. 7 – for copper 0.0014 $\mu m \cdot a^{-1}$, for silver 0.0062 $\mu m \cdot a^{-1}$ and for lead 0.0261 $\mu m \cdot a^{-1}$. The corrosion rate of copper is slightly reduced due to forming of patina layer. The corrosion rate of lead was significantly higher than for other metals which shows the effect of organic acid pollution.

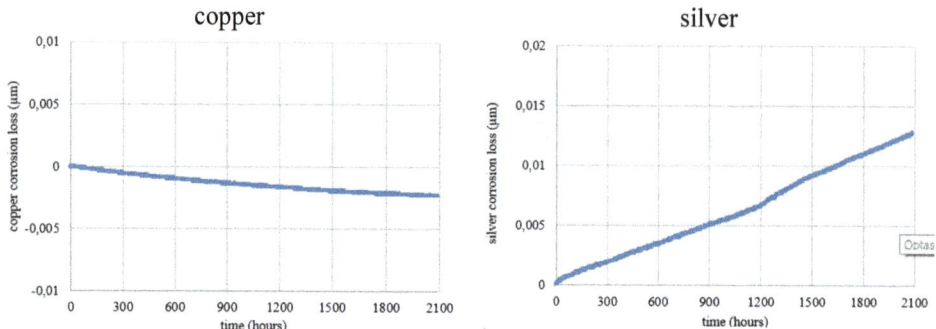

Figure 7: The AirCorr sensors' corrosion loss (copper, silver).

4 CONCLUSION

In both storage and display, the processes of corrosion may threaten the stability of metal artefacts. Organic acids, mainly acetic and formic, are known to be significant pollutants in the museum and similar indoor environments. Higher concentrations are usually observed in confined spaces, such as display cases or storage enclosures, due to low air-exchange rates. The Klementinum Baroque library hall is large space, but the exchange of air is reduced as there is limited access to this hall. The library was closed for public since the 1920s till the year 2000.

The acetic acid is known to affect metals (particularly lead), calcareous materials (shell, limestone, calcium-rich fossils), soda-rich glass, and cellulose. Examples of deterioration due to acetic acid include the corrosion of lead-rich organ pipes in churches [13], discoloration of pigments [14], and depolymerisation of paper [15]. The acetic acid has been shown to not effect paper lifetime significantly at values below 250 $\mu g \cdot m^{-3}$ (100 ppb) [16], but in Klementinum Baroque library hall this limit values had been exceeded during the half of year. Bronze and zinc are also affected by organic acids but to a lesser extent [17].

The measurement of their concentration by commercially available simple methods cannot give correct values due to very low concentration and detection limits of these instruments. Passive samplers are more precise and give better values.

Although the values of corrosion mass changes (increasing and loss) depend on used methods (coupons' size, evaluation), in principle the atmospheric indoor corrosivities of Baroque library hall for tested metals are the same. The complex measurement gives information about indoor corrosivity and prevention means which are necessary for long-term protection of exposed culture heritage artefacts.

Results of determination of corrosivity category in 17 localities of culture heritage objects show that lead is the most sensitive metal from others (zinc, copper, silver) [18]. In many exposure localities the yearly corrosion loss r_{corr} of lead was higher than maximal value for corrosivity category IC5 according to revised ISO 11844-1 (1,600 $mg \cdot m^{-2} \cdot a^{-1}$), although for other standard metals the corrosivity categories were only IC1 to IC3. Baroque library hall Klementinum belongs between these localities, too.

ACKNOWLEDGEMENT

This article was created with the support of NAKI Project No. DG18P02OVV050 from the Ministry of Culture of the Czech Republic.

REFERENCES

[1] Brimblecombe, P., The composition of museum atmospheres. *Atmospheric Environment B*, **24**, pp. 1–8, 1990.

[2] Tétreault, J., *Airborne Pollutants in Museums, Galleries, and Archives: Risk Assessment, Control Strategies, and Preservation Management*, Canadian Conservation Institute: Ottawa, 2003.

[3] Balocco, C., Petrone, G., Maggi, O., Pasquariello, G., Albertini, R. & Pasquarella, C., Indoor microclimate study for cultural heritage protection and preventive conservation in the Palatina Library. *Journal of Cultural Heritage*, **22**, pp. 956–967, 2016.

[4] Saheb, M. & Dubus, M., Indoor corrosivity in museums and archives assessment: Standards and recommendations. *7th Indoor Air Quality*, 2006.

[5] Kreislova, K., New development of indoor corrosivity classification. *11th International Conference Indoor Air Quality in Heritage and Historic Environments*, 2014.

[6] Prosek, T. et al., Real-time monitoring of indoor air corrosivity in cultural heritage institutions with metallic electrical resistance sensors. *Studies in Conservation*, **58**(2), pp. 117–128, 2013.

[7] Dubus, M. & Prosek, T., Standardized assessment of cultural heritage environments by electrical resistance measurement. *e-Preservation Science*, **9**, pp. 67–71, 2012.

[8] Angellini, E. et al., Atmospheric corrosion of facts in museum indoor environments. *EUROCORR 2019*, 2019.

[9] Lafuente, D., Cano, E., Llorente, I., Crespo, A., Künne, J. & Schieweck, A., The effects of organic pollutants on metals in museums: Corrosion products, synergistic effects and the influence of climatic parameter. *Proceedings of METAL 2013*, 2013.

[10] Gibson, L.T. & Watt, C.M., Acetic and formic acids emitted from wood samples and their effect on selected materials in museum environments. *Corrosion Science*, **52**, pp. 172–178, 2010.

[11] ISO 8407, Corrosion of metals and alloys – Removal of corrosion products from corrosion test specimens. 2020.

[12] ISO 11844-2, Corrosion of metals and alloys – Classification of low corrosivity of indoor atmospheres – Part 2: Determination of corrosion attack in indoor atmospheres.

[13] Niklasson, A. et al., Air pollutant concentrations and atmospheric corrosion of organ pipes in European church environments. *Studies in Conservation*, **53**, pp. 24–40, 2008.

[14] Oikada, T. et al., Volatile organic compounds from wood and their influences on museum artifact materials I. Differences in wood species and analyses of causal substances of deterioration. *Journal of Wood Science*, **51**(4), pp. 363–369, 2005.

[15] Dupont, A.L. & Tetreault, J., Cellulose degradation in an acetic acid environment. *Studies in Conservation*, **45**(3), pp. 201–210, 2000.

[16] Tennent, N.H. & Baird, T., The identification of acetate efflorescence on bronze antiquities stored in wooden cabinets. *The Conservator*, **16**, pp. 39–43 and 47, 1992.

[17] Menart, E., de Bruin, G. & Strlič, M., Effects of NO_2 and acetic acid on the stability of historic paper. *Cellulose*, **21**(5), pp. 3701–3713, 2014.

[18] Kreislova, K., Fialova, P., Strachotova, K., Svadlena, J. & Majtas, D., Indoor corrosivity classification based on lead coupons, in press.

SIMPLIFIED PROCEDURE FOR STRUCTURAL INTEGRITY EVALUATION OF COMPLEX WALLS THROUGH SIGNAL ENERGY ANALYSIS BASED ON VIBRATIONAL DATA

IVANO ALDREGHETTI, GIOSUÈ BOSCATO, MAURO FIORIN,
LORENZO MASSARIA & VINCENZO SCAFURI
Laboratory of Strength of Material LabSCo, University Iuav of Venice, Italy

ABSTRACT

The Structural Health Monitoring based on vibrational data offers an effective, reliable, and non-destructive methodology to assess the structural integrity of existing buildings. This aspect is particularly appreciated for the architectural heritage because enables to characterise the global structural condition of a monument in respect of its preservation. On the other hand, the structural assessment by dynamic identification is an onerous procedure, both in terms of costs of monitoring and data processing time. For this reason, the procedure proposed in this paper offers a simplified solution for structural integrity assessment. The goal is to detect, localise, classify, and assess the alteration of the global stiffness and weak areas through the comparison of the signal energy based on vibrational data and related pattern. The procedure was applied on a case study to better understand structural integrity procedure and to evaluate approach reliability. The time histories recorded on complex historical façade through "input–output" and "output only" methods were elaborated and compared. A fixed grid of excitation and recording points was created to map the distribution of signal energy values. The lower signal energy values are able to detect the loss of stiffness marking the vulnerability of structure, validated by experimental modal analysis. The comparison between experimental and numerical dynamic parameters with the results of signal energy analysis enables to identify the degree of connection between the sub-parts defining the potential macro-element mechanisms.
Keywords: SHM, signal energy, structural integrity, experimental modal analysis, dynamic identification.

1 INTRODUCTION

The conservation of cultural heritage depends on preventive control and monitoring of structures [1]–[4]. The methodology that guarantees the global control of structure aimed at verifying vulnerability is based on the analysis of vibrational response to ambient and mechanical actions. As the type of input varies, known or unknown, the EMA (Experimental Modal Analysis) and OMA (Operational Modal Analysis) are defined in input–output and only-output methodologies, respectively [5]–[9].

Although the reliability of the dynamic identification procedures is now proved [10]–[15], the methodology remains onerous in terms of time both in operational phase and in data processing phase. In this sense, this research proposes a simplified methodology based on vibrational analysis and relative signal energy to identify the critical areas that affect the global structural integrity.

Through a mapping of signal energy distribution of vibrational data building with respect to survey points and stressed areas, it is possible to identify the critical points. The procedure was applied to a case study through the excitation induced both by ambient noise and mechanically by means of the instrumented hammer.

In the first case, the structure was affected globally with an input that consistently involved the building; while in the second case the signal energy of locally induced vibration was suitably normalised with respect to the known input entity to compare the amplitudes of different signals.

WIT Transactions on The Built Environment, Vol 203, © 2021 WIT Press
www.witpress.com, ISSN 1743-3509 (on-line)
doi:10.2495/STR210121

The results obtained were compared with the modal shapes identified experimentally, demonstrating the good reliability of proposed procedure.

2 SIMPLIFIED PROCEDURE

The disastrous events of the last decades that have affected the national and international cultural heritage have confirmed the need to monitor the structures to define a prevention approach for safeguarding.

The control/monitoring procedure enables a simple and immediate detection of crucial points that present discontinuity on structural integrity by analysing the vibration propagation on the structure. These points will be validated identifying modal shapes that reveal macroelements and/or sub-parts responses than global behaviour.

The proposed simplified procedure aims to interpret the values obtained from the analysis of the acceleration intensity of vibrations. The vibrations are induced mechanically by the impact hammer according to the input–output procedure. The method organises the signal energy values in a graduated scale from zero to the maximum value of the hammering, comparing the data collected in a given time lapse, the procedure characterizes the structure into coloured zones representing the value of signal energy.

The process to obtain this kind of graphics rendering in the procedure requires a vector file containing the drawing of the perimeter of the wall, the openings (see the left rectangle, Fig. 1) the positions of accelerometer sensors (circles 3, 5, 7, 6 of Fig. 1) and hit points (filled circles B1 and B2 of Fig. 1), the vibrations and related signal energy values.

Then, the procedure will create a model where different channels are combined into adjacent triangles having three channels as vertices. It divides the sides of these triangles based on the signal energy dispersed on that distance (Fig. 2(a)) and then joins the points with the same value of the signal energy by means of a smooth curve (Fig. 2(b)).

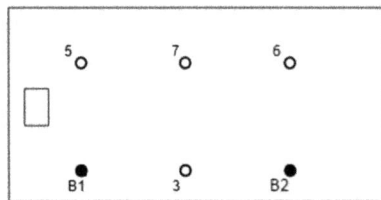

Figure 1: Generic façade: Data needed for the analysis with two hits (filled circles, B1 and B2) and channel points (circles 3, 5, 7, 6).

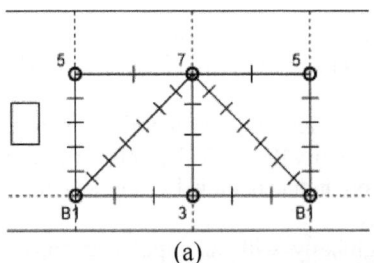
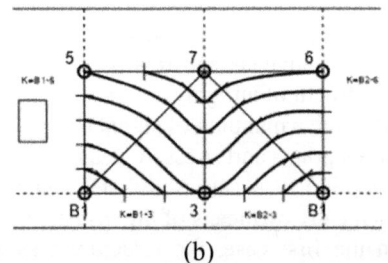

(a)	(b)

Figure 2: (a) Division by energy values of the initial scheme; and (b) Subsequent combination by means of a smooth curve.

For the part of the façade that cannot be analysed with experimental data, the procedure applies a method where the signal energy for every point is based on:

$$k = \frac{\text{Hit's Energy} - \text{Point's Energy}}{distance}$$

where k is the coefficient of energy dispersion on the façade and it depends on the position of the recording point compared to the hitting point and is calculated according to the closest signals (Fig. 3), and d is the distance between the point and the reference hit point. In case of openings between the two points, the distance is calculated according to the minimum path between the recording point and hit point (Fig. 4).

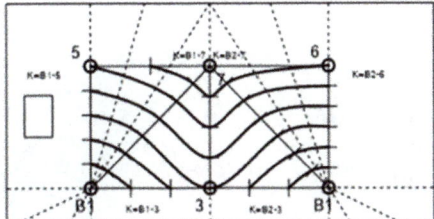 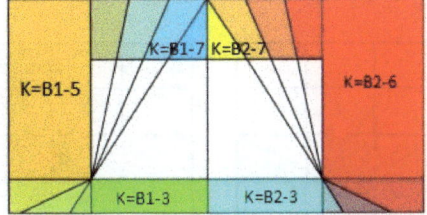

Figure 3:   Division of the external part of the façade into areas with the same energy dispersion coefficient k, and areas where k changes gradually in a radial manner with respect to the hit.

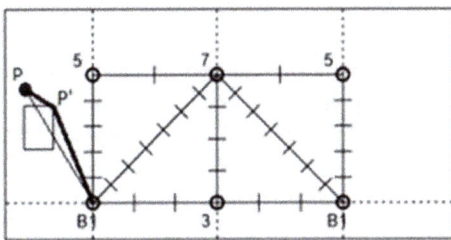

Figure 4: The distance between P and B1 is the sum of PP' and P'B1.

It is thus obtained the subdivision of façade that will be composed of uniform curved lines in the areas among measured channels, in circumference arcs on all constant dispersion coefficient zones that over the windows (Fig. 5).

An immediate understanding would be obtained by analysing the propagation of the vibration power by comparing each pixel calculated to the adjacent ones and reassigning a colour to their sum that highlights the energy absorbed by the façade. An energy value is assigned to each pixel, comparing this value with those of adjacent pixels, it is possible to assign the value of the dispersion coefficient (Fig. 6(a) and 6(b)). Using a colour scale for the values of k, it is possible to obtain another graphic which highlights the different parts of the façade according to their response to the signal. In Fig. 6(c), red colour highlights the stiffest part of façade with a lower k value, whereas dark green colour characterises the most deformable part of the wall.

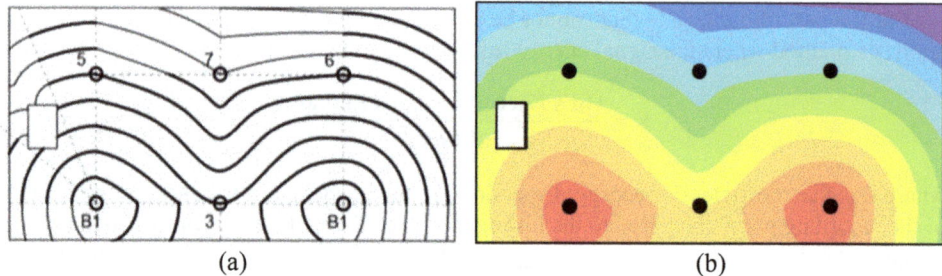

Figure 5:   Contour plot and related colour map of unscaled signal energy over a generic façade.

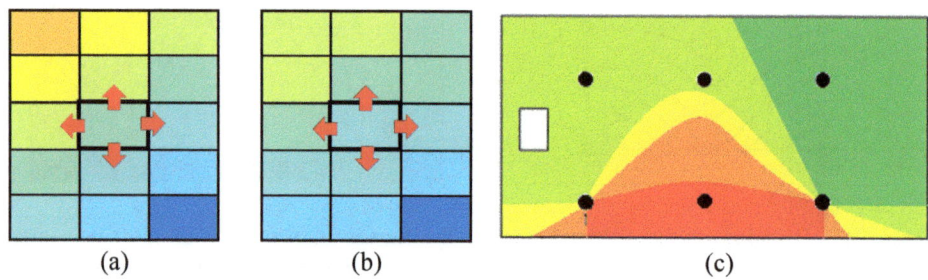

Figure 6:   Pixel coloration in a deformable situation (a) or a rigid one (b) energy dispersion colour map over a generic façade (c).

Another type of analysis can be done with a different methodology, by comparing each pixel of Fig. 7(a), obtained using the vibration values measured in situ over the generic façade, with respect to the 'ideal' behaviour of the same façade (Fig. 7(b)). Fig. 8 shows the differences from the ideal model, expected in the case of a façade made of a homogeneous material and uniform thickness, characterised by an energy propagation equal to the average dispersion or that of standard masonry. The figure shows in shades of red in the areas with defective energy compared to the ideal situation and in blue those that are in excess. Fig. 9

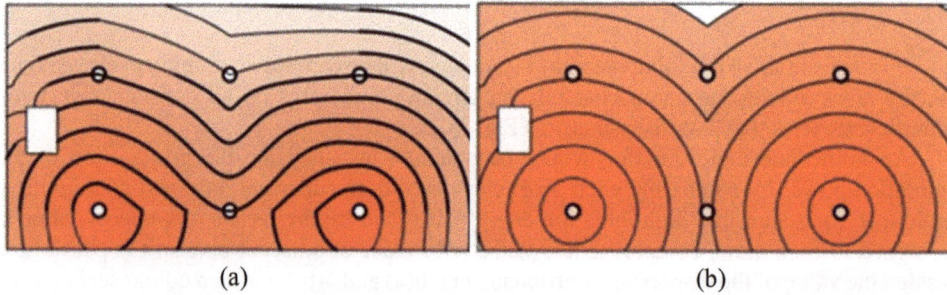

Figure 7:   Vibration energy over a generic façade. (a) Real measured values; and (b) 'Ideal' values.

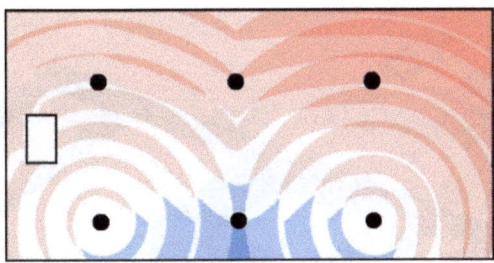

Figure 8: Difference between the real situation and ideal situation. In red the zones that are in defects and in blue the zones in excess of unscaled signal energy.

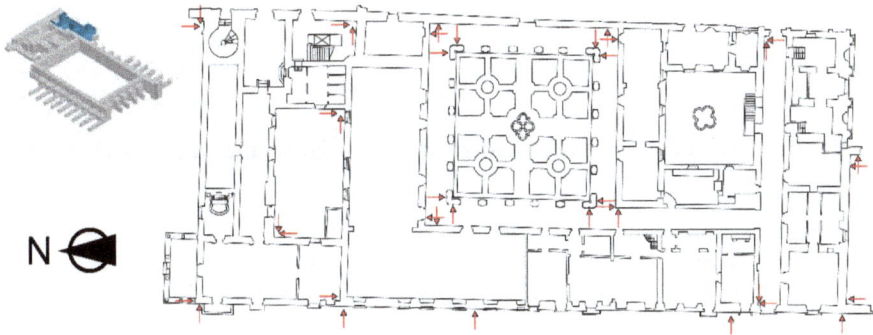

Figure 9: Scheme of the accelerometer directions, plan and 3D.

identifies areas having larger stiffness with blue colour and areas having an excessive deformability with a red colour.

3 APPLICATION OF PROCEDURE TO CASE STUDY

The proposed structural survey system is analysed through its application to a specific case study subject to monitoring.

The macro-volume Western Museum of Lucania and the House of the Priore of the Certosa di San Lorenzo in Padula (Salerno), Italy, see Fig. 9, were analysed, considering them as a structurally homogeneous area, since they both belong to the same period of construction. The sensor configuration for the dynamic monitoring was based on the planovolumetric composition of the structure. The vibration measurements were made using the accelerometer directions positioned according to the details in Fig. 9.

The modal shapes determined experimentally present large displacements corresponding to the macro-elements already identified through the planovolumetric characteristics of the building. The first mode of vibration (frequency 1.86 Hz and damping coefficient 1.51%) shows that the most vulnerable part of the building is the western wall. This part is characterised by two internal floors and a basement, it has an irregular thickness along its height, and it is not characterised by orthogonal stiffening masonry walls, that causes the façade to be slender and vulnerable for the out-of-plane flexural mechanisms, Fig. 10.

The other vibration modes identified experimentally concern other sub-parts of the macro-volumes analysed.

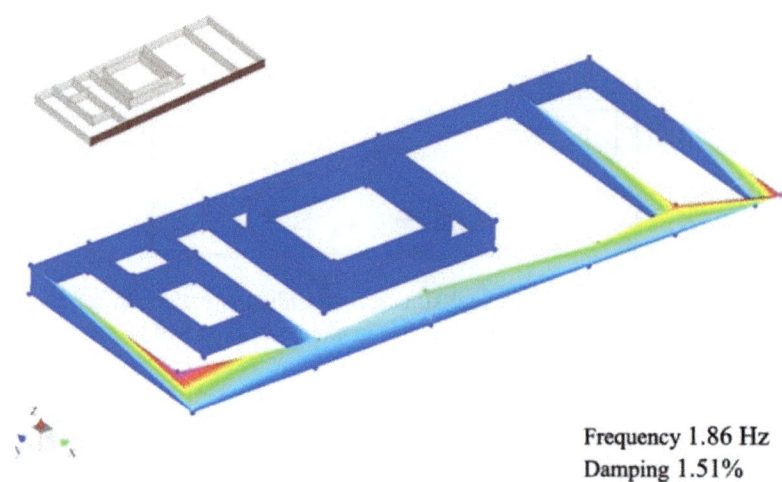

Frequency 1.86 Hz
Damping 1.51%

Figure 10: First modal shape of macro-volume that triggers the western wall response.

The western wall identified in the first modal shape of macro-volume (Fig. 9) is considered as a macro-element and it is analysed below (Fig. 11). The structural interaction between western wall and transversal wall was analysed by channels CH2, CH6, CH12, CH4, CH9, CH3 and CH7 (see Fig. 11, red box).

Figure 11: Front and related plan with sensors scheme of western wall.

4 DYNAMIC IDENTIFICATION INPUT/OUTPUT AND OUTPUT ONLY

Fig. 12 shows the sensors, the hit position and the model used for the dynamic identification of the western wall.

The 16 channels registered both the vibration induced by environmental noise and the one generated through the impact hammer from 10 hits (Fig. 12, blue rows). It is necessary to point out that for hit 5 it was necessary to eliminate the signal recorded by channel 15, since the signal saturated because of excessive proximity to hit point.

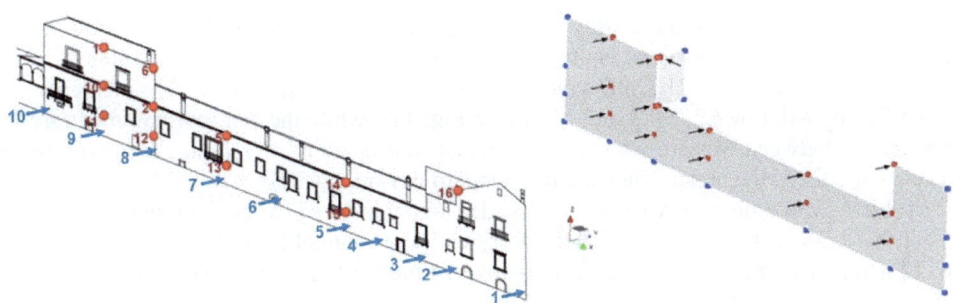

Figure 12: D scheme of accelerometer position, direction (red circles) and hit points (blue rows).

The values obtained were normalised in function of different entities of hits and then processed to calculate the signal energy at each excitation. The data obtained from the channels were sorted according to each analysed wall panel (see Fig. 11) and coloured based on a representative scale for an immediate visual interpretation, Tables 1 and 2.

Table 1: Hit input, signal energy values (gravitational acceleration, g).

| | Accelerometer positions | | | | | | | | | | | | | | |
| | ME1 | | | | | | ME2 | | ME3 | | ME4 | Connection between western wall and transversal wall | | | |
	1	10	11	2	6	12	5	13	14	15	16	3	7	4	9
Hit 1	0.0022	0.0004	0.0004	0.0019	0.0003	0.0007	0.0004	0.0008	0.0006	0.0008	0.0392	0.0006	0.0002	0.0003	0.0003
Hit 2	0.0022	0.0007	0.0004	0.0021	0.0004	0.0008	0.0006	0.0025	0.0014	0.0022	0.1119	0.0009	0.0004	0.0004	0.0004
Hit 3	0.0022	0.0036	0.0009	0.0019	0.0004	0.0007	0.0007	0.0010	0.0026	0.0018	0.0174	0.0009	0.0008	0.0004	0.0003
Hit 4	0.0022	0.0021	0.0004	0.0020	0.0009	0.0009	0.0019	0.0021	0.0226	0.0087	0.0212	0.0017	0.0009	0.0006	0.0004
Hit 5	0.0025	0.0011	0.0003	0.0022	0.0038	0.0011	0.0138	0.0075	0.2379	8.6534	0.0034	0.0023	0.0028	0.0008	0.0008
Hit 6	0.0031	0.0014	0.0006	0.0032	0.0073	0.0023	0.0497	0.0319	0.0399	0.0097	0.0033	0.0047	0.0046	0.0021	0.0017
Hit 7	0.0052	0.0032	0.0009	0.0059	0.0432	0.0081	0.3310	1.5385	0.0261	0.0062	0.0021	0.0076	0.0172	0.0037	0.0082
Hit 8	0.0034	0.0023	0.0021	0.0052	0.0407	0.0486	0.0148	0.0079	0.0073	0.0013	0.0014	0.0035	0.0076	0.0033	0.0041
Hit 9	0.0069	0.0283	2.0861	0.0073	0.0150	0.0078	0.0034	0.0042	0.0013	0.0008	0.0009	0.0054	0.0114	0.0048	0.0045
Hit 10	0.0032	0.0013	0.0017	0.0019	0.0006	0.0007	0.0002	0.0012	0.0002	0.0005	0.0012	0.0015	0.0011	0.0008	0.0003
Mean value	0.0029	0.0018	0.0089	0.0034	0.0019	0.0019	0.0045	0.0034	0.0057	0.0019	0.0056	0.0029	0.0047	0.0017	0.0021
SD %	0.1	0.11	0.06	0.16	0.27	0.25	0.62	0.29	0.93	0.33	0.73	0.23	0.57	0.17	0.27

Table 2: Ambient vibration input, signal energy values (gravitational acceleration, g).

| | Accelerometer positions | | | | | | | | | | | | | | |
| | ME1 | | | | | | ME2 | | ME3 | | ME4 | Connection between western wall and transversal wall | | | |
	1	10	11	2	6	12	5	13	14	15	16	3	7	4	9
Signal energy	0.201	0.039	0.037	0.168	0.072	0.055	0.060	0.094	0.027	0.066	0.215	0.058	0.026	0.026	0.036

Tables 1 and 2 show the energy values (expressed in gravitational acceleration, g) for each channel based on different input, represented by the hits with the instrumented hammer and ambient vibration. The red values show overflow data. Four macro-elements are identified by ME1, ME2, ME3 and ME4 (see Fig. 11); while the last four columns refer to connection between western wall and transversal wall (see Fig. 11). In the last two rows of Table 1 are listed the mean value and the standard deviation (SD).

Table 2 shows the energy values of recorded signal induced by ambient vibrations.

Table 1 allows to immediately note the criticality of channel 11, followed by channels 10, 6, 12 and 5. These channels are placed in areas of the wall that can be considered less stiff, since the signal energy is exhausted quickly before being detected. These channels therefore consist mainly of low values except for the presence of extremely high sudden peaks detected in the case in which the hit is close to recording channel.

As well known, a façade portion simply subjected to ambient vibration is characterised by larger signal energy values in stiff areas and smaller energy values in deformable areas.

In Fig. 13 channels 10 and 11 are in points characterised by structural discontinuity that does not transmit efficiently the accumulated accelerations unlike channels 1 and 2 which are located in the stiffest points of the wall.

The same approach was adopted with vibration induced by impact hammer. For this analysis, along each column of channels, the values obtained when the hit gets too close to the channels had to be removed because they can invalidate the average recorded value. Hit number 9 is eliminated from channels 1, 10 and 11; Hits 7 and 8 from channels 6 and 12 (channel 2 is considered sufficiently distanced thanks to the irregularity in elevation of the façade from Zone A to Zone B, see Fig. 11); Hits 6 and 7 from the channels 5 and 13; Hits 4, 5 and 6 from the channels 14 and 15; Hits 1, 2 and 3 from channel 16. Even in this case, if represented graphically as in Fig. 13, it is evident that channels 6, 10, 11 and 12 are in areas characterised by structural discontinuity, differently than channels 1 and 2. In this type of analysis, it can be noted that wall stiffness close to channel 15 seems to be smaller than that determined close to channel 14. This result confirms the analysis of recorded signals from ambient vibrations.

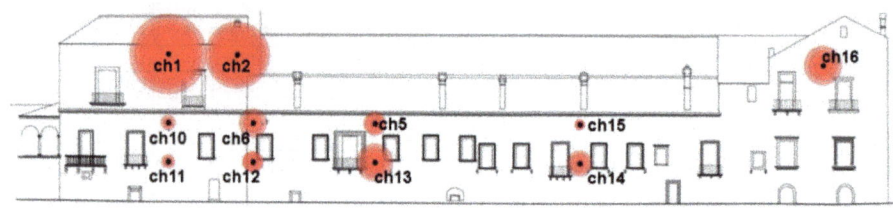

Figure 13: Scheme of signal energy values in case of ambient vibration.

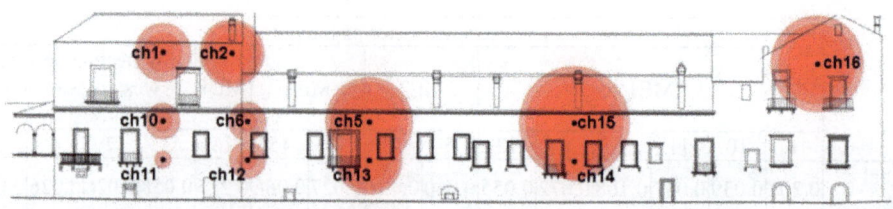

Figure 14: Scheme of signal energy values in case of vibration induced by hits.

Figs 15 and 16 shows the map of signal energy over the entire western wall and highlights the problem of vibration transfer over the façade. This image was obtained by adding together the signal energy values measured for each channel for hits 10, 8, 6, 4 and 3. Dividing the signal energy values in different colour zones can represent the ability to dissipate or accumulate the acceleration of the induced vibrations.

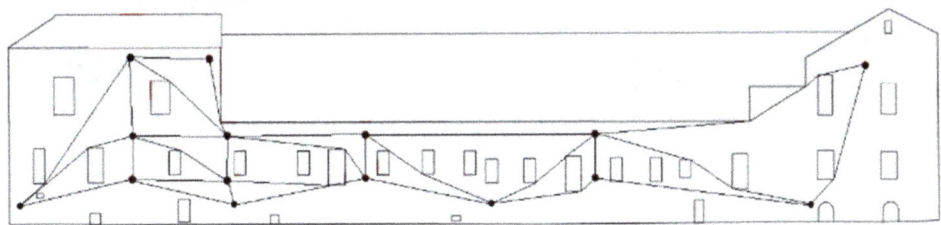

Figure 15: Outline of the distances between channels and hits on the façade.

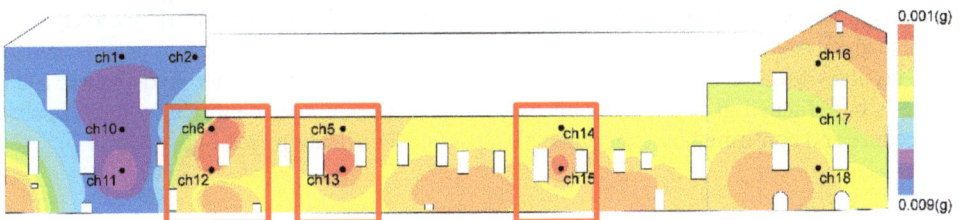

Figure 16: Coloured graphic scheme of the façade behaviour with hits 10, 8, 6, 4 and 3.

The distance between one channel and the second one is highlighted by a segment (Fig. 15), which is divided in proportion to the energy values of the channels and adopting a coloured reference scale. A contour map is then generated, with contour lines perpendicular to the points along the segments connecting adjacent channels and by dividing. the façade in areas having the same energy range, obtaining an approximated pattern of the signal energy.

Reading this result, it is necessary to pay attention to variations in the signal energy values between the points, that is, when two points close to each other have a significant difference in the value of the signal energy. An example is the area close to point 12, where great dispersion of energy can be observed. The same chromatic difference is found for example from channel 16 to hit 3 in more than 13 m. The same trend can be seen in channel 15 compared to channel 14, where one is probably influenced by the presence of a flue and the other one is stiffened by the presence of the roof (Fig. 12).

Comparing the results of Figs 14 and 15 it is clear that the vulnerability zones of the façade are localised close to points 11, 12, 13, 15 and 16.

What is interpreted through the analysis of the signal energy is confirmed by the experimentally identified modal shapes, showed in Fig. 17. It is evident from the triggered modes that the most vulnerable zones mainly concern points 11, 12, 13, 15 and 16 (Figs 11, 12 and 18).

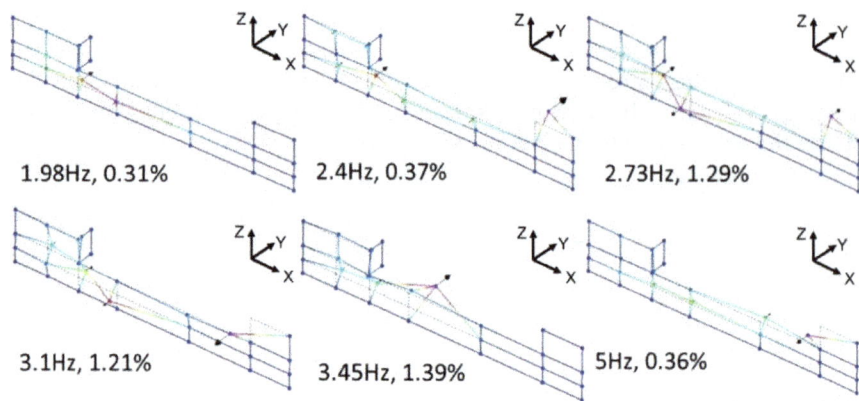

1.98Hz, 0.31%   2.4Hz, 0.37%   2.73Hz, 1.29%

3.1Hz, 1.21%   3.45Hz, 1.39%   5Hz, 0.36%

Figure 17: Dynamic identification of West wall.

Fig. 18 schematises the main experimental modal shapes that trigger the flexural horizontal and vertical mechanisms of macro-elements M1–M2 and M4 respectively. The displacements entities are not significant then were omitted. Out-of-plane vertical mechanism that involves ME4 is well known, while the combined mechanism that involves ME1 and ME2 is investigated in Section 5. This potential mechanism was identified by the proposed procedure and confirmed by experimental dynamic identification (see Fig. 19).

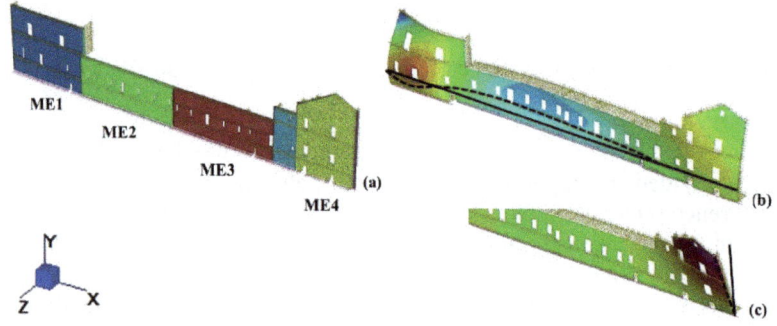

Figure 18: Macro-elements and potential mechanisms. (a) Macro-elements; (b) Out-of-plane horizontal mechanism of ME4; and (c) Out-of-plane vertical mechanism of ME1-ME2.

5 LOCAL MECHANISM

The macro-elements vulnerability can be evaluated through the analysis of the dynamic behaviour of structure correlating the modal shapes to local mechanisms.

The case study highlighted some local mechanisms, which involve both parts of macro-volumes triggered by the irregularity in elevation. The local mechanism was triggered by vertical bending of wall; nevertheless, the structural interaction between walls is affected by: (i) the degradation of structural interaction among west wall and normal walls; (ii) the irregularity along wall elevation; and (iii) the geometric slenderness of wall.

The modal shapes show the structural interaction between West-wall and perpendicular wall between ME1 and ME2 (see Fig. 19), in particular no interaction is found up to 4 m. Despite the irregularity along wall height, the effect of the roof guarantees structural continuity between walls. At a height of 8 m, the modal deformations coincide for both modes, confirming a good structural interaction between walls conferred by the corner connection. Fig. 19 shows the wall mechanism that enables to analyse the structural interaction between adjacent buildings, connected or in adherence, which constitutes the complex and/or aggregate system.

Despite the well-known out-of-plane mechanism, the one analysed is instead characterised by the upper stiff plane or volume which triggers an overturning mechanism generated by stiff overlapping volumes. Furthermore, the presence of effectively connected horizontal floor and corners with an effective degree of constraint, and the lack of connection with the orthogonal walls influence the potential mechanism (Fig. 19(a)), especially for a slender wall such as the west wall.

The second mode identified (Fig. 19(b)) can be defined as an out-of-plane mechanism triggered by the loss of interaction with the transverse wall; the hammering effect of the perpendicular wall and the loss of effectiveness of the interaction contribute to the overturning mechanism of the macro-element (Fig. 19(c)).

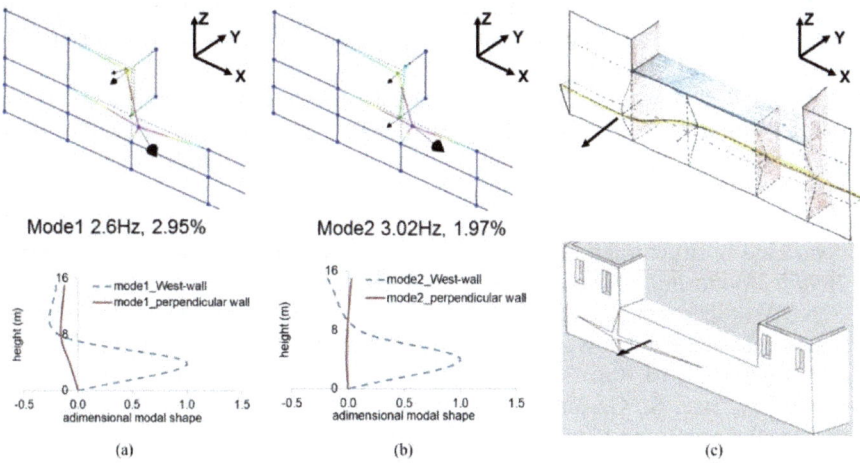

Figure 19: Main modal shape of west wall, potential failure local mechanism of wall.

6 CONCLUSIONS

The proposed procedure enables to evaluate the structural integrity through acceleration measurements induced by mechanical and/or ambient vibrations. The simplified procedure enables to identify the critical points of façade defining the potential failure mechanisms.

The recorded signal energy values are mapped through a graduated scale from 0 to the maximum value normalised by greater hit values to design the graphic model.

The procedure steps are:

- knowing in advance the geometric measurements of the wall fixing the recording points and return them in vector format;
- recording the vibrations on the wall assuring an acceptable noise-signal relationship;

- filtering and cutting the significative recorded signal;
- the tests should be carried out by combined procedure, input–output and output-only; and
- the data should be processed to calculate the normalised energy signal as a function of the input.

In this phase the procedure is reliable in case of façades represented as two-dimensional macro-elements; further developments are needed to evaluate other applications to more complex three-dimensional mechanisms.

REFERENCES

[1]   Farrar, C.R. & Worden, K., An introduction to structural health monitoring. *Philosophical Transactions of the Royal Society A*, **365**(1851), pp. 303–315, 2007.

[2]   Syrmakezis, C.A., Seismic protection of historical structures and monuments. *Structural Control and Health Monitoring*, **13**, pp. 958–979, 2006. DOI: 10.1002/stc.89.

[3]   De Stefano, A. & Ceravolo, R., Assessing the health state of ancient structures: The role of vibrational tests. *Journal of Intelligent Material Systems and Structures*, **18**, pp. 793–807, 2007.

[4]   D'Ambrisi, A., Mariani, V. & Mezzi, M., Seismic assessment of a historical tower with advanced numerical model tuned on ambient vibration data. *Advanced Material Research*, **133–134**, pp. 617–622, 2010.

[5]   Ceravolo, R., Pistone, G., Zanotti Fragonara, L., Massetto, S. & Abbiati, G., Vibration-based monitoring and diagnosis of cultural heritage: A methodological discussion by way of three examples. *International Journal of Architectural Heritage*, 2014. DOI: 10.1080/15583058.2013.850554.

[6]   Podestà, S., Riotto, G. & Marazzi, F., Reliability of dynamic identification techniques connected to structural monitoring of monumental buildings. *Structural Control and Health Monitoring*, **15**, pp. 622–641, 2008. DOI: 10.1002/stc.219.

[7]   Sepe, V., Speranza, E. & Viskovic, A., A method for large-scale vulnerability assessment of historic towers. *Structural Control and Health Monitoring*, **15**, pp. 389–415, 2008. DOI: 10.1002/STC.24.

[8]   Ceroni, F., Sica, S., Garofano, A. & Pecce, M., SSI on the dynamic behaviour of a historical masonry building: Experimental versus numerical results. *Buildings*, **4**, pp. 978–1000, 2014.

[9]   Diaferio, M., Foti, D., Giannoccaro, N.I. & Ivorra, S., Optimal model through identified frequencies of a masonry building structure with wooden floors. *International Journal of Mechanical*, **8**, pp. 282–288, 2014.

[10] Gallipoli, M.R. et al., Empirical estimates of dynamic parameters on a large set of European buildings. *Bulletin of Earthquake Engineering*, **8**, pp. 593–607, 2010.

[11] Ramos, L.F., Alaboz, M. & Aguilar, R., Dynamic identification and monitoring of St. Torcato Church. *Advanced Material Research*, **133–134**, pp. 275–280, 2010.

[12] Ramos, L.F., Marques, L., Lourenço, P.B., De Roeck, G., Campos-Costa, A. & Roque, J., Monitoring historical masonry structures with operational modal analysis: Two case studies. *Mechanical Systems and Signal Processing*, **24**, pp. 1291–1305, 2010.

[13] Gentile, C. & Saisi, A., Ambient vibration testing of historic masonry towers for structural identification and damage assessment. *Construction and Building Materials*, **21**, pp. 1311–1321, 2007.

[14] Foti, D., Diaferio, M., Mongelli, M., Giannoccaro, N.I. & Andersen, P., Operational modal analysis of an historical tower in Bari. *Conference Proceedings of the Society for Experimental Mechanics Series*, IMAC XXIX, vol. 7, Jacksonville, pp. 335–342. DOI: 10.1007/978-1-4419-9316-831.

[15] Aras, F., Krstevska, L., Altay, G. & Tashkov, L., Experimental and numerical modal analyses of a historical masonry palace. *Construction and Building Materials*, **25**, pp. 81–91, 2011.

SECTION 4
VULNERABILITY ASSESSMENT

NOVEL RISK ASSESSMENT METHODOLOGY FOR CULTURAL HERITAGE SITES

FABIO GARZIA[1,2,3]
[1]Safety & Security Engineering Group – DICMA, SAPIENZA – University of Rome, Italy
[2]Wessex Institute of Technology, UK
[3]European Academy of Sciences and Arts, Austria

ABSTRACT
Cultural heritage sites are places subjected to certain risks embodied, for example, by theft, vandalism, damaging, terrorism which could harm equally persons and cultural heritage. For this reason, it is necessary to activate suitable countermeasures to avoid the above risks and to defend against them by means of intrusion detection, access control, video surveillance, communication systems, security personnel and procedures suitably combined to achieve an integrated system or solution. In the present work a new risk assessment method for cultural heritage sites (RACHS) is shown, illustrating as a case study, with no loss of its broad applicability, its usage with a museum. The suggested risk assessment method permits of finding the precise quantity of physical security defences (intrusion detection system, access control, video surveillance, communication devices, security personnel, etc.) which a given cultural heritage site requires and the correlated features depending on the potential targets that can be damaged. It also permits to avoid of overrating the risk as in the case of considering superfluous defensive countermeasures which occasionally can be not necessary, thus decreasing the associated additional costs.
Keywords: risk assessment, risk analysis, security, safety, cultural heritage sites.

1 INTRODUCTION
Cultural heritage sites are places subjected to certain risks embodied, for example, by theft, vandalism, damaging, terrorism which could arm equally persons and cultural heritage.

For this reason, it is necessary to activate suitable countermeasures to avoid the above risks and to defend against by means of intrusion detection, access control, video surveillance, communication systems, security personnel and procedures suitably combined to achieve an integrated system or solution [1]–[3].

Considering the devices and installations prospective, it is also essential them to be suitably powered and to be able to transmit the data and information necessary for security management. This implies that power providers and communication tools and networks have to be correctly shielded to prevent potential attacks versus them which could damage the performances of integrated technologies employed and therefore let the entire site to be subjected to excessive risks [4].

For this reason, it is essential to evaluate all the potential risks to select the correct countermeasures that have to be implemented versus every possible malicious action. If security systems are already present, their correctness must also be assessed every time the risk framework varies [5]–[7].

In the present work a new risk assessment method for cultural heritage sites (RACHS) is shown, illustrating as a case study, without no loss of its broad applicability, its usage with a museum.

The suggested risk assessment method permits of finding the precise quantity of physical security defences (intrusion detection system, access control, video surveillance, communication devices, security personnel etc.) which a given cultural heritage site requires and the correlated features depending on the potential targets that can be damaged. It also

WIT Transactions on The Built Environment, Vol 203, © 2021 WIT Press
www.witpress.com, ISSN 1743-3509 (on-line)
doi:10.2495/STR210131

permits to avoid of overrating the risk as in the case of considering superfluous defensive countermeasures which occasionally can be not necessary, thus decreasing the associated additional costs.

It is also a new method with respect to other security risk assessment methods for heritage sites [7]. As a matter of fact, it utilizes a suitable initial risk analysis to continue further, estimating the level of defence of every target associated to every threats. So, it gives additional valuable knowledge as demonstrated afterwards.

2 DESCRIPTION OF THE METHODOLOGY

The suggested method of risk assessment applied to cultural heritage sites (RACHS) characterizes a particular application obtained from the Physical Security Adapted Layer of Protection Analysis (PSA-LOPA) method [8], [9]. It allows of obtaining the correct quantity of physical security protections (video surveillance, access control, intrusion detection system, etc.) which a certain location needs and the related features. It even supports the specialist in preventing risk overestimate, avoiding of including superfluous protective countermeasures that occasionally results to be futile, thus reducing any needless expense.

For these reasons, the right application of the RACHS methodology means to use an easy and useful examination technique to fix not only which physical security protections (PSPs) the given cultural heritage site necessitates to be classified as properly protected but mainly whether the existing PSPs are essential and adequate.

The LOPA method is divided into diverse subsequent phases:

1. Identification of the physical security risk scenario.
2. Analysis of the severity of the consequences of the above scenario and distribution of a specified Target Factor score.
3. Identification of the initial trigger (Initiating Event).
4. Assessment of the occurrence of incidence of the Initiating Event.
5. Identification of any other elements (Enabling Factors) that, joined with the Initiating Event, activate the scenario.
6. Assessment of the certain time in which the risk is revealed (Time at Risk).
7. Identification of independent defences (Independent Protection Layers, IPLs).
8. Assessment of the probability of failure of the physical security protections (Probability of Failure on Demand, PFD).
9. Assessment of credits.
10. Assessment of the suitability of risk and associated improvement activities.

To achieve the risk assessments, it is necessary to fix how the existing PSPs can diminish the probability of occurrence of the scenario, establishing the notion of 'credit'. The sense of credit is associated to the probability of failure, linked to every precise PSP_i, according to the next equation [10]:

$$credits(IPL)_i = -log(PFD)_i. \tag{1}$$

After that the various credits have been computed, the PSA-LOPA evaluation [8] is achieved with the evaluation of the risk coefficient, correlated to the k scenario, utilizing the equation [10]:

$$R_k = TF_k - F_k^I - F_k^E - I_k^T - \sum_i credits(IPL)_i, \tag{2}$$

where:

TF is the Target Factor.

F^I is the opposite of the logarithm of the incidence of the Initiating Event.

F^E is the opposite of the logarithm of the rate of incidence of the Enabling Factor.
I^T is the indicator of the Time at Risk.
IPLs are the Independent Protection Layers tath, in the contemplated situation, represents the physical security defences, or levels of protections, which IE triggers.

Since LOPA method [10] was firstly thought to estimate industrial risk, the above-mentioned expression required to be accustomed. By adjusting it to the physical security risk, it gave extremely precise and useful results where the PSA-LOPA methodology [8], [9] has been employed. The same results were obtained when the derivative RACHS method was applied to cultural heritage sites.

Physical security is ordinarily applied using the concept of layers of protection since every intruder face various layers of shields as perimeter protection, video surveillance, technological barriers, sensors etc., before reaching the desired goal. This justify the appropriateness of LOPA when the considered flow of risk is reversed.

In fact, LOPA evaluates the different layers of shields beginning from the target and advancing through different levels of protection which could gradually generate harms.

In the PSA-LOPA the progressions are suitably inverted, seeing the various layers of protection as a sort of successive defences to avoid that an aggressor could reach a particular target, causing the expected damages (Fig. 1).

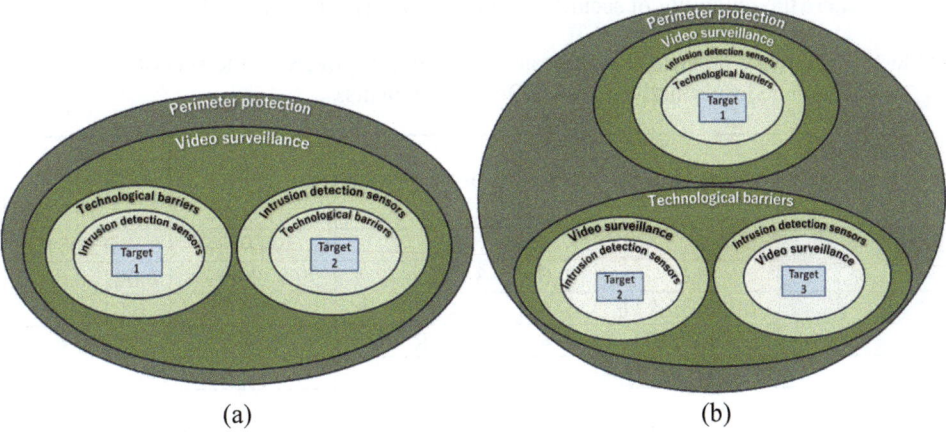

(a)   (b)

Figure 1:   Graphic illustration of various kind of layers of protections. (a) Two targets; and (b) Three targets.

In this work technological defences are solely contemplated even if the technique can be extended by considering not only technological protections but even physical defences and human factor [11] that are crucial parts for security management.

Some assertions were set to make simpler the technique indicated. These assertions can certainly be adjusted to achieve a better grade of evaluation even if they are not contemplated in this basic estimation. They are embodied by:

- The F^I factor was supposed to be equal to 1, as the intrusion security system is calculated when the incursion has already occurred.
- The I^T factor is not contemplated to simplify even if the times of exposure to a security risk can be identified in some situations.

- The F^E factor is not contemplated to simplify even if in some security incidents it is conceivable to distinguish 'enabling' factors or factors that enable the progress of a security incident.

As PSA-LOPA intends only technological protections and devices whose failure rate is lesser than one and since the failure rate is applied in eqn (1) as probability of failure, the sign the logarithm calculation is adjusted since the outcome of the argument (failure rate, lesser than one) delivers already a negative outcome.

Evidently, the greater the reliability of a specific security shield and the lesser its failure rate. This indicates that the number expected with eqn (1) significantly decreases, diminishing the correlated R factor of the connected risk scenario. This is an obvious consequence because it signifies that the shield used is reliable, ensuring an enhanced grade of security shield and a resultant reduction of the connected risk. The interesting advantage of the proposed method is embodied by its capacity of estimating the various level of defences from the semi-quantitative point of view, leading to an estimation of whether extra levels of protection are required. This additionally provides all the needed evidence to improve the cost/benefit ratio.

In the starting it is needed to categorize the damage levels and the needed level of performances of the security defences for the associated physical security risks and these activities are peculiar of a certain site of a particular organization. An instance is shown in Table 1 where the reliability of security solution (RSS) is shown as well.

Table 1:   Summary table of the requested performance levels of the security system, the damage levels, and the PSA-LOPA coefficients.

Requested level of performance	Damage	TF	R (PSA-LOPA)	RSS	PFD
5	SEVERE	9–10	$R < -3$	>99.99%	<0.0001
4	HIGH	7–8	$-3 < R < -2.1$	99.9–99.99%	0.001–0.0001
3	MODERATE	5–6	$-2 < R < -1.1$	99–99.9%	0.01–0.001
2	LIMITED	3–4	$-1 < R < 0$	90–99%	0.1–0.01
1	NEGLIGIBLE	1–2	$R > 0$		

The performance level of the security shield is associated to the degree of damage that the security incident can generate. The five levels of damage have been obtained by a standard classification of a general organization, that connects every level to the calculation of economic, physical, company's reputation, legal, expenses etc. damages. So, the PSA-LOPA technique has been adapted and customized to a plenty of situations and it is valid for every sort of organization.

The TF target factor (attained through an appropriate initial risk assessment prepared by the studied organization) has been associated to every level of damage. For instance, a potential target of the greatest strategic and economic importance for the studied organization (data centre, vaults etc.) is correlated with the greatest damage level, and the designated score can vary from 9 to 10, and so on for the levels considered as lower risk.

Different sort of initial evaluations can be done applying, for example, risk matrixes such as: interaction matrix accesses – target/security protections (Table 2), interaction matrix targets-security protections (Table 5), interaction matrix impact on targets-threats (Table 6), etc., which provide significant data to calculate the level of damage of every target of the given site of the studied organization required for the successive PSA-LOPA semi-quantitative assessment.

WIT Transactions on The Built Environment, Vol 203, © 2021 WIT Press
www.witpress.com, ISSN 1743-3509 (on-line)

Table 2: Instance of table of interaction matrix accesses: Targets/security protections.

Access	Target i							Target i + 1					⋯⋯
	Intrusion detection	Access control	Video surveillance	Security personnel	⋮	Number of level of protections	⋮	⋮	⋮	⋮	Number of level of protections	⋮	⋮
Access j													

The R factor, i.e. the valuation of the risk factor achieved from eqn (2), is linked with the damage levels of considered organization, thus associating each of them to the associated levels of total reliability of the security solution (RSS) and its likelihood of failure on demand (PFD).

Consequently, the PSA-LOPA method is applicable to all possible targets T_i inside the considered site of the studied organization that are exposed to physical security risks. The choice of the objectives comprised in the assessment is performed pondering the cruciality of them for the organization, considering the information on the exposure to the physical security risk which a fitting initial risk assessment can guarantee.

The suggested risk assessment method for cultural heritage sites (RACHS), shown in the following, embodies an appropriate development and adjustment of PSA-LOPA.

3 DESCRIPTION OF RACHS METHODOLOGY

In cultural heritage sites there are definite risks embodied, for example, by theft, vandalism, damaging, or terrorism which could damage people and the cultural heritage equally.

Thus, suitable measures are required for risk prevention and protection, such as: intrusion detection, access control, video surveillance, communication systems, security personnel and procedures aptly merged to achieve an integrated system or solution [1]–[3]. Further, these technologies can be appropriately integrated to guarantee the safety distance between people, when necessary, during pandemic and post pandemic periods.

It is essential to keep in mind that it is vital that security countermeasures are as non-invasive as feasible. In this way, cultural heritage sites are no subjected to aesthetics and architectural impacts, but their safety and security are always ensured.

It is also particularly vital that devices and installations are appropriately powered and can transmit all the data and information required for security management. This means that power providers, communication tools and networks have to be aptly protected. This is to prevent that a possible assault versus them could generate a reduction or a breakdown of the performances of the integrated technologies being utilized, consequently exposing the whole location to higher risks [4].

Moreover, in cultural heritage sites there can be security personnel endowed by radio communication tools through the day but nobody, or a decreased guarding, through the night. This involves that to be confident the targets to be suitably safeguarded, two distinct assessments must be made, one for day and one for night conditions.

The key elements that are usually present in a cultural heritage site, such as, for example, a museum, and which can be imaginable objectives of intentional attacks are embodied by: external space around the site, entrance hall, ticket office, coffee shop, toilets, shop, luggage depot, internal exhibit rooms with different works of art, offices, control room, data centre, warehouse, main electrical power room, generator set, uninterruptible power supply – UPS, air conditioning central device, external electrical power delivery point, external data network delivery point.

After the potential threats and the possible targets of a cultural heritage sites are focused, RACHS method continues with the construction of a suitable interaction matrix impact on targets-threats. In this matrix all the targets are associated with the respective and different impacts of the diverse threats, introducing, in each link box, a numerical value between 1 and 10, dependent on the impact generated by each threat on each target (0: absent; 1, 2: negligible; 3, 4: limited; 5, 6: moderate; 7, 8: high; 9–10: severe). In this way it becomes possible to determine a mean value of threats for each target focused.

An instance of the considered matrix, for the targets of the case study of a museum considered in the following, is shown in Table 6, while the impact scale is summarized in Table 3.

Table 3: Impact scale with related numerical values.

Impact	Numerical values
SEVERE	9–10
HIGH	7–8
MODERATE	5–6
LIMITED	3–4
NEGLIGIBLE	1–2
ABSENT	0

The levels of physical security shields that can be thought for the RACHS methodology are symbolized by video surveillance, access control, intrusion detection and radio communication devices utilized by security personnel. They are labelled P_1, P_2, P_3, P_4, respectively, even if the method warrants of considering numerous levels of defences, as indicated before, not restricted to technological defence systems since it is certainly extensible to physical barriers and human factor reliability and errors [11].

The likelihood of failure of protection levels P_1, P_2, P_3, P_4, are achieved from the failure levels of every sort of used means. About video surveillance, the failure rate is multiplied by the percentage of visual coverage of the area evaluated (i.e. equal to 1 if all the considered area is comprised). Similar issues, with suitable variation, are applicable for security personnel endowed by radio devices.

After that all the targets of the site have been appropriately focused, it is possible to assess for every target T_i, applying the associated level of defences P_{1i}, P_{2i}, P_{3i}, P_{4i}, through the prior equations, the connected PSA-LOPA risk factor R_i, i.e. achieving the actual physical security level of defence of all the targets of the location.

It is now possible to produce a concise table (as shown in Table 4) where:

- the first column expresses the different targets.
- the second and the third columns express the possible damages confrontable by the actual level of protection determined via PSA-LOPA (using eqns (1) and (2)), through the day and through the night, respectively, converted using Table 1.
- the fourth column expresses the expected damage calculated by means of the results of initial analysis, converted using Tables 1 and 3.
- the fifth and the sixth columns express the actual level of performance of the security protections through the day and through the night, respectively, computed via PSA-LOPA (by means of eqns (1) and (2)), converted using Table 1.

- the seventh column expresses the required level of performance of the security protections necessary to face the expected damages calculated by means of the results of initial analysis, converted using Tables 1 and 3.

Table 4: Instance of a summary table of the RACHS method outcomes.

Target	Damage confrontable by the actual level of protection (day)	Damage confrontable by the actual level of protection (night)	Estimated damage	Actual level of performance (day)	Actual level of performance (night)	Requested level of performance
Target i						
.....						

The damage confrontable by the actual level of protection and the actual level of performance has been judged in a different way for the day from the one for the night since the number of defence levels could be different in the two circumstances. For example, there could be greater quantity of security personnel components endowed by radio communications tools through the day and their number could be decreased through the night. If it is planned the absence of security personnel during the night, there can be other sort of protection defences activated.

The outcomes attained allows of evaluating rapidly if the performance level of protections of every target, and consequently of the whole site, are fitting or the current layers of protection of each target need reinforcement (increasing their reliability, for example) or augmenting their number to scope the required performance level. Then, a proper decision is made for the level of protection of each target as a function of the probability of the related threat, and therefore of the associated risk.

Table 4 can also be condensed using a suitable histogram graph to obtain suddenly a clear view of the state or in a further manner via a radar graph. Both are shown in the next general case study of a museum.

4 EXAMPLE OF APPLICATION TO A MUSEUM

In the following, a museum is considered as a general case study, without any loss of generality with respect to other kind of cultural heritage sites.

For our purposes we presume that in the museum all the targets earlier focused are present, and that external and internal video surveillance, access control, intrusion detection and security personnel endowed by radio are used to protect them. This excludes now other countermeasures which can utilized as supplementary security defences, if needed. For the consequent analytic calculation, these security defences are judged as being categorized by mean technical/operative characteristics of commercial devices (that are not shown here for briefness). An outline of the context for day and night is shown in Table 5 (interaction matrix targets-security protections), where "internal exhibit room" represents the ith room of the museum and "work of art(j) of exhibit room(i)" represents the jth exposed element of room(i). Since there can be different works of art in the different exhibit rooms, these two targets must be repeated in Table 5 a number of times equal to the number of different elements protected by different levels of protection, if they are interested by different layers of protections. If the levels of protections are the same for all the exhibition rooms and related works of art, they must be considered only once. In this way it is possible to reach a great detail in the analysis

Table 5: Interaction matrix targets – security protections in the studied museum ('X' signifies the presence, '–' signifies the absence).

Target	Kind of protection				
	External video surveillance (day/night) [X/–]	Internal video surveillance (day/night) [X/–]	Access control (day/night) [X/–]	Intrusion detection (day/night) [X/–]	Security personnel equipped with radio (day/night) [X/–]
External space around the site	X/X	–/–	–/–	–/–	X/–
Entrance hall	–/–	X/X	–/–	–/–	X/–
Ticket office	X/X		X/X		X/–
Coffee shop	X/X	–/–	–/–	–/–	X/–
Toilets	X/X	–/–	–/–	–/–	–/–
Shop	–/–	X/X	–/–	–/–	X/–
Luggage depot	–/–	X/X	X/X		
Internal exhibit room (i)	–/–	X/X	–/–	–/–	X/X
Work of art (j) of exhibit room (i)	X/X	–/–	–/–	X/X	X/X
Offices	X/X	–/–	X/X	–/–	–/–
Control room	X/X	–/–	X/X	–/–	X/X
Data centre	X/X	X/X	X/X	–/–	–/–
Warehouse	X/X	X/X	X/X	X/X	X/–
Main electrical power room	X/X	X/X	X/X	–/–	–/–
Generator set	X/X	X/X	X/X	–/–	–/–
Uninterruptible Power Supply – UPS	X/X	X/X	X/X	–/–	–/–
Air conditioning central device	X/X	X/X			
External electrical power delivery point	X/X	–/–	–/–	–/–	X/–
External data network delivery point	X/X	–/–	–/–	–/–	X/–

since it is conceivable to consider the levels of protection of each work of art that can different according to their value.

All the required information to carry out an initial assessment have been obtained through open-source data available on the Internet. In this manner it has been conceivable to develop the interaction matrix impact on targets – threats for the studied site whose outcomes are displayed in Table 6, where mean values of each target are rounded to the upper integer to use a precautionary approach.

Table 6: Table of interaction matrix impact on targets – threats for the studied location.

Target	Vandalism	Physical violence against people and/or objects	Damage	Sabotage	Espionage	Theft	Arson	Robbery	Explosive device	Terrorist attack	Mean value
External space around the site	7	6	7	7	0	7	7	6	8	8	7
Entrance hall	8	8	7	7	0	7	8	8	10	10	8
Ticket office	8	8	7	7	6	8	8	9	8	8	8
Coffe shop	8	8	7	7	0	6	8	6	8	8	7
Toilets	7	8	7	7	0	6	8	3	8	8	7
Shop	8	8	7	3	0	8	8	6	8	8	7
Luggage depot	4	8	6	3	0	7	8	8	8	8	6
Internal exhibit room(i)	9	9	9	8	4	9	10	10	10	10	9
Work of art(j) of the exhibit room(i)	10	10	10	8	4	10	10	10	10	10	10
Offices	7	8	7	7	7	8	7	7	8	8	8
Control room	8	9	8	8	8	4	9	7	9	8	8
Data centre	8	9	8	8	8	8	9	7	9	8	9
Warehouse	8	9	8	8	4	8	10	8	9	9	9
Main electrical power room	8	9	8	8	4	8	8	8	9	9	8
Generator set	8	9	8	8	4	8	8	8	9	9	8
Uninterruptible Power Supply (UPS)	8	9	8	8	4	8	8	8	9	9	8
Air conditioning central device	7	9	8	7	6	8	8	6	7	6	8
External electrical power point delivery	8	8	9	9	6	9	8	8	9	9	9
External data network delivery point	8	8	9	9	6	9	8	8	9	9	9

It is now conceivable to continue with the computation, according to what specified previously, bearing in mind that mean values of every target of Table 6 are considered as the associated target Factors (TF) and they embody the estimated damage and the related requested level of performance in Table 7, after suitable numerical translation by using Tables 1 and 3. Results of Table 7 are shown in Figs 2 and 3.

As it is possible to see from Figs 2 and 3, except for targets 7, 9, 11, 12, 13, 14, 15, 16, targets are categorized by an actual level of performance (equally during the day and the night or just during one of them), that are lesser with respect to the demanded level of performance. In some situations, the night decrease depends on the lack or the reduction of

Table 7: Resuming table of RACHS methodology results for the considered site.

Target	Damage confrontable by the actual level of protection (day)	Damage confrontable by the actual level of protection (night)	Estimated damage	Actual level of performance (day)	Actual level of performance	Requested level of performance
External space around the site	HIGH	NEGLIGIBLE	HIGH	4	1	4
Entrance hall	MODERATE	NEGLIGIBLE	HIGH	3	1	4
Ticket office	SEVERE	MODERATE	HIGH	5	3	4
Coffe shop	HIGH	NEGLIGIBLE	HIGH	4	1	4
Toilets	NEGLIGIBLE	NEGLIGIBLE	HIGH	1	1	4
Shop	HIGH	NEGLIGIBLE	HIGH	4	1	4
Luggage depot	SEVERE	SEVERE	MODERATE	5	5	3
Internal exhibit room(i)	LIMITED	LIMITED	SEVERE	2	2	5
Work of art(j) of the exhibit room(i)	SEVERE	SEVERE	SEVERE	5	5	5
Offices	MODERATE	MODERATE	HIGH	3	3	4
Control room	SEVERE	SEVERE	HIGH	5	5	4
Data centre	SEVERE	SEVERE	SEVERE	5	5	5
Warehouse	SEVERE	SEVERE	SEVERE	5	5	5
Main electrical power room	SEVERE	SEVERE	HIGH	5	5	4
Generator set	SEVERE	SEVERE	HIGH	5	5	4
Uninterruptible Power Supply (UPS)	SEVERE	SEVERE	HIGH	5	5	4
Air conditioning central device	MODERATE	MODERATE	HIGH	3	3	4
External electrical power point delivery	LIMITED	NEGLIGIBLE	SEVERE	2	1	5
External data network delivery point	LIMITED	NEGLIGIBLE	SEVERE	2	1	5

security personnel endowed by radio. This implies that is required to augment of one or more the levels of security protection for them. This be done by introducing, for example, a suitable intrusion detection system, thermal camera, motion detection, video analysis or other type of countermeasures. If high quality strengthening countermeasures are used, it is possible to verify, repeating the computation process, if the upgraded actual level of protection (equally in the day and in the night) scopes, or in some cases exceeds, the demanded level of protection, guaranteeing the apt security defending of every targets of the studied site.

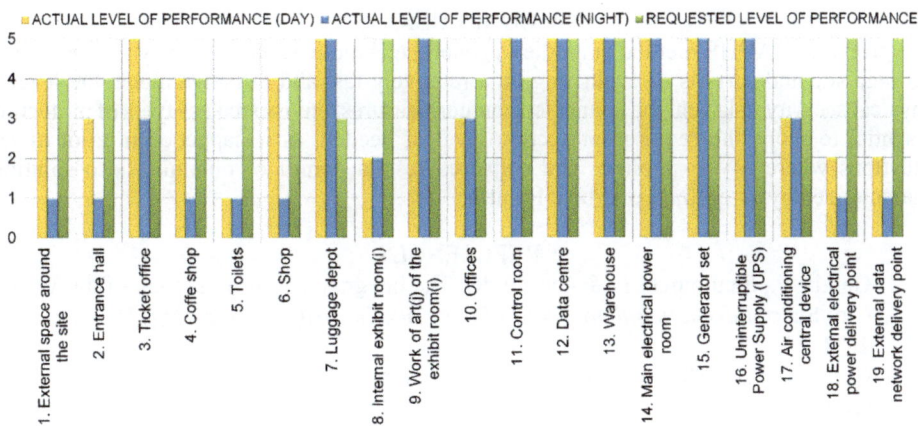

Figure 2: Histogram graph of the obtained results.

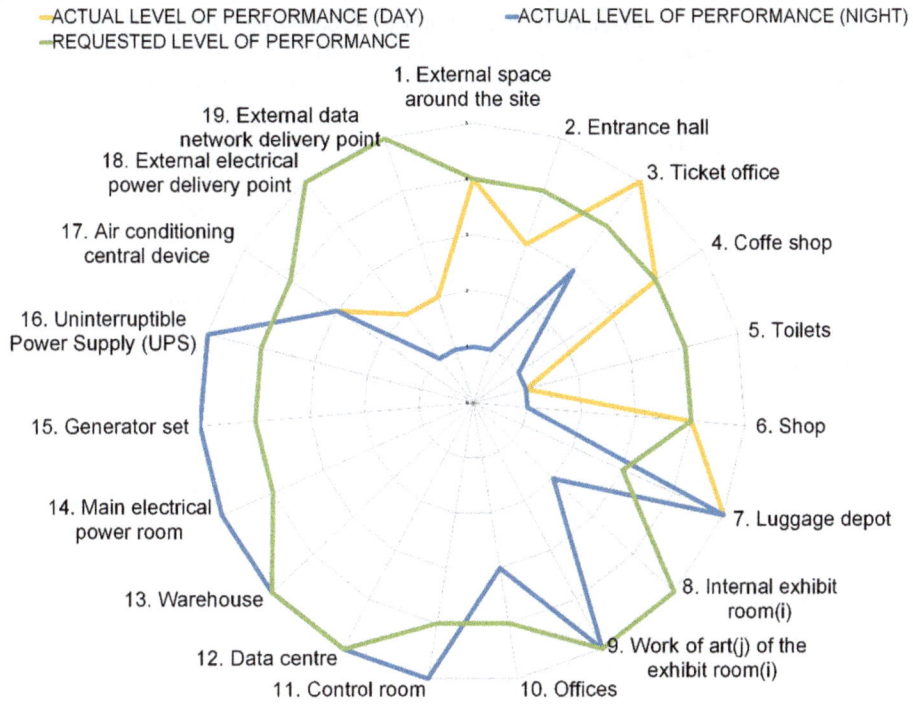

Figure 3: Radar graph of the obtained results.

So, it is conceivable to evaluate and definitively achieve all the needed and extra defences, carefully pondering the cost/benefit ratio, to guarantee the desired level of protection to the various targets, also considering the related likelihood of the threats.

WIT Transactions on The Built Environment, Vol 203, © 2021 WIT Press
www.witpress.com, ISSN 1743-3509 (on-line)

5 CONCLUSIONS

The proposed RACHS method embodies a general technique suitable for any sort of cultural heritage site and permits of evaluating, in a relatively fast and effective mode, the level of physical security risks and the connected countermeasures, envisioned as layers of protection, essential to scope the required protection level, if needed, as it happened in a lot of real situations where it was applied. The delivered results permit of obtaining also solutions characterized by an optimal cost/benefit ratio.

REFERENCES

[1] Garzia, F., Sammarco, E. & Cusani, R., The integrated security system of the Vatican City State. *International Journal of Safety and Security Engineering*, **1**(1), pp. 1–17, 2011.

[2] Garzia, F., The Internet of Everything based integrated system for security/safety/general management/visitors' services for the Quintili's Villas area of the Ancient Appia way in Rome, Italy. *WIT Transactions on the Built Environment*, vol. 174, WIT Press: Southampton and Boston, pp. 261–272, 2018.

[3] Garzia, F., Lombardi, M. & Ramalingam, S., An integrated Internet of Everything – Genetic algorithms controller – Artificial neural networks based framework for security systems management and support. *Proceedings of IEEE International Carnahan Conference on Security Technologies*, 2017.

[4] Garzia, F., *Handbook of Communication Security*, WIT Press: Southampton and Boston, 2013.

[5] Broder, J.F. & Tucker, E., *Risk Analysis and the Security Survey*. Butterworth-Heinemann: New York, 2012.

[6] Norman, T.L., *Risk Analysis and Security Countermeasure Selection*, CRC Press, 2010.

[7] CCI/ICC & ICCROM, *The ABC Method: A Risk Management Approach to the Preservation of Cultural Heritage*, Canadian Conservation Institute, 2016.

[8] Garzia, F., Lombardi, M., Fargnoli, M. & Ramalingam, S., PSA-LOPA – A novel method for physical security risk analysis based on LOPA – Layers of protection analysis. *Proceedings of IEEE International Carnahan Conference on Security Technologies*, pp. 187–191, 2018.

[9] Garzia, F., Sammarco, E., New risk analysis methodology for religious buildings. *WIT Transactions on Engineering Sciences*, vol. 129, WIT Press: Southampton and Boston, pp. 215–227, 2020.

[10] Willey, R., J., Layer of protection analysis. *Proceedings of 2014 International Symposium on Safety Science and Technology, Procedia Engineering*, vol. 84, pp. 12–22, 2014.

[11] Borghini, F., Garzia, F., Borghini, A. & Borghini, G., *The Psychology of Security, Emergency and Risk*, WIT Press: Southampton and Boston, 2016.

BUILT HERITAGE IN THE 2020 EARTHQUAKES IN ZAGREB AND PETRINJA, CROATIA: EXPERIENCE AND CONSEQUENCES

JOSIP GALIĆ, DAVOR ANDRIĆ, LUCIJA STEPINAC & HRVOJE VUKIĆ
Department of Architectural Technology and Building Science, University of Zagreb, Croatia

ABSTRACT

Croatia is a seismically active country, with recorded devastating earthquakes – among which is the Zagreb earthquake of 1880, when many buildings have been destroyed. Since that time 140 years have passed, and little has been done to prepare the city and its buildings for the return of a similar event, last year's quake found Zagreb rather unprepared. The same year an even more massive earthquake hit Petrinja. An overview of observed damage caused directly and by the aftermaths of the earthquakes in Croatia: Zagreb in March and Petrinja in December 2020 is given. Most of the damage is recorded on historical buildings with brick walls and wood floors and roofs, but some damage on contemporary structures is also noted. Most of the damage is found on old buildings; most of them are part of Zagreb historical centre and are in the protected heritage area or are listed as protected heritage buildings. These include damage like cracks in walls, collapse of unsupported walls, and parts of architectural details, chimneys, gable walls as well as damaged or collapsed vaulted ceilings, wood floor structures and roofs. Buildings of the modern period have sustained damage as well. Several typical types of damage stand out: infill walls of concrete frames damaged due to different flexibility, columns damaged mostly due to insufficient reinforcement, and steel to concrete connection details failure. Main problems for reconstruction and retrofitting are of financial nature, cultural asset status of large number of damaged buildings, scale, and complexity of some buildings mostly in Zagreb, and sustainability of settlements in Sisak-Moslavina County in case of Petrinja earthquake. Based on these observed phenomena, a set of appropriate approaches to the post-earthquake renewal is proposed including techniques for retrofitting depending on the desired level of quake proofing.
Keywords: earthquake, Zagreb, Croatia, Petrinja, masonry, damage, heritage.

1 INTRODUCTION

Croatia has a long history of intense earthquakes all across the country, with some notable being the Dubrovnik earthquake in 1667, Međimurje earthquake in 1738, the great Zagreb earthquake in 1880, Turjak earthquake in 1898 and Vinodolski earthquake in 1916 [1]. Recent history records earthquakes of magnitude 6 throughout the 20th century, most of them on the Croatian south, like M 6.2 earthquake in Imotski in 1942, M 6.1 in Makarska, in 1962, and M 6.0 in Ston-Slano in 1996 [2]. Other notable earthquakes in countries of former Yugoslavia include devastating earthquakes in Skopje in 1963, Banja Luka in 1969, and Montenegro in 1979 that severely impacted Dubrovnik city in the Croatian south.

Regulations regarding earthquakes are usually enacted only after the event takes place – when people become more aware of the problem. For instance, after the earthquake in Zagreb in 1880 and the earthquake in Ljubljana in 1895, residential and public buildings are built as brick masonry, instead of two-layer stone walls, obeying the maximum height limit set at five floors (20 m). The walls are also braced by metal clamps at the floor plane. Ordinance on temporary technical regulations for construction in seismic regulations [3] was enacted after the Skopje earthquake in 1963, and after the Ulcinj earthquake in 1979, Ordinance on technical standards for the construction of buildings in seismic areas [4] was enacted. The most notable Croatian codes relating the rules of seismic design are given in Table 1.

 WIT Transactions on The Built Environment, Vol 203, © 2021 WIT Press
www.witpress.com, ISSN 1743-3509 (on-line)
doi:10.2495/STR210141

Table 1: Important Croatian codes relating the rules of seismic design.

Year	Codes and regulations
1947	Interim technical regulations
1964	Ordinance on temporary technical regulations for construction in seismic regulations [3]
1981	Ordinance on technical standards for the construction of buildings in seismic areas [4]
2007–2009	Technical regulations for concrete [5] and masonry [6] structures
2012	Modern technical regulations – EN 1998

The latest earthquakes in Zagreb and Petrinja, in 2020 and 2021, have revealed many problems. Three stand out: (1) the cities and the state of Croatia are unprepared to react promptly to the disaster; (2) the army no longer has the efficient construction operative, and (3) large number of buildings are of low seismic resistance – about 30% of the building stock predates 1963, until when residential and public buildings were mostly built as masonry.

Most of the buildings were poorly maintained and the upgrades were not comprehensive – they were mostly performed without seismic improvements. Until 1941 the construction of the buildings was of acceptable quality, and number of storeys was appropriate but inadequate layout of the load-bearing structures was present. After the World War II, construction significantly worsened, building structures with walls in only one direction, and often surpassing the set limits for masonry buildings. There was also a significant problem of illegal construction in Croatia – that was later legalised – which was not accompanied by appropriate technical documentation and technically sound construction. Another problem that helped generate and further worsen the situation, especially the neglect for older buildings, was inappropriate criteria for heritage preservation of historical buildings. Too often, everything was considered a protected heritage asset that needed to be preserved as much as possible, blocking the potential upgrades that could have brought necessary seismically improvements with them.

1.1 The Zagreb earthquake

The city of Zagreb stands at the contact of large tectonic units: Alps in the northwest, Pannonian Basin in the east, and Dinarides in the south. Due to the pushing and/or subsiding of individual tectonic units, the lithosphere breaks, and faults are created as the source of seismic activity. The area of the city of Zagreb is influenced by the zone of the Žumberak-Medvednica-Kalnik fault. The strongest known earthquake occurred on November 9, 1880, at 07:03. Estimated magnitude was 6.0–6.5, and less intense earthquakes shook the area in 1905 and 1906 [7]. Zagreb had 30,000 inhabitants and about 2,500 residential buildings of which 1,400 were damaged or demolished.

On Sunday, March 22, 2020 at 6:24 am M 5.5 earthquake at a depth of 10 km and intensity VII according to the EMS scale with the epicentre in Markuševac, approx. 9 km to the centre of Zagreb, hit the city killing one person – a 15-year-old girl. In the period from March 22 to November 22, 2020, there were over 200 earthquakes of magnitude greater than or equal to M 1.3 [8]. The largest of these aftershocks was of M 4.9 about 40 minutes after the main quake.

According to the data of the Seismological Service, accelerographs at a distance of 8.2 km and 11.7 km from the epicentre recorded the peak acceleration of the ground soil $a_{max, A} = 0.22$ g and $a_{max, B} = 0.20$ g for the M 5.5 earthquake, while the $a_{max, A} = 0.07$ g and $a_{max, B} = 0.04$ g were recorded for the M 4.9 earthquake [9].

In the period from March 22, 2020, till July 01, 2020, rapid damage inspections were conducted according to EMS classification under the guidance of Croatian Centre for Earthquake Engineering [10]. The Same Centre created a manual and a form – by adapting the existing form used in Italy – for conducting the inspections [11]. Buildings were issued a green, orange (yellow), or red label according to the estimated scale of sustained damage (Fig. 1).

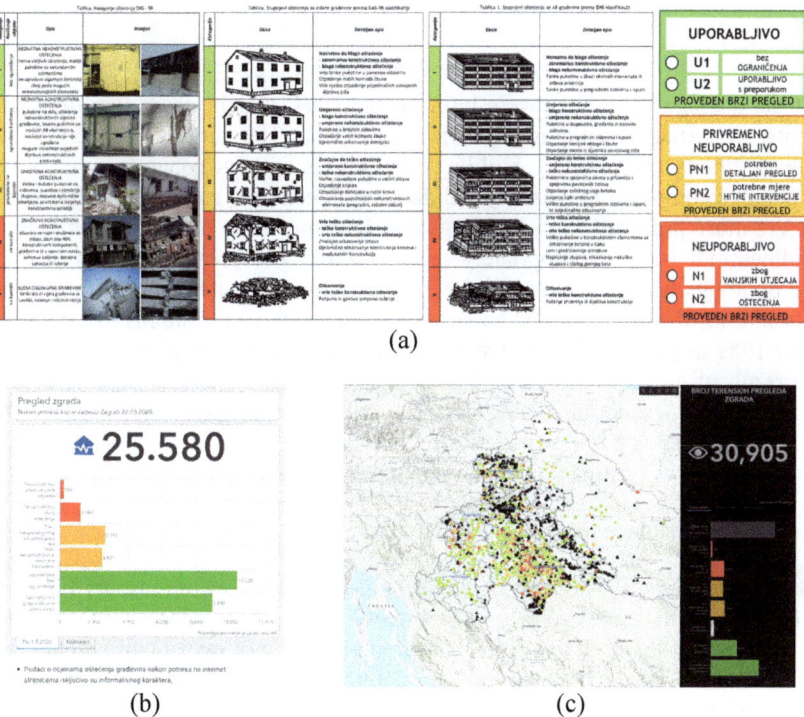

(a)

(b)   (c)

Figure 1:   Damage inspections. (a) Manual for rapid damage assessments [11] and total number of inspections (b) after the Zagreb earthquake, and (c) after the Petrinja earthquake [10].

1.2 The Petrinja earthquake

The tectonics of the Pokupsko–Petrinja–Sisak area is caused by the continuous movement of the Adriatic microplate to the north, due to which there is great stress and activation of individual faults at the contact of the Dinarides and the Pannonian Basin. Since the acquisition of the first seismographs in Croatia in 1909, up until 2019 1,364 earthquakes were detected in the vicinity of Petrinja. The strongest known earthquake of 5.8 magnitude took place on October 8, 1909 in Pokupsko. By analysing that particular earthquake, seismologist and geophysicist Andrija Mohorovičić made one of the greatest discoveries in seismology – the boundary between the Earth's crust and the mantle – the Mohorovičić discontinuity [12].

On December 28, 2020, at 6:28 am the first M 5.0 earthquake occurred with the epicentre in Petrinja (about 50 km SE from ZG). On the same day, there were two more earthquakes, an M 4.7 at 07:49 am, and an M 4.1 at 07:51, with a series of weaker ones. On the next day, December 29, 2020, at 12:19 pm, the strongest earthquake M 6.2 occurred with the epicentre 3 km southwest of Petrinja at a depth of 10 km. It was the strongest earthquake recorded by instruments in Croatia since 1909. Seven people were killed, 5 of them in the village of Majske Poljane [13]. Thus, the earthquake of Petrinja activated the system of faults in the underground of the wider Sisak, Petrinja and Glina area. After the main earthquake until January 28, 2021, 622 more earthquakes of magnitude greater than 2.0 were recorded.

2 CHARACTERISTIC DAMAGE

Most of the damaged buildings were built before 1964, which means that they were built before the first true seismic regulations in Croatia. The largest number of damaged buildings is among old masonry buildings. Buildings without concrete ceilings sustained greater damage compared to masonry buildings with concrete ceilings. Collapses and other damage are more pronounced in buildings that have not been properly maintained and on the buildings that were inappropriately reconstructed. The greatest amount of damage in concrete buildings was present in buildings with frame structures, mostly to non-structural elements like infill walls, but also to structural elements like columns. Modern buildings built after 1981 in general did not sustain significant damage except in some isolated cases – as the result of errors in design or inadequate execution of works. Steel structures were mostly undamaged, except in some cases where the bracings gave way. Wooden structures would have had breakage if they had already been damaged or had inappropriately executed connections details or unsuitable structural systems (roofs).

2.1 Problems and damage to masonry structures

Several main deficiencies and reasons for damage and collapses of masonry buildings were noted. First one was the extremely poor quality of construction materials and thus low lateral resistance of load-bearing walls. Shear strength is not realistically higher than $f_{v0} = 0.10$ MPa. Second one is that the structural integrity of the buildings was insufficient, causing local failures, separation, and collapse of structural elements. Generally, the connections between the ceiling structure and the walls do not exist or is extremely weak – acting only through friction. The walls are usually not held transversally by other walls (out of plane problems), and the connection of longitudinal and transversal walls is exclusively done by a masonry connection which is inadequate due to its weak properties. The only additional connection is sometimes made by the built-in ties that were often found to be damaged. The roofs of the buildings were mostly dilapidated, and their structures were not properly connected with the masonry structure. Thirdly, inadequate load-bearing wall layout and inappropriate geometry is often the issue – which in most cases manifests in openings that are too large, buildings that have too many storeys, and systems with one-way ceilings that leave lateral walls unladen by the weight of the ceilings further increasing the number of cases of damage.

Masonry structures sustained several characteristic types of damage (Fig. 2). These include damage and collapse of unsupported protrusions like chimneys, gables, facades, towers and alike. Further, damage to brick vaults and wooden ceilings were often present as well as separation, cracking and collapse of the walls. Common damage to masonry was in form of slipping or diagonal tensile/shear failure.

Figure 2: Damage to masonry structures. *(Source: Authors.)*

2.2 The damage to reinforced concrete structures

Damage to concrete structures (Fig. 3) was less spread, but still there were some reoccurring types of damage that they have sustained. Concerning the reinforced concrete frame structures, damage manifested mostly as partition and infill walls failing due to the flexibility of the frames and their horizontal displacements. Typical column damage was caused by sparsely spaced or missing reinforcement ties. Damage to the joints of steel and concrete girders on concrete columns and walls was also observed.

Figure 3: Damage to RC structures. *(Source: [10] and authors.)*

2.3 Geotechnical problems

Soil liquefaction was present in the Sisak-Moslavina area during the Petrinja earthquakes. The decrease in the strength of sandy soil caused the structures to crack where the ground under their foundation sunk. The expelling of soil with fluid-like behaviour and water to the surface in the form of geysers was observed, as well as the occurrence of large sinkholes due to the fact that below the alluvial sediments there are limestone rocks, eroded by the long-term action of groundwater. In this contact zone, the finer fraction is washed away, and cavities are formed, notably in places like Mečenčani and Donji Kukuzari. Landslides – mainly displacement of previously active landslides – also occurred where loss of shear strength was caused due to inertial forces. Cracks and subsidence of the embankment was present as well [14].

3 THE RECONSTRUCTION OF THE BUILDINGS IN ZAGREB AND PETRINJA AREAS

Many buildings were damaged or demolished in the Zagreb and Petrinja earthquakes and costly and time-consuming renovations followed. The Law on Reconstruction of earthquake-affected areas [15] has been passed, which envisages: that all public buildings and protected cultural asset buildings would be completely renovated with the costs to be borne by the owner; demolished private houses were to be renovated by the state covering the 100% of costs – but only the necessarily "usable" portion of the houses; and for the damaged private residential/business buildings, the state and the cities would co-finance up to 80% of the costs of structural retrofitting and improvements that would reach up to the 50% of the HRN EN 1998 norm.

3.1 Main problems of reconstruction

Several problems emerged after the post-earthquake assessments of damage. First one was how to secure an amount of money needed for renovation? Further, in Zagreb, a large number of damaged buildings are in the historically protected area or listed as individual cultural asset. Further on, in Zagreb case, some of the buildings are relatively large, and their structural improvement is quite complicated to implement because it requires the evacuation of residents and high associated costs. In Sisak-Moslavina County on the other hand, the main problem is how to ensure the sustainability of settlements since many of them are facing decades-long depopulation.

The Faculty of Architecture has issued two manuals in the field of seismic improvement of old masonry buildings in order to fill the informational void that emerged after the earthquake: Techniques for the Repair and Strengthening of the Masonry Buildings [16], and Manual for Seismic Retrofitting of the Existing Masonry Buildings [17]. Subsequently, a third manual called Urban Renewal was also published [18].

In the Techniques for the Repair and Strengthening of the Masonry Buildings, an overview of main causes of damage and failure of masonry structures along with the overview of conventional and special techniques for retrofitting and reinforcement of masonry buildings is given. It also includes an overview of the most commonly used techniques for retrofitting and reinforcement of reinforced concrete buildings. It is supplemented with information on the process of creating a project for the retrofitting and strengthening of the load-bearing structure of existing buildings (Fig. 4).

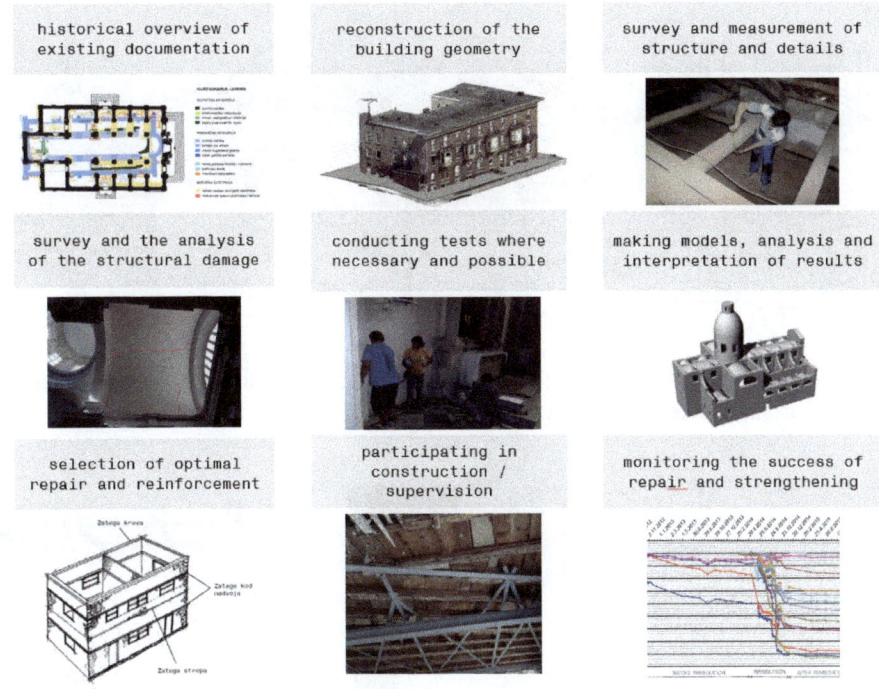

Figure 4: The process of making a project for the retrofitting and strengthening of a building [16].

In the Manual for Seismic Retrofitting of the Existing Masonry Buildings a concrete case of a typical Zagreb historical lower-town building with masonry structure and wood ceilings (Fig. 5) was used to demonstrate different levels of interventions in accordance with the conclusions of the expert group at the Faculty of Civil Engineering, University of Zagreb, organised in cooperation with the Croatian Chamber of Civil Engineers. Levels range from basic retrofitting (Fig. 6), basic strengthening (Fig. 7), overall strengthening of existing structure (Fig. 8) all the way to the more invasive and overall reconstruction (Fig. 9). These are coupled with the cost estimate for each level with all the prices for necessary works – total estimated prices are given in Table 2.

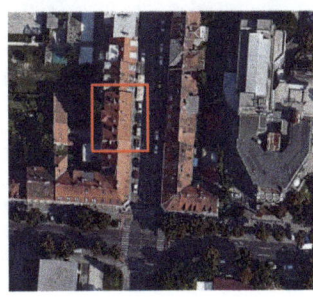

Figure 5: Case building for the simulation of retrofitting and reinforcement levels, Kačićeva 22, Zagreb, Croatia [17].

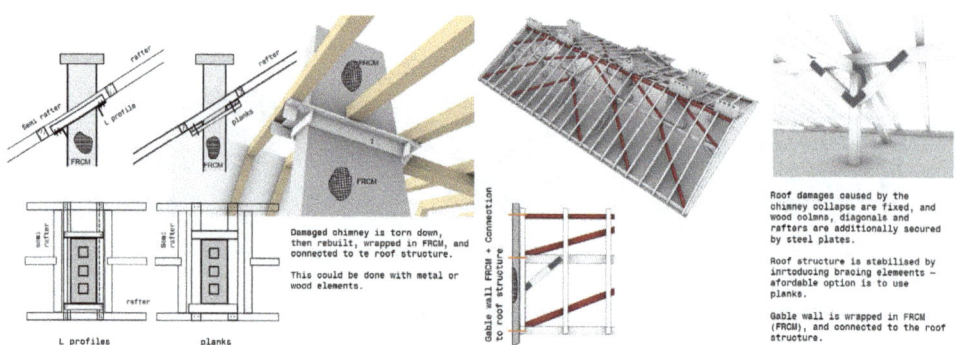

Figure 6: First level – basic retrofitting: chimneys, gable walls are rebuilt, roof structure is stabilised, and roof floor is rigidised and connected to the walls [17].

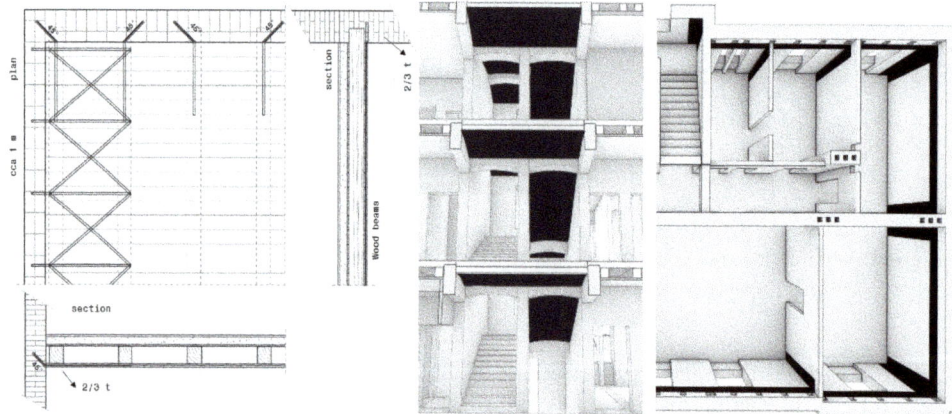

Figure 7: Second level – same as first, but with addition of all floors being connected to the walls that are partially reinforced with FRCM [17].

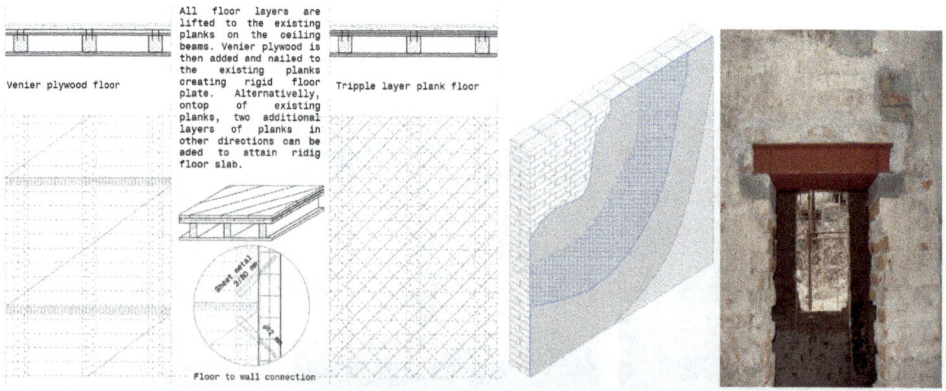

Figure 8: Third level – all floors are rigidised and connected to the walls that are completely reinforced with FRCM and steel lintels [17].

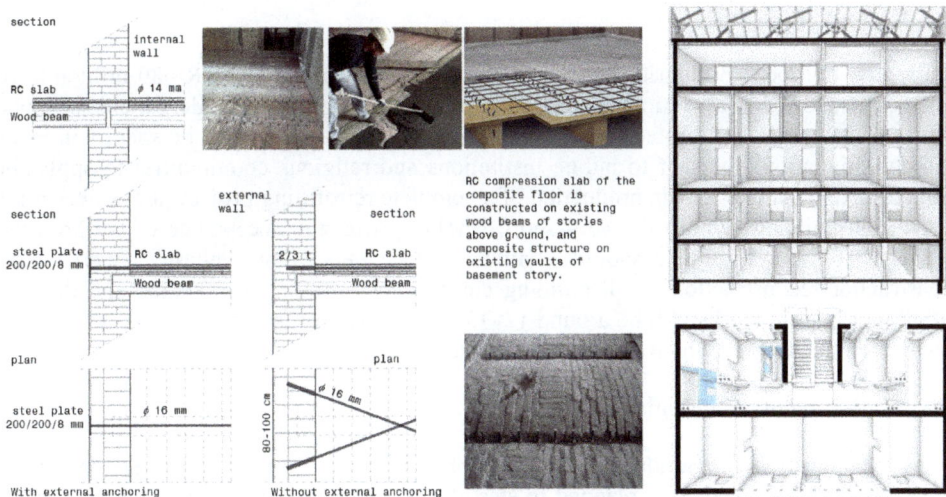

Figure 9:   Fourth level – Reinforced concrete walls are inserted into floor plan and all floors are fitted as composite floors [17].

Table 2:   Cost estimate for case building in downtown Zagreb with prices in Euro at the time of the earthquake [17].

	Level	Description	Total cost in €	€/m²
	1	Return to the state before the earthquake	122,600	75
	2	Economic raising of resistance	218,450	135
	3	Further raise of resistance	533,650	335
	4	Execution of additional RC walls(not preferred option)	430,900	270

The question "how to improve old masonry buildings" could be addressed in several ways, depending on the state of individual case. Application of appropriate interventions on the load-bearing structure in accordance with the individual aim of retrofitting can span anywhere from: ensuring the stability of cantilever protrusions such as chimneys, gables, etc.; stabilising the roof and connecting it with the walls; stiffening the ceiling structures in order to attain the rigid disk behaviour of the ceiling planes and connecting them to the walls; interconnecting the walls, and ensuring adequate walls to ceilings connections; increasing the shear strength of masonry walls; adding new rigid elements in the form of new masonry walls, concrete walls, steel or concrete frames; introducing the seismic insulation at the foundation level to prevent vibration input, etc.; demolition and construction of new modern structure etc.

3.2 Renewal of Zagreb

Currently, European Solidarity Fund funds in the amount of about EUR 680 million have been activated for interventions on public buildings like hospitals, kindergartens, schools, colleges, institutes, and protected heritage buildings. The funds should be spent within 18 months. Invitation was sent to public institutions and religious communities to apply for projects. Applications for the production of a complete retrofitting projects are expected for about 500 buildings within the very short deadline. After the acceptance of the projects, tenders for execution should initiate. In the first phase, works on construction/reconstruction (shell building construction) should be executed and the real deadline for these works will be around 12–15 months. After that, the works on the second phase from the funds of the Recovery Fund would continue.

3.3 Renovation of the Petrinja area

An EU application for Solidarity Fund funds has been sent, and a significant amount is expected to be received. It is planned to start the reconstruction as soon as possible, as in Zagreb, with the proviso that family houses will most likely be built with prefabricated construction (about 4,000 houses) due to the speed of construction and the shortage in construction companies. The design and renovation of public buildings will start in parallel.

It is currently expected that around HRK 10 billion (about €1.3 billion) will be available in the next two years for projects and construction as part of the renovation of these two areas.

4 CONCLUSION

After Zagreb and Petrinja earthquakes, several things were experienced and noted: needs of many people had to be taken care of at the critical moment; information and the response of the cities and the state were not immediate, and too much time was spent preparing and planning instead of acting; even things that presented immediate danger were sometimes not promptly addressed; the organisation of recovery actions goes slowly and there is a risk of missing the deadlines for financial support.

Authors would therefore like to propose this three-phase approach to the repairs, retrofitting, and reconstruction of old buildings in the earthquake-stricken areas.

Phase 1 – instead of waiting for action plans, and new laws to be enacted, damage repair works should be executed as fast as possible in order to return the buildings into the state of operation. This is the most effective way to have the city up and running. However, it also lacks the improvements to structures that would raise their reliability since their basic characteristics are roughly the same as they were before the earthquake.

Phase 2 – careful valorisation of existing buildings can then help determine what is valuable, and to what extent, and what is not of particular value. This issue is haunting Zagreb for decades in the form of practices that make sure that everything is conserved as much as possible regardless of its true cultural, technical, or practical value – which has caused more assets to fall into disrepair instead of being improved.

Phase 3 – strategic implementation of integral renovation needs to be carried out, including the synergistic urban and architectural renewal. It is important to design a renovation scheme that will allow building upgrades that can raise their value and the quality of spaces in order to (financially) facilitate the implementation of the projects that can bring the critical elements of buildings architecture and technology up to modern standards.

REFERENCES

[1]   Herak, D., Dasović, I. & Ivančić, I., Crtice iz (geofizičke) povijesti [notes from (geophysical) history]. https://www.pmf.unizg.hr/geof/popularizacija_geofizike/ crtice_iz_povijesti#DBK1667. Accessed on: 2 Mar. 2021.

[2]   Ivančić, I., Seizmičnost Hrvatske [Seismicity of Croatia]. https://www.pmf.unizg.hr/ geof/seizmoloska_sluzba/seizmicnost_hrvatske#. Accessed on: 2 Mar. 2021.

[3]   Pravilnik o privremenim tehničkim propisima za građenje u seizmičkim područjima. Sl. list SFRJ 39/64.

[4]   Pravilnik o Tehničkim normativima za izgradnju objekata visokogradnje u seizmičkim područjima. Sl. list SFRJ 31/81, 49/82, 29/83, 20/88, 52/90.

[5]   Tehnički propis za betonske konstrukcije, p. 3381, NN 139/2009.

[6]   Tehnički propis za zidane konstrukcije, p. 4, NN 1/2007.

[7]   Miklin, Ž., Potresi na zagrebačkom području. http://omagi.hgi-cgs.hr/Zagreb_ potresi.html. Accessed on: 17 Jan. 2021.

[8]   Pola godine od Zagreboačkog potresa. https://www.pmf.unizg.hr/geof/ seizmoloska_sluzba/o_zagrebackom_potresu_2020/pola_godine_od_potresa. Accessed on: 28 Jan. 2021.

[9]   Šavor Novak, M. et al., Zagreb earthquake of 22 March 2020 – Preliminary report on seismologic aspects and damage to buildings. *Građevinar*, **72**(10), pp. 869–93, 2020. DOI: 10.14256/JCE.2966.2020.

[10] Hrvatski centar za potresno inženjerstvo. https://www.hcpi.hr/. Accessed on: 6 Feb. 2021.

[11] Uroš, M. et al., Procjena oštećenja građevina nakon potresa – postupak provođenja pregleda zgrada [Post-earthquake damage assessment of buildings – procedure for conducting building inspections]. *Građevinar*, **72**(12), pp. 1089–1115, 2021. DOI: 10.14256/JCE.2969.2020.

[12] Markušić, S. et al., Destructive M6.2 Petrinja earthquake (Croatia) in 2020 – preliminary multidisciplinary research. *Remote Sensing*, **13**(6), p. 1095, 2021. DOI: 10.3390/rs13061095.

[13] Civilna zaštita RH – Potres kod petrinje. https://civilna-zastita.gov.hr/vijesti/potres-kod-petrinje-3357/3357. Accessed on: 20 Jan. 2021.

[14] Bačić, M. & Terzić, J., Pregled nekih geotehničkih problema na području Banovine pogođenim potresom M6.4, 29.12.2020. *Preliminarno izvješće-*, 2021.

[15] Zakon o obnovi zgrada oštećenih potresom na području Grada Zagreba, Krapinsko-zagorske Županije, Zagrebačke Županije, Sisačko-moslavačke Županije i Karlovačke Županije, NN 10/2021 (5.2.2021.), p. 191, 2021.

[16] Galić, J., Vukić, H., Andrić, D. & Stepinac, L., *Tehnike Popravka i Pojačanja Zidanih Zgrada*, Arhitektonski fakultet Sveučilišta u Zagrebu: Zagreb, 2020.

[17] Galić, J., Vukić, H., Andrić, D. & Stepinac, L., *Priručnik za Protupotresnu Obnovu Postojećih Zidanih Zgrada*, Arhitektonski fakultet Sveučilišta u Zagrebu: Zagreb, 2020.

[18] Jukić, T., Mrđa, A. & Perkov, K., *Urbana obnova: Urbana regeneracija Donjega grada, Gornjega grada i Kaptola/Povijesne urbane cjeline Grada Zagreba*, Arhitektonski fakultet Sveučilišta u Zagrebu: Zagreb, 2020.

DIGITAL SURVEY AND PARAMETRIC 3D MODELLING FOR THE VULNERABILITY ASSESSMENT OF MASONRY HERITAGE: THE BASILICA OF SAN DOMENICO IN SIENA, ITALY

ANGELO MASSAFRA, CARLO COSTANTINO, DAVIDE PRATI,
GIORGIA PREDARI & RICCARDO GULLI
Department of Architecture, Alma Mater Studiorum, University of Bologna, Italy

ABSTRACT

This research focuses on the combined use of digital survey tools and parametric 3D modelling procedures for supporting the vulnerability analysis of masonry heritage. The behaviour of masonry structures is strongly related to the geometrical properties of macro-elements, their connections, and material characteristics. The collapse of these buildings occurs mainly due to the loss of equilibrium of the macro-elements than the breaking of the materials. Moreover, the numerical computing of structural element stresses requires an in-depth knowledge of material properties that could be reached mainly through destructive tests, which have high costs and could damage the architectural figurativeness of protected buildings. Besides, many tests are required because of the high variability of mechanical properties within an ancient masonry building due to the various transformations and construction techniques that usually characterise its construction history. A quick and non-destructive approach is proposed as a preliminary support tool. It is based on the observation of high-resolution unmanned aerial vehicles (UAV) orthophotos combined with terrestrial laser scanning (TLS) point cloud geometric information analysis through visual programming generative algorithms. The case study is the Basilica of San Domenico, one of the most important churches in Siena, Italy, built in the 13th century, completed in the 15th century, and repeatedly modified until the beginning of the 20th century. The Basilica has construction features (the large volume without side chapels and high slenderness) that make it particularly vulnerable to horizontal actions. These inherent vulnerabilities were joined by the effects of an earthquake in 1798, whose damages were partially remedied. The workflow application allowed highlighting the hypothetical displacements that the transept and the bell tower have undergone over time, improving their behaviour comprehension. It also helped to point out the critical issues of the whole building, supporting the vulnerability analyses based on the macro-elements and orienting the intervention hypothesis for its preservation.
Keywords: cultural heritage, Basilica of San Domenico in Siena, masonry structures, terrestrial laser scanning, UAV photogrammetry, generative algorithms, parametric 3D modelling, structural systems reverse engineering, displacement analysis.

1 INTRODUCTION

The importance of preserving masonry heritage cannot be disregarded since most of the historical buildings in Europe were built with this construction technique. Effective ways of retaining these architectures are needed, based on appropriate knowledge of their structural systems. Minimally invasive approaches are required, especially on protected buildings, to carry out an aware conservation process without damaging them.

In general, to evaluate the seismic risk assessment, a simplified qualitative procedure is implemented to have a first overview of the main vulnerability of damaged heritage buildings after an earthquake. This evaluation, specific for protected buildings, is valid as a preliminary analysis but not helpful in designing targeted interventions for seismic improvement that require more complex quantitative assessment [1].

WIT Transactions on The Built Environment, Vol 203, © 2021 WIT Press
www.witpress.com, ISSN 1743-3509 (on-line)
doi:10.2495/STR210151

In the case of churches, the simplified procedure is reduced to defining a single parameter i_v (vulnerability index) [2]. Great importance is given to the assessments of out-of-plane mechanisms (1st mode) with linear kinematic analysis, which is typically done by decomposing the structure into macro-elements. In fact, studies related to earthquake damages undergone by churches, carried out during the last decades' main Italian seismic events, have highlighted how their seismic behaviour is characterised by an autonomous structural response of the macro-elements rather than of the church as a whole structure. This happens because, as for most historic masonry buildings, disconnections of the construction elements occur and do not guarantee structural continuity, causing the earthquake action to further increase these disconnections.

Although these qualitative assessments help deepen the analyses of the construction and the cracking framework of masonry buildings, which are still the most reliable path towards evaluating numerical assessments, they strongly depend on the technician's judgment validating the interpretation of the activated kinematic chains.

Therefore, it has been decided to exploit the point cloud data obtained with on-site complementary terrestrial laser scanning (TLS) and aerial photogrammetry (UAV) techniques to support and make more effective evaluations and increase the objectivity value of the qualitative analyses. In particular, the use of generative algorithms already developed to analyse other construction elements [3], [4] has allowed deepening the automation of the analysis of the out-of-plane of the transept façades and bell tower of the Basilica of San Domenico in Siena, Italy. Similar studies that combine digital survey techniques and out-of-plane analysis are not widespread in scientific literature and often require a complex and dedicated survey campaign [5], [6].

2 THE CASE STUDY

The presented case study is the Basilica of San Domenico, a masonry church built in the 13th century on Camporegio Hill in Siena, Italy, near the medieval fountain of Fontebranda. On the one hand, the prime location ensures the church a landmark role over the city along with the Palazzo Pubblico tower and the Cathedral bell tower. On the other hand, it has been the leading cause of many structural issues since its construction.

Although the Dominican Order settled in Siena in 1221, the documentary sources indicate that San Domenico's nave construction began only around 1246, and it was completed around 1300. Some documentary sources suggest that the construction incorporated the ancient church of San Gregorio, but this has never been confirmed [7], [8]. The first expansion of the Basilica began even before the nave was completed and concerned the Beato Ambrogio Chapel's construction. The location of the chapel, demolished in 1479, was at the corner between the nave, the current bell tower, and the transept [8]. The construction of the Lower Church between 1306 and 1352 turned out to be one of the most critical phases for the contemporary church shape; in fact, it would have been employed as a foundation for the future transept. However, the plague interrupted the public and private donations for the construction progress, and therefore it was suspended for over thirty years. Indeed, the transept was completed only between 1380 and 1480 through many difficulties and interruptions, using the chapels owner families' funds [9].

The bell tower is documented since 1321, but the current construction is a later work dating between 1489 and 1517 when the statue of St. Caterina from Siena was placed on the top spire. Later on, the spire and the statue were removed due to the excessive danger caused by lightning bolts.

Between the 16th and 18th centuries, many interventions were carried out: the reconstruction of the roof, changes on the openings, altars addition and other Baroque decorations in the nave, the consolidation of the masonry. In 1798, the church was severely damaged by an earthquake to the point that its demolition and reconstruction were considered. Only after a lot of debate, the Sienese population opted for its restoration.

Finally, extensive renovation work was carried out during the 20th century to uncovered as far as possible the medieval frescos and restore the original natural illumination coming from the large arched windows.

3 THE INVESTIGATION PROTOCOL WORKFLOW
The research allowed defining a procedure to systematise the assessment of masonry structures' behaviour in terms of displacement and deformations over time (Fig. 1). This research protocol includes an on-site instrumental acquisition phase using UAV photogrammetry and TLS survey techniques, followed by the digital rendering of the acquired data, extracting high-resolution orthophotos and generating 3D point clouds. The orthophotos are carefully observed to identify the building conservation state, cracks, and damages. Consequently, the vulnerabilities and the associated damage mechanism (e.g., the overturning of the façade or diagonal cracks on the transversal wall) can be evaluated starting from the crack patterns. At the same time, parametric models from the TLS point cloud permit highlighting the hypothetical displacements that macro-elements have undergone during the lifespan.

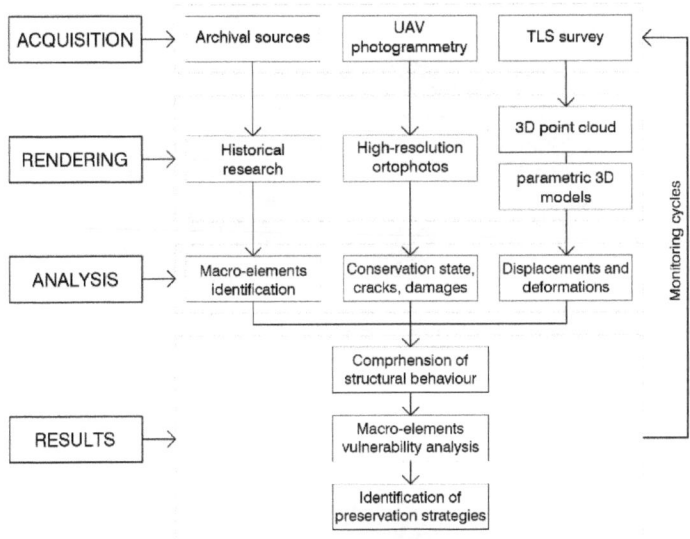

Figure 1: Investigation protocol workflow. *(Source: Authors' graphic elaboration.)*

These methodologies allow examining in detail building portions that would have been impossible to investigate with traditional methods because of their heights (up to about 40 m), the obstacles hiding them from the ground, or the occlusion generated by other buildings or vegetation. By combining these different analyses with historical data, it is possible to understand the investigated buildings' effective structural behaviour. The protocol allows to precisely recognise the crack pattern and connect it to the most probable collapse kinematics of masonry elements, which is the key for numerically assessing the vulnerability macro-elements analysis, orienting the intervention hypothesis for the building preservation.

Finally, future repetitions of the whole protocol would permit efficient monitoring cycles regarding the displacements and the conservation state of the buildings over time.

3.1 The TLS and UAV photogrammetric survey

The TLS and UAV photogrammetric survey campaign took five days. It was conducted employing a FARO CAM2 FOCUS 3D® and a DJI Mavic Mini, an open category UAV of C0-class identification label (MTOM < 250 g), using a targetless approach on both types of equipment. The photogrammetry needed, at least a 50% overlapping of the pictures with a proper resolution, to ensure an ideal outcome. These images were elaborated in Agisoft Metashape® 1.6.5 to make the photogrammetric 3D model and orthophotos of both the external and internal elevation (Fig. 2). Overall, the total airtime was about 8 h, allowing the shooting of 708 pictures.

On the other hand, for the entire TLS survey of the building (Fig. 3), it was necessary to shoot 130 scans with a medium resolution of 7.67 mm/10 m and a quality filter of 3×. With a medium overlapping between the scans of 36%, despite the building complexity and size, it was possible to achieve an extremely accurate alignment, with an average 1.5 mm standard deviation between corresponding points and a maximum 2.6 mm deviation.

Figure 2:   The photogrammetric mesh model of the exterior of the Basilica of San Domenico in Siena. *(Source: Authors' graphic elaboration.)*

Figure 3:   TLS point cloud of the Basilica of San Domenico in Siena: identification of the investigated masonry walls and the reference system. *(Source: Authors' graphic elaboration.)*

The remarkable height of the Basilica, which reaches 42.85 m in the transept and 48.25 m in the bell tower, the difficulties in surveying near the steep slopes of the hillside, and in getting precise information of the upper part of the inner transept were the main challenges of the survey campaign.

3.2 High-resolution orthophotos and crack patterns analysis

The use of high-resolution orthophotos, made from UAV photogrammetry techniques, consists of a valuable and effective way to analyse the masonry building crack patterns and compare the cracks present on the inside wall with the outside ones. Having the exact correspondence between the external and the internal cracks allows to accurately value their locations, their width, and if they concern the entire thickness of walls.

The orthophotos of the Basilica of San Domenico have been used to divide the cracks according to their width into three categories: small cracks (<2 mm); medium cracks (2 mm ÷ 1 cm); dangerous cracks (>1 cm). Simultaneously, those walls characterised by mortar absence were detected, and those areas where the previous cracks have been reintegrated. Based on all these considerations, the mechanisms associated with their corresponding crack patterns have been defined as follow: (1) Out-of-plane flexures; (2) Diagonal shear cracks; (3) Anchor-rod punching; (4) Do-not-restrained arch trusts; (5) Crushing cracks; (6) Cracks related to different construction phases; and (7) Foundation subsidence.

The crack patterns analysis identified vertical cracks due to out-of-plane overturning at the corners of the transept heads (façades A and B). Those on façade B are particularly visible, as they reach 5–6 cm in width. This mechanism is also confirmed inside the church by similar vertical cracks, albeit with a smaller width.

Furthermore, in the B façade, there are numerous cracks (Fig. 4), which follow a sub-vertical direction in the upper part of the elevation and become diagonal near the base. Other minor damages are attributable to the roof trusses' concentrated loads and the masonry's poor quality and condition in correspondence with the tympanum. The crack pattern in the inner transept surface is complex and referable to multiple factors as the out-of-plane overturning of the chapels, the instability of the corner section of the wall, the shear cracks, the mid-height-out-of-plane overturning, the do-not-restrained arch thrust of the choir. Furthermore, it cannot be excluded that these mechanisms are also caused by local foundation subsidence.

3.3 The model generation procedure

The assumption behind the algorithmic 3D model generation procedure is to take advantage of the large amounts and accuracy of spatial information provided by TLS devices, such as point clouds. These data allow a highly detailed analysis of the masonry elements and, therefore, the elaboration of specific and comparative information on their static behaviour and their state of preservation. The usual modelling methodologies, such as vectorisation of orthophotos or 3D modelling using CAD or BIM software, heavily depend on the operator's choices and are affected by low repeatability and accuracy. For these reasons, it was decided to transform the point cloud into 3D models using Grasshopper® generative algorithms. These tools, which are continuously updated, provide advantages over manual modelling methods, both in the speed of execution and the elimination of inaccuracies.

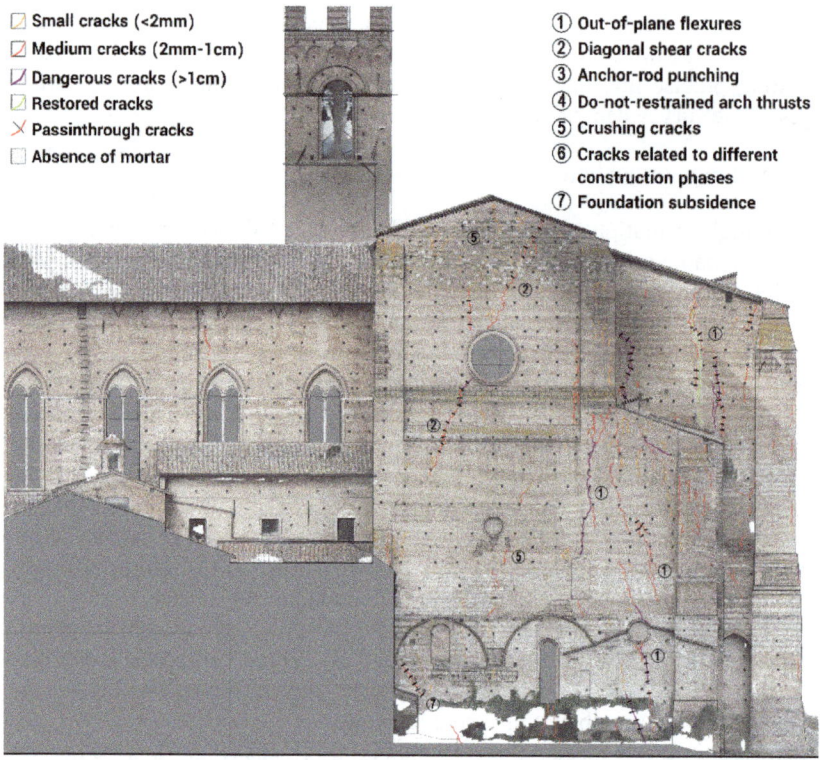

Figure 4:   Façade B of the Basilica. Orthophoto highlighting the crack patterns and their interpretation and analysis. *(Source: Authors' graphic elaboration.)*

3.3.1 Parametric 3D modelling

The parametric modelling workflow was applied to each façade identified in Fig. 2, considering the reference system shown in the same figure. The x-axis is parallel to the nave with the positive direction towards the valley, the y-axis is parallel to the transept with the positive direction towards the bell tower, the z-axis is zenith with the positive direction towards the sky.

The segmented point cloud of the façade, which is also called the "Basic 3D Model" as it represents the structure's current condition, was transformed into the "Ideal 3D Model" through generative algorithms. This model represents the wall's hypothetical undeformed configuration, excluding the out-of-plane deformations that have presumably occurred over time. In particular, the Ideal 3D Model coincides with the "average vertical plane" of the façade, in other words, the plane that represents the ideal condition of a perfectly flat and vertical façade.

The operations for generating the Ideal Plane are illustrated in Fig. 5. Firstly, the point cloud of the façade was segmented and cleaned up in Geomagic Control®; it was divided in the inner and outer surface, horizontal bands of points were selected, exported into Rhino®, and linked to the Grasshopper® algorithm (Fig. 5). The algorithm vectorised the horizontal cross-sections of the wall (in this case, it was chosen to consider a section-curve every 2 m of height). Then the algorithm selected the portion of the curves that best represented the planar condition of the façade, removing pilasters, buttresses, protrusions, recesses, etcetera.

If the façade had different main planes, which means that it presents different section thicknesses, only the curves relating to the lowest plane of the façade were selected. These curves were divided into a certain number of points (e.g., 100 points) and, for each group of points, the "linear regression line" was calculated. This line is the one-degree polynomial curve, lying in the horizontal plane, that best interpolates the points group. A similar procedure was carried out for calculating the "vertical linear regression plane", which is the plane that best interpolates the best-fit lines, forced to be perfectly vertical (Fig. 5).

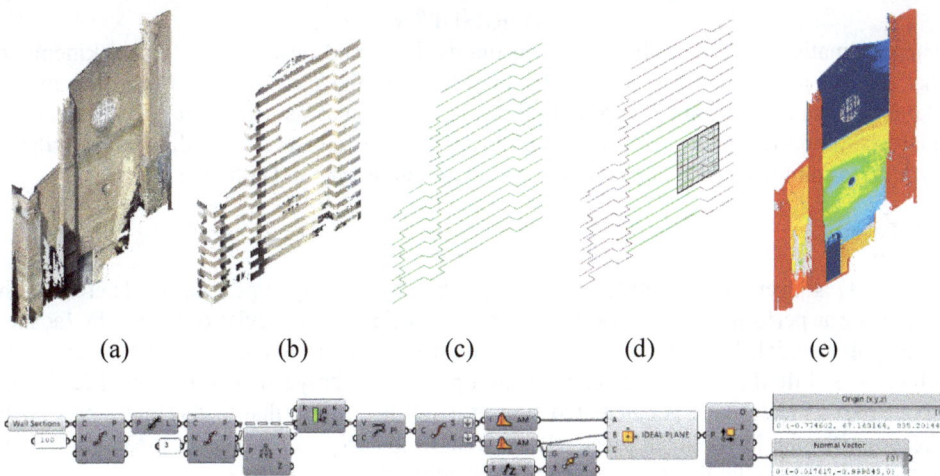

(a)   (b)   (c)   (d)   (e)

Figure 5:   The model generation procedure and grasshopper algorithm. (a) TLS point cloud; (b) Point cloud segmentation; (c) Vectorised cross-sections; (d) Ideal Plane; and (e) TLS point cloud-Ideal Plane comparison. *(Source: Authors' graphic elaboration.)*

This plane is the Ideal Plane that represents the ideal condition of the wall. It is important to note that this plane is not properly representative of the façade's condition at the time of its original construction since it is often difficult to find an absolute reference to attribute the original deformation state of the façade, especially in such complex buildings. However, it consists of a useful approximation to evaluate the façade areas that have lost flatness and could be considered a relative reference for assessing its out-of-plane deformation state. It permits knowing how much a point of the cloud deviates from the mean plane and identifying areas of the façade that show typical deviation values.

3.3.2 3D models and point cloud comparison

Once the Ideal 3D Model definition was reached, a comparative analysis was performed using Geomagic Control® to evaluate the deviations between the Ideal 3D Model (the ideal plane) and the Basic 3D Model (the TLS point cloud). For all the investigated masonry elements, the minimum and maximum reference deviation threshold were manually set in a chromatic scale to better indicate the extent of the movements that occurred between the surveyed configuration and the ideal one. The graphic representation through the most appropriate chromatic scale clarifies the entities of the displacements. It is possible to select any position on each portion of the masonry to highlight the local deviations' numerical value using the graphical representation (Figs 6–9). This representation provides qualitative and quantitative information and facilitates an accurate and widespread understanding of all the displacement and deformations that each wall has undergone.

Comparisons can be made in a 3D space and on 2D projection planes user-selected according to the deviation types that need to be highlighted. The software also allows for the precise quantification of deviations using control points. With the "annotation" tool, it was possible to report the exact value of the distances measured between the two compared objects' corresponding points and their subdivision along the x, y, and z axes.

4 RESULTS

The chromatic scale highlights deviation trends that could describe the typical kinematics that ancient masonry structures usually manifest and the interaction between the macro-elements. The integration of the proposed displacement analysis with the usual crack patterns analysis, supported by the historical research and the in-situ inspections, allowed formulating hypotheses on the church's behaviour, producing interesting and coherent results.

4.1 Façade A

Façade A, adjacent to the bell tower, presents thickness change in elevation. Therefore, the analysis was performed firstly for the lower part, evaluating its deviations from the façade's mean plane model. The upper part was analysed with reference to the lower part's ideal plane, which was shifted in correspondence to the upper part's barycenter. More specifically, the upper part's mean plane is calculated first in a similar way to that of the lower part of the façade. Then the lower mean plane is translated to the origin of the upper mean plane. This translated plane is the Ideal 3D Model of the upper part of the façade.

The analyses show that the lower portion of the façade's exterior surface exhibits a relative offset between its base and top of about 9 cm. The top is projecting outward from the building in the positive y-direction. The upper portion of the façade follows the same trend as the lower part. Moreover, it is rotated with respect to its mean xy-plane so that the side adjacent to the bell tower is projected towards the outside of the building (Fig. 6).

It is important to note that this rotation does not mean that the façade left side is rotated towards the inside of the building. More likely, this means that there is a delta deviation from the mean plane equal to about 9 cm caused by the interaction with the bell tower movement. The combined interpretation of both façade analyses suggests a simple tilting of the façade with some interference with the bell tower's movements (Fig. 7).

4.2 Bell tower

The bell tower's analysis was carried out on its two main elevations facing the outside of the building. As far as the tower elevation adjacent to façade A is concerned, the analyses show the displacements' compatibility since the bell tower also presents a relative displacement between its base and its top equal to about 7 cm, with the top part projected towards the outside (similar to façade A value).

Observing the other elevation, parallel to the y-axis, the deviation between the top and the base is even more marked, equal to about 19 cm (Fig. 7). Therefore, it is possible to hypothesise a tilting of the bell tower in the direction towards the outside of the building. The top is about 7 cm displaced from the base in the positive y-direction and about 19 cm in the negative x-direction. Moreover, observing the façade's cracking pattern, a vertical lesion is evident at the border between façade A and the bell tower, which shows a separation between the two macro-elements.

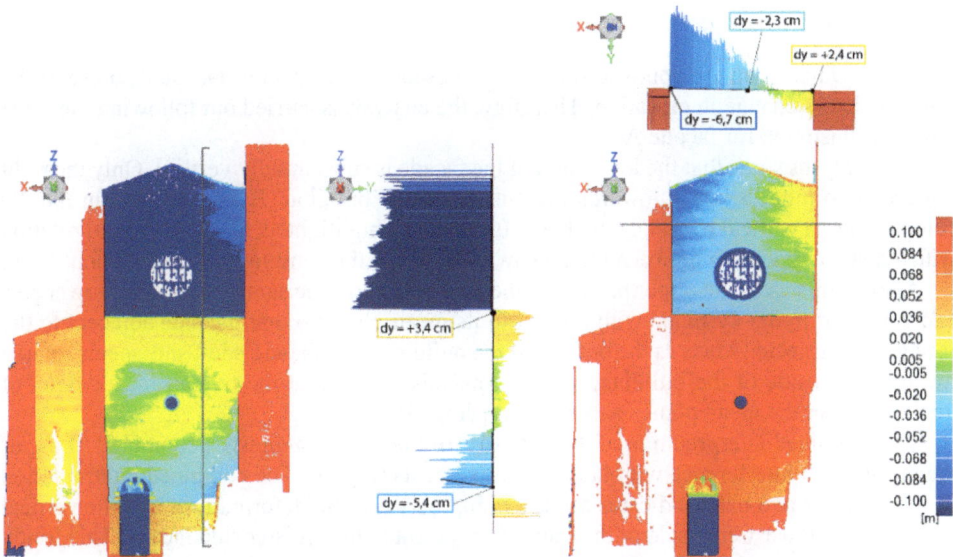

Figure 6:   Displacement analysis of the façade A. The green points lie on the mean plane of the façade. The red points deviate from the mean plane in positive y-direction towards the outside of the building, the blue ones in the negative y-direction towards the inside of the building. The deviation values are represented on a scale 100 times larger to allow greater readability in the sections. *(Source: Authors' graphic elaboration.)*

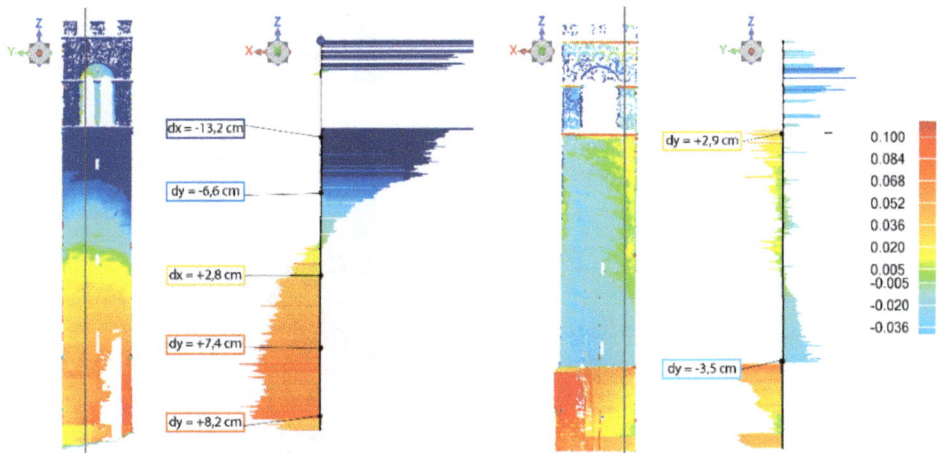

Figure 7:   Displacement analysis of the bell tower. On the left, the red points deviate from the mean plane in the positive x-direction towards the inside of the building, the blue points in the negative x-direction towards the outside of the building. On the right, the red points deviate from the mean plane in the positive y-direction towards the outside of the building, the blue points in the negative y-direction towards the inside of the building. *(Source: Authors' graphic elaboration.)*

4.3 Façade B

Façade B, which is on the opposite side of the transept with respect to façade A, presents the same thickness change in elevation. Therefore, the analysis is carried out following the same methods illustrated for façade A.

The analyses show that the lower part of the façade is substantially vertical. Only the right top seems to move slightly towards the outside of the building. By going deep in the 2D comparisons, a vertical overhang of about 10 cm can be highlighted, and a horizontal rotation with a delta of deviation between the downward side and the symmetrical one of about 5 cm.

Also, in this case, the upper portion of the façade follows the same trend as the lower part and is rotated in the xy-plane with respect to its mean plane so that the side adjacent to the façade C is projected towards the outside of the building. The façade's left side seems rotated towards the inside of the building, but this analysis means that the overall façade delta of deviation from the mean plane is about 10 cm (Fig. 8).

The combined interpretation of the two parts of the façade suggests a combined tilting of the façade with interference with façade C movements (Fig. 9). The combined interpretation of the façade's two parts indicates a relationship between the deformations and the vertical lesion on the downstream side of the transept, adjacent to the cloister side façade. This lesion shows a detachment of the façade in its upper part, resulting in the façade's overturning.

4.4 Façade C

Since it was not possible to survey the façade C accurately, due to the high hill slope and the presence of a tree curtain, the internal surface was used for the analysis.

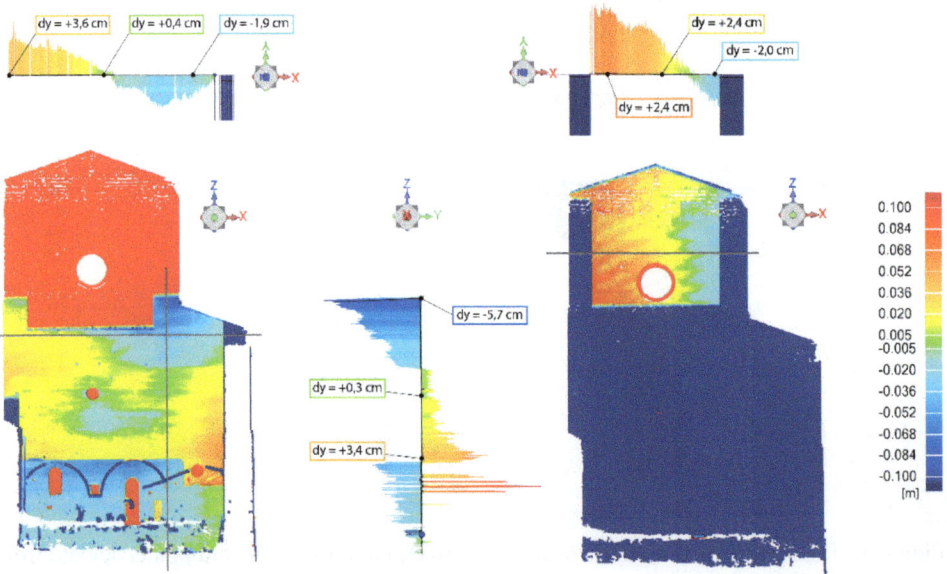

Figure 8: Displacement analysis of façade B. The red points deviate from the mean plane in the positive y-direction towards the inside of the building, the blue points in the negative y-direction towards the outside of the building. *(Source: Authors' graphic elaboration.)*

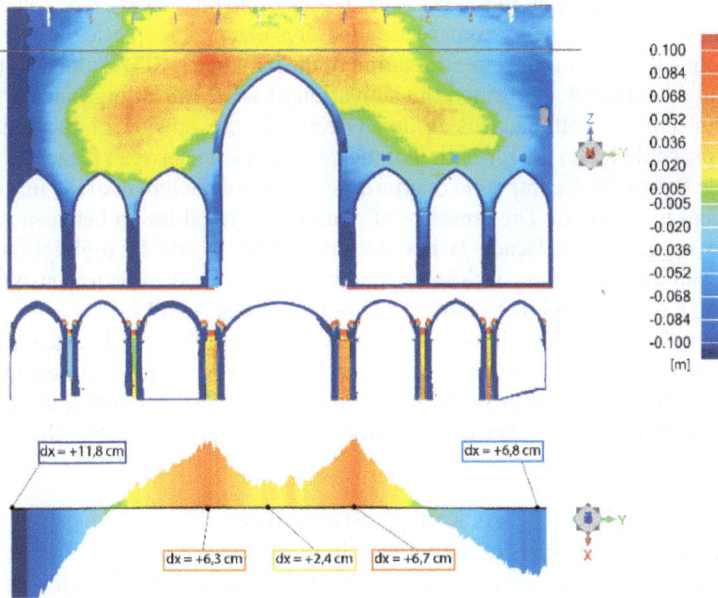

Figure 9:   Displacement analysis of façade B. The red points deviate from the mean plane in the negative x-direction towards the inside of the building, the blue points in the positive x-direction towards the outside of the building. *(Source: Authors' graphic elaboration.)*

From a qualitative point of view, the deformations of this part of the façade show how the lateral zones are offset from the mean façade plane towards the outside of the building, while the central zones are offset towards the inside of the building. Unlike all other façades, there is no particular offset in the vertical direction (Fig. 9).

Analysing the 2D comparison, the maximum deviation between the lateral and the central parts is equal to about 18 cm. Moreover, the transition from the blue-coloured points (on the side) to the yellow-coloured ones (on the centre) occurs according to the so-called "deformation isolines" with about 45° inclination. This supports the hypothesis of activation of a kinematic chain of combined overturning on both transept sides, with the interference of the deformations of façades A and B.

5 CONCLUSION

This research proposes a quick and non-destructive approach as a preliminary tool for supporting the vulnerability analysis of masonry heritage, based on the observation of the crack pattern through high-resolution UAV orthophotos combined with the analysis of TLS point cloud geometric information through visual programming generative algorithms.

The method could contribute to developing a more accurate and objective macro-elements analysis that, even if it strongly depends on the subjectiveness of the technician's judgment, is the most reliable path towards evaluating numerical assessments for masonry buildings. In fact, many seismic events that occurred in the last decades in Italy demonstrated that the collapse of masonry structures occurs mainly due to the loss of equilibrium of the macro-elements than the breaking of the materials.

The workflow was applied to the Basilica of San Domenico's case study in Siena, Italy, a 13th-century church with many construction features that make it particularly vulnerable to horizontal actions. The combined interpretation of the parametric 3D results with the analysis of the cracking framework confirmed the activation of some prevailing kinematic chains: a simple overturning of both façades A and B towards the outside of the building. This overturning probably led to the formation of the vertical lesion between façade A and the bell tower. Façade B presents a compound overturning with the detachment of the masonry wedge placed adjacent to façade C. The presence of a marked vertical lesion between the two wall curtains also suggests that façade B has detached from façade C, highlighting a simple overturning pattern. Moreover, façade C presents a compound overturn on both masonry wedges on the sides, highlighting relationships with façades A and B.

The protocol's application allowed highlighting the hypothetical displacements that the transept and the bell tower have undergone over time, improving the comprehension of their behaviour and the critical issues of the whole building. It also supported the vulnerability analyses based on the macro-elements and oriented the intervention hypothesis for church preservation.

ACKNOWLEDGEMENT

This research is part of a series of activities jointly carried out between the Department of Architecture research group and the Soprintendenza Archeologia Belle Arti e Paesaggio per le province di Siena, Grosseto e Arezzo, which is currently underway.

REFERENCES

[1] Direttiva del Presidente del Consiglio dei Ministri 9 febbraio 2011. *Valutazione e riduzione del rischio sismico del patrimonio culturale con riferimento alle Norme tecniche per le costruzioni di cui al D.M. 14/01/2008*, 2011.

[2] Linee Guida per la Valutazione e Riduzione del Rischio Sismico del Patrimonio Culturale Allineate alle Nuove Norme Tecniche per le Costruzioni (D.M. 14 Gennaio 2008). http://www.beniculturali.it/mibac/multimedia/MiBAC/documents/ 1295444865088_LINEE.pdf. Accessed on: 15 Feb. 2021.

[3] Massafra, A., Prati, D., Predari, G. & Gulli, R., Wooden truss analysis, preservation strategies, and digital documentation through parametric 3D modelling and HBIM workflow. *Sustainability*, **12**, p. 4975, 2020. DOI: 10.3390/su12124975.

[4] Prati, D., Zuppella, G., Mochi, G., Guardigli, L. & Gulli, R., Wooden trusses reconstruction and analysis through parametric 3D modelling. *International Archives of the Photogrammetry, Remote Sensing and Spatial Information Sciences*, **XLII-2/W9**, pp. 623–629, 2019. DOI: 10.5194/isprs-archives-XLII-2-W9-623-201.

[5] Capra, A., Bertacchini, E., Castagnetti, C., Dubbini, M., Rivola, R. & Toschi, I., Rilievi laser scanner per l'analisi geometrica delle torri degli Asinelli e Garisenda. *INARCOS*, **LXVI**(4), pp. 35–42, 2011.

[6] Bertacchini, E., Boni, E., Capra, A., Castagnetti, C. & Dubbini, M., Terrestrial laser scanner for surveying and monitoring middle age tower. *XXIV FIG International Congress 2010*, Sydney, Australia, TS4D-n. 4445, pp. 1–13, 2010.

[7] Lusini, V., San Domenico in Camporegio. *Bullettino Senese di Storia Patria*, volume tredicesimo, Tip. lit. dei sordomuti di L. Lazzeri: Siena, 1906.

[8] Riedl, P.A. & Seidl, M., *Die Kirchen von Siena, 2: Oratorio della Carità – S. Domenico*, Brückmann Verlag: Munchen; [poi] Polistampa: Firenze, 1985.

[9] Romagnoli, E., Biografia cronologica de' bellartisti senesi 1200–1800. *Firenze SPES*, 1976.

TOWARDS A SUSTAINABLE *RE-CONSTRUCTION* METHOD FOR SEISMIC-PRONE HERITAGE SETTLEMENTS OF GUJARAT, INDIA, BASED ON ADVANCED RECORDING TECHNOLOGIES

BERNADETTE DEVILAT[1], JIGNA DESAI[2], ROHIT JIGYASU[3],
MOHAMED GAMAL ABDELMONEM[1], FELIPE LANUZA[1] & MRUDULA MANE[2]
[1]Centre for Architecture, Urbanism and Global Heritage, Nottingham Trent University, UK
[2]Center for Heritage Conservation, CEPT Research and Development Foundation, India
[3]International Centre for the Study of the Preservation and Restoration of Cultural Property ICCROM, Italy

ABSTRACT

Post-earthquake reconstruction of housing in heritage settlements confronts challenges such as inadequate damage assessment and replacement, displaced population and loss of heritage significance. Recording the built fabric and ways of life within heritage settlements are key for conserving their historic value, which is increasingly possible with the evolution of digital technologies. This paper presents an ongoing research project developing a novel methodology for heritage conservation and post-disaster *re-construction* using state-of-the-art 3D-laser-scanning (LiDAR) technologies to enable comprehensive damage assessment and design of solutions for repair, retrofitting, reuse and disaster risk mitigation, facilitating community empowerment, while virtually preserving the living heritage of vernacular settlements in Kutch, Gujarat. Through the development of a scalable method of *re-construction*, the aim is to break the unsustainable cycle of buildings' replacement, dereliction and subsequent loss of heritage, advancing from previous research in Chile. Key aspects of vernacular architecture in Kutch are presented alongside the critical evaluation of previous experiences of post-earthquake reconstruction in the region, introducing the challenges from the LiDAR data capture done so far. The early findings show the potential of the record as fast and accurate documentation of complex living settings, incorporating the technical information required of each house within the contextual information of its surroundings – even when working remotely – where complementary social, archival and historical information can be placed and analysed.
Keywords: earthquakes, reconstruction, heritage conservation, 3D-laser-scanning, LiDAR, Kutch, India, risk mitigation, vernacular architecture, re-construction.

1 INTRODUCTION

Vernacular heritage is at risk of damage due to various natural and human-induced hazards everywhere in the world. However, heritage located in seismic areas faces even more challenges due to the recurrent threat earthquakes pose. Despite the periodic recurrence of earthquakes, most actions are predominantly focused on post-event response and there is little emphasis on developing proactive risk reduction measures.

The recovery of vernacular heritage is thus not often a priority after earthquakes and is usually left unattended for a long time until specific solutions are developed. This may take several years, as in the cases of the earthquakes in L'Aquila in 2009, Amandola in 2016 – both in Italy; or in Zúñiga, Chile, in 2010 [1]. In India, after the Marathwada 1993 and Bhuj 2001 earthquakes, the historic villages of Killari and Adhoi, amongst many others, were abandoned, and sometimes even destroyed, and their population was relocated to new settlements that were culturally incompatible [2]. The social problems produced reveal the importance of the vernacular built environment in sustaining specific ways of living and the cultural heritage of vulnerable inhabitants.

WIT Transactions on The Built Environment, Vol 203, © 2021 WIT Press
www.witpress.com, ISSN 1743-3509 (on-line)
doi:10.2495/STR210161

Regarding housing in affected areas, damaged dwellings cannot be immediately repaired and retrofitted to continue their inhabitation. Repairs are usually costly and large numbers of affected structures reduce the chances of adequate damage assessments. These conditions make the conservation of vernacular heritage difficult in the aftermath of disasters, as efforts and resources tend to focus on urgent humanitarian action. Current governmental reconstruction, with non-specific approaches, results in small new houses, sometimes relocated in a new settlement.

In response to those challenges, the project '3D for Heritage India' (https://3d4heritageindia.com) seeks to advance towards a sustainable *re-construction* [1] method for seismic-prone heritage settlements of Kutch, Gujarat, in the north-west of India, based on advanced recording technologies – mainly terrestrial 3D-laser-scanning (also known as LiDAR). A LiDAR record captures the physical environment in one of the most accurate, fast and comprehensive ways currently possible, resulting in a digital, measurable and coloured 3D point cloud with a precision of millimetres. Once the digital model of reality is obtained, it will offer a basis for damage assessments before and after a seismic event, improving the chances of a rapid and effective response, as well as offering a technical platform for risk mitigation measures. The visual outcomes of the record will also enable working with the community, increasing resilience.

The proposed methodology will take the form of a Guidance Document, applicable in similar cases, which can be adapted to different seismic areas in India, and possibly beyond. Aiming to embed it in current institutions of India, using the technology as a participatory method of engagement, a key aspect is integrating different stakeholders within the process of the research project itself: academia, local authorities, NGOs – such as the Hunnarshala Foundation – and inhabitants. Knowledge and skills transference on 3D-laser-scanning documentation for heritage structures locally is also considered, in the form of training workshops with students from the Center for Heritage Conservation (CHC) CEPT Research and Development Foundation (CRDF), as part of a complementary training project (see details in Section 6).

2 *RE-CONSTRUCTION*: A VISION FOR POST-EARTHQUAKE RECOVERY IN HERITAGE VILLAGES OF CHILE

This current study is informed by previous research on post-earthquake reconstruction in heritage villages of Chile, which are built using mainly vernacular building techniques such as adobe and quincha. Devilat [1] indicates that despite earthquakes occurring regularly, specific approaches for housing in heritage settlements are created afterwards and are insufficient to avoid large-scale damage and disruption. She proposed *re-construction* as an alternative vision to current reconstruction approaches mainly based on uncritical replication. By using 3D-laser-scanning for documenting the heritage settlements of San Lorenzo de Tarapacá (Fig. 1), Zúñiga and Lolol, affected by the 2005 and 2010 earthquakes, she introduced an in-depth understanding of contextual information of the whole village, recorded each in just three days. This was an experiment to see how much data could be captured in a short period, resembling a post-earthquake emergency situation, with extra days for collecting social information. Based on the resulting comprehensive and measurable 3D scanning data of what is available in each site, she proposed design strategies to increase reparation and re-use of damaged vernacular dwellings, discussed the implications of heritage reconstruction with analysis throughout different periods of time and use, and argued for the introduction of the technology institutionally. This informs a more inclusive and sustainable method for risk mitigation, *re-construction* and emergency actions, relevant to the case of India with similar challenges to avoid people's relocation.

Figure 1: San Lorenzo de Tarapacá-Chile damaged after the 2005 earthquake (left) and after governmental reconstruction from the terrestrial 3D-laser-scanning obtained on-site in 2013 (right). *(Source: Bernadette Devilat.)*

Vernacular architecture in Chile and India is a sustainable response to their climatic conditions and cost-effective in its use of resources, with materials available locally and building techniques that have been improved from the direct observation of previous seismic behaviour. Current building techniques – with their associated social perception of status, ready-made housing provided by the State and external contractors in previous reconstruction processes, have contributed to the loss of this vernacular knowledge. This results in inhabitants not engaging in the maintenance and repair of their houses, a key aspect to secure proper seismic resistance.

Not all housing in heritage settlements can be adequately retrofitted or re-built using governmental subsidies, posing a potential danger for human life in future earthquakes, as occurred in Chile after the 2005 and 2010 earthquakes [1], and in India, after the 1993 earthquake in Maharashtra, and Bhuj 2001 [2]. Reusing and rehabilitating buildings is a sustainable mode of conservation. Constant maintenance can increase their resistance and is significantly cheaper if compared to demolishing and building anew. However, Pereira-Roders and van den Brand [3] indicate that the tools available for rehabilitation lack development in comparison with those available for new constructions; a lack that also involves trying to integrate inhabitants and relevant stakeholders in the process. LiDAR technology will be employed as a tool for rehabilitation and damage assessment, providing a better basis for analysis and design. It will be a digital platform to engage with the affected community and work in the repair of their affected dwellings – when technically possible – to inform new constructions based on digital data and to plan risk mitigation measures, building from its previous application in Chile.

3 VERNACULAR HERITAGE IN KUTCH, GUJARAT

The district of Kutch sits northwest of the State of Gujarat, west of India. It is a sort of peninsular territory extending from the Indian landmass, bordering Pakistan on the north, the Arabian Sea on the west and the Gulf of Kutch on the south. A harsh semi-arid climate predominates, and a great portion of the northern area corresponds to the Rann of Katch, which are low, flat and salty wetlands that get flooded during the monsoon season, and are mostly uninhabited. Over 900 villages are scattered across the rest of the territory, which is divided into 10 different Talukas or Tehsils. Villages corresponding to each Taluka have recognisable traits giving each their character, according to the interactions of different communities that have settled in the area and their own crafts, which in turn shape their layout – as is the case of Nirona [4].

The different housing typologies present across Kutch can therefore be categorised according to the communities that build them [5]. The houses of the Darbar, traditionally a martial community, are gated spaces with clearly segregated areas for men and women. Large families live in a house with a *Dela* (gated open space) with a shared kitchen. There is usually a semi-open *Osri* (a deep veranda) occupied predominantly by men. The walling material of the house could be adobe, cob or stone masonry, depending on the local resources, whilst roofs are usually timber construction with country tiles. The houses are located on high plinths to indicate the social stature of the community. Ahirs were traditionally involved in animal husbandry and their houses reflect their occupation. The houses of this community have smaller built-up areas and large open spaces to accommodate animal sheds, storage of fodder and other livelihood related activities. The houses, constructed in locally available materials, typically had a small room (*Ordo*) and a kitchen. The room was a store for family valuables. This nomadic pastoral community lived in wattle and daub *Bhunga* (circular houses with conical thatch roof) in their native villages whilst building rudimentary tents or no shelters when migrating with their animals. Most of the Jats in this region are involved in camel herding and are not economically prosperous. This community continues to live in the traditional *Pakkhas* – made of reed grass, coir ropes and wood. The construction of these houses is akin to weaving and is usually done by the women of the community and can be easily moved to a different place. Their houses are usually rebuilt after every monsoon.

It must be noted that these categories are described in general terms only, as each of these communities has subcastes. Typically, a single town or a village will contain multiple community settlements, to which various artisan communities present across the region also contribute to building. They are bearers of the traditional knowledge of the region's vernacular architecture. For example, the Meghwals and the Salats are known for their stone masonry construction, Kolis specialise in thatch roofs, whereas Prajapatis were potters who also cast the terracotta roof tiles. Most of their own homes were made of wattle and daub and arranged in informal clusters on the outer edges of the villages.

Figure 2: Moti Khakhar village in Kutch, Gujarat, India. *(Source: Mrudula Mane, 2021.)*

The communities living in villages across Kutch have adapted their built environment over the years to locally available materials, the climate, as well as to their changing lifestyle. The local characteristics of vernacular architecture thus vary across the different Talukas, with each having its own character. An example of one of the recurrent traditions across Kutch, in terms of materials and construction techniques, determines that several houses are built with 50 cm thick stone walls and timber framing; they tend to be flat buildings with small openings, conferring resistance both to earthquakes and the hot semi-arid climate. In Moti Khakar, houses built on plinths configure a traditional streetscape (Fig. 2), resembling the continuous facades and seats that feature in villages across the desertic climate of Chile, like San Lorenzo de Tarapacá.

4 POST-EARTHQUAKE RECONSTRUCTION EXPERIENCES IN KUTCH, GUJARAT, INDIA

Kutch sits on a series of seven geographical faults, which makes this territory prone to earthquakes. The last great earthquake was the Bhuj earthquake, also known as Gujarat earthquake, in 2001 (7.7 Mw). Its magnitude combined with the fact that the epicentre was just 16 km deep produced great destruction and casualties. At least 100,000 houses were destroyed and more than 300,000 were severely damaged [6]. Many of these damaged houses, made using traditional construction techniques and representing the vernacular heritage of the region, are still lying vacant, in damaged condition, due to the post-earthquake outmigration of their population. The Bhuj earthquake triggered a huge response in terms of reconstruction of the destroyed and damaged areas. Different stakeholders were involved in each specific village coordinated by the State authority; hence a variety of reconstruction programmes took place instead of a monolithic response to the situation. Berenstein [7] groups these approaches in five distinctive categories: (1) owner-driven reconstruction with governmental assistance;(2) subsidiary housing approach, which is similar to the first but involves further assistance of an NGO; (3) participatory housing approach with an NGO assuming a leading role; (4) contractor-driven reconstruction of the affected village; and (5) contractor-driven reconstruction in the form of a new settlement or neighbourhood, apart from the affected area.

Most of the reconstruction initiatives followed the people's preference to stay in the same place instead of being relocated to a different site [8]; of which many were driven by NGOs that also worked with the local communities, actively involving them in the decision-making. The positive effects of community involvement in the reconstruction process were already acknowledged in the UNDP [6] and the World Bank [8] reports, informing reconstructions strategies. After the 1993 Marathwada and 2001 Bhuj earthquakes, the problems were the lack of technology transfer to the locals and how people returned to their original damaged settlements since 'city-like' plans for relocated villages did not consider their lifestyles [2]. The subsequent Gujarat Disaster Management Policy sought to address these problems as well [9]. Whilst these documents mention the need to train local masons to build seismically safe houses using new construction techniques, the urgency of focusing on reconstruction and rehabilitation prevented the development of an understanding of the merits of traditional construction knowledge in the region. Heritage considerations were not a part of the agenda either.

An example of a more culturally sensitive approach is the work of the Kutch Navnirman Abhiyan, a network of NGOs working in Kutch, in which the Hunnarshala Foundation takes part. Abhiyan played a key role in the reconstruction after the Bhuj earthquake, working in the villages of Reha, Bhojay and Kotdi Mahadevpuri. Their approach consisted of enabling the local people to construct their own houses and thus transfer their knowledge into the new structures, whilst integrating innovative solutions for water harvesting and sustainable construction. They also intended to demonstrate best practices in owner-driven housing reconstruction that could be emulated by development institutions and government authorities for larger-scale projects. In this context, Hunnarshala's particular approach to the design of new houses encompassing traditional values has been referred to as synthetic vernacular [10].

While this paper argues for a greater degree of rehabilitation of existing structures, new designs are sometimes unavoidable and play a key role in sustaining people's cultural heritage and relating to their traditional surroundings, whilst including communities' current needs and vision. Along with the importance of community engagement in the reconstruction processes in India, a significant challenge is to scale up considering the vernacular village as a whole, each house in relation to others around it and to its context, and to increase coverage to diminish relocation and dereliction. The LiDAR record offers the capability to map the existing built fabric comprehensively and accurately, to enable integral approaches that can tackle these challenges, following Desai's [11] understanding of heritage conservation essentially as an inclusive and participatory practice towards long-term sustainable development of the affected communities.

5 RECONNAISSANCE TRIP: THE CURRENT SITUATION

A reconnaissance trip across the district of Kutch was done to evaluate the state of different villages affected by the Bhuj earthquake, informing the selection of a case study for the LiDAR survey. The local team visited twelve villages in four days, located in the northeast, south and west areas of the district: Bela, Vrajvani, Moti Reha, Nani Reha, Nani Tumbdi, Depa, Moti Khakkar, Bhojay, Kotdi Mahadevpuri, Roha, Tera, and Jadva. The selection of villages visited responds to previous knowledge of the region in the research team (in both academic and professional capacity), and a preliminary trip by Aditya Singh from Hunnarshala, to look across a diversity of places with a significant presence of traditional architecture at risk and greater potential of collaboration with local people.

After 20 years, the memories of the 2001 earthquake are still strong. Villagers recall how traditional houses resisted the quake, although resulting in different levels of damage. The region has a strong culture of social work and community participation, key factors for effective recovery after disasters such as earthquakes. Therefore, the traditional houses that were repaired and regularly maintained since then, are still standing today. Whilst some areas recovered to different degrees, the destruction produced by the earthquake forced individuals and communities to migrate, with the negative associated social changes that result from the abandonment of many neighbourhoods and in some cases even entire villages. People considered that a collaborative initiative of 3D-laser-scanning the traditional constructions in the area can be significant in supporting their post-disaster response and appreciate the opportunity of being involved in this research project.

The villages visited feature diverse aspects of vernacular heritage worth recording for purposes of studying traditional construction systems and conservation, although not counting with legal recognition and protection. The selection criteria are summarised in Table 1. Important considerations were the potential official heritage recognition of the village considering its immediate vicinity to an already recognised heritage asset, and the existence of previous records to better understand the changes and persistence of the built fabric over the years.

Table 1: Comparative summary of villages surveyed in the Reconnaissance Trip, with two selected alternatives. (*Source: Mrudula Mane, 2021.*)

Sr. No.	Village	Estimated number of houses reconstructed	Estimated number of potential houses	Is it located in a cluster?	Previous records of the village	Other heritage assets	Is community still residing the village?	Accessibility
1.	**Bela in Rapar**	**20–30**	**100–110**	**Yes**	**Yes**	**Yes**	**Yes**	**Yes**
2.	Vrajvani in Rapar	250–270	20–30	No	Yet to search	No	Yes	No
3.	Moti Reha in Bhuj	50–60	100–120	Yes	Yes	No	No	No
4.	Nani Reha in Bhuj	100–150	20–30	No	No	No	Yes	Yes
5.	Nani Tumbadi in Mundra	20–25	60–70	Yes	No	No	Partially	No
6.	Depa in Mundra	40–45	10–15	No	No	No	Partially	Yes
7.	Moti Khakkar in Mundra	25–30	70–80	Yes	Yet to search	No	Partially	No
8.	Kotdi Mahadevpuri in Mandvi	100–150	20–30	No	Yes	Yes	Partially	Yes
9.	**Roha in Nakhatrana**	**10–15**	**100–110**	**Yes**	**Yes**	**Yes**	**Yes**	**Yes**
10.	Tera in Abdasa	300–350	50–70	Yes	Yes	Yes	Yes	Yes
11.	Jadva in Lakhpat	120–130	10–12	No	No	No	Yes	Yes

The project will focus on Bela (Figs 3–5), situated in the northeast of Kutch between its major fault lines. Bela is characterised by a coherent built fabric of continuous facades and porticoes. Most houses were damaged by the 2001 earthquake but did not completely collapse. Bela's traditional constructions conforming the area to scan are arranged around a Darbargadh or small fortification, placed in the centre of the village. There are approximately 30 abandoned residential units within the fortification, surrounded by several inhabited houses.

Figure 3: Aerial Photograph of Bela in Kutch, Gujarat, India, May 2020. *(Source: Google Earth.)*

Figure 4: Bela village in Rapar, Kutch, Gujarat, India. *(Source: Mrudula Mane, 2021.)*

Bela's location responds well to practical aspects of the LiDAR data capture as a training workshop for students and staff at CEPT University, as well as other stakeholders in the context of the related training project (see next section). Its officially recognised heritage asset, the Darbargadh, sits within the village and not outside of it – as is the case of Roha, facilitating the overall data capture. Bela has a cohesive and numerous resident community to follow on to the next stage of community engagement using the captured data, which means that more people may benefit directly from the LiDAR record in terms of assessment compared to other potential case studies.

Figure 5: Bela village in Rapar, Kutch, Gujarat, India. *(Source: Mrudula Mane, 2021.)*

6 LIDAR SURVEY OF HOUSES IN AHMEDABAD:
SHAPING OUR WAY FORWARD

Much of the effort in digital technologies for heritage conservation over the past two decades had primarily been directed towards accurate documentation, recording and representation of historic sites and buildings with not much emphasis on the socio-spatial patterns of human aspects of city life [12], [13]. The proliferation of the use of 3D modelling techniques, nonintrusive imaging, geophysics and augmented reality cameras has offered a multiplicity of platforms to simply store, archive and communicate vast amounts of information on cultural heritage sites, traditions and contents [14]. Whilst current advanced recording technologies offer unprecedented capabilities for documenting and visualising heritage, the innovation of this project is not in the technology itself but in its application for developing a scalable *re-construction* method. For this, the survey of Bela using LiDAR technologies and questionnaires to local people, and the further analysis relating archival and historical information, constitutes a pilot case study for similar settlements.

For the technical aspects of LiDAR documentation, it is relevant to embed the skill in the new generation of conservation professionals, which has been done through a parallel training project titled: 'Surveying heritage buildings in Ahmedabad, India: empowering local action and skills for heritage conservation' (https://ntu3dscanlibrary.com/). This project aims to generate local capacities for recording, surveying and protecting vernacular heritage buildings, as a knowledge transfer to improve their maintenance and tackle deterioration, by giving access to a Faro m70 3D scanner and providing training to tutors who will then transfer the skills to students. One of the activities of this project consisted of a LiDAR workshop of two residential buildings in Ahmedabad. This workshop included an introductory session to explore the uses of this technology in the conservation of heritage buildings and planning the on-site scanning using previous records; on-site capturing sessions over two days; and a remote final session on post-processing, combining the scans, visualising and discussing the results. Each day on-site, one building was scanned by the leading tutor Mrudula Mane and seven students, following a similar training scheme tested before in London, UK [15].

The Ahmedabad Municipal Corporation has identified and listed 2,236 residential structures situated within the World Heritage Site of the walled city as heritage assets. The two houses documented (Figs 6 and 7) are listed heritage buildings of Grade III status. The scanned buildings are situated in the densely populated traditional neighbourhood known as Zaveri Wad in Kalupur ward, custodied by the religious organisation Jinagya Awas Trust.

Figure 6:   Ground floor plan and elevation of house 361, Ahmedabad. *(Source: Bernadette Devilat, 2021, using 13 3D scans captured by Mrudula Mane and workshop participants, see acknowledgements.)*

The narrow houses (2.5 m to 4 m) have steep wooden staircases, presenting an intricate and challenging setting for a LiDAR survey. Nonetheless, the scans were successfully combined into a three-dimensional model with less than 5 mm of maximum error in only a couple of hours. The data captured shows the buildings' context in rich detail, which corroborates the potential of this tool to understand vernacular heritage within a wider scope.

Figure 7: Aerial view and section of house 429, Ahmedabad. *(Source: Bernadette Devilat, 2021, using 11 3D scans by Mrudula Mane and workshop participants, see acknowledgements.)*

It will offer better visualisation – images, videos and architectural drawings – to provide technical assessments and be used as unprecedented and accurate information of the context to develop risk mitigation and *re-construction* strategies considering the re-use, rehabilitation and repair of existing constructions.

7 CONCLUSIONS AND PROJECTIONS

Documenting buildings, especially those with historic value, is more than just capturing the physical condition of their structure. It is about recording the history of a place and the life around it. The LiDAR documentation process, as demonstrated through the Ahmedabad example, records the complexity of the place as found in a short period, which can be of extreme relevance for documentation after disaster and conflicts by offering a platform for action research. This record also has the potential of bringing out the nuanced differences in the way members of a community live and how they practice their culture. If studied carefully, it can potentially suggest which human associations with the built environment are enduring and which are temporal, besides the technical information about building systems, earthquake damage, distortions and use.

The associated training scheme not only served as a knowledge transfer experience but facilitated working remotely, based on what was captured on-site by the local team, which is relevant in the context of restrictions posed by the COVID-19 pandemic and opens new forms of site-specific working in challenging contexts.

The *re-construction* method under development in this project goes beyond the documentation. The record allows for the in-depth study of material culture and ways of living of the domestic typologies presented, but only if used as a combination of sources such as social, archival and historical information. Following that, it is expected to use the LiDAR record as a mapping basis to analyse, measure and visualise the vernacular housing elements identified in this paper with occupancy and insights provided by its inhabitants, construction systems and traditional techniques that withstood the earthquake, among other aspects. In this way, it will contribute to enhancing the number and quality of heritage buildings conserved, reducing risks to these buildings and human lives. As a potential tool for improving public policies, it could also enable a more culturally sensitive approach to the conservation of vernacular heritage, particularly to its *re-construction* after earthquakes as an alternative to the complete replacement of buildings.

ACKNOWLEDGEMENTS

This project is funded by the UKRI Arts and Humanities Research Council (AHRC) and the Department for Digital, Culture, Media and Sport (DCMS). Project Reference: AH/V00638X/1. With thanks to Faro Technologies for their support and Jinagya Awas Trust in Ahmedabad for facilitating access to the buildings captured during the LiDAR training workshop. Thanks to the participating students: Anushka Mital, Kanchi Chaudhari, Neha Chandel, Satyajeet Chavan, Anagha L., Bhanumati V., Sneha Anand, and the Teaching Assistants Juhi Bafna and Zeus Pithawala. With thanks also to Aditya Singh from Hunnarshala Foundation, our Project Partner, for his constant support and proactive participation. Thanks to Sandeep Virmani and Mahavir Acharya, also from Hunnarshala. During the reconnaissance trip across Kutch and the data capture in Ahmedabad, members of this project carefully observed COVID-19-related preventive measures to protect staff students and surrounding communities and will continue to do so in the following fieldwork and engagement activities. The training project "Surveying heritage buildings in Ahmedabad, India: empowering local action and skills for heritage conservation", which is complementary to 3D for Heritage India, is funded by Nottingham Trent University.

REFERENCES

[1]   Devilat, B., Re-construction and record: Exploring alternatives for heritage areas after earthquakes in Chile. PhD thesis, The Bartlett School of Architecture, University College London, 2018.

[2]   Jigyasu, R., From Marathwada to Gujarat – Emerging challenges in post-earthquake rehabilitation for sustainable eco-development in South Asia. *First International Conference on Post-disaster Reconstruction: Improving Post-Disaster Reconstruction in Developing Countries*, Montreal, pp. 1–22, 2002.

[3]   Pereira-Roders, A.R. & van den Brand, G.J.W., Sustaining rehabilitation: A call to strengthen the building rehabilitation knowledge base. *Proceedings of the CIB W70 Trondheim International Symposium: "Changing User Demands on Buildings – Needs for lifecycle planning and management"*, Trondheim, pp. 128–138, 2006.

[4]   Desai, J., Conservation of a craft habitat; mapping spatial and temporal networks of the crafts practices of Nirona, Kutch. Doctoral thesis, CEPT University, 2014.

[5]   Hunnarshala Foundation, People in centre, Buildaur, & thumb impressions. *Socio Technical Facilitation of IAY in Gujarat: Housing Typology Study*, Commissionerate of Rural Development, pp. 32–40, 2014.

[6]   From Relief to Recovery: The Gujarat Experience, United Nations Development Programme (UNDP), p. 9, 2001. https://www.in.undp.org/content/india/en/home/library/environment_energy/from-relief-to-Recovery.html. Accessed on: 7 Apr. 2018.

[7]   Berenstein, J.D., Housing reconstruction in post-earthquake Gujarat. A comparative analysis, The Humanitarian Practice Network at the Overseas Development Institute, London, 2006.

[8]   Gujarat Earthquake Recovery Program Assessment Report, World Bank and the Asian Development Bank, p. 3, 2001. https://reliefweb.int/sites/reliefweb.int/files/resources/788AADD8C64A0D16C1256A1C00461C35-worldbank-indannexes-14mar.pdf. Accessed on: 7 Apr. 2018.

[9]   Gujarat State Disaster Management Authority, Gujarat earthquake reconstruction and rehabilitation policy. 2001.

[10]   Gillick, A., Synthetic vernacular – The coproduction of architecture. PhD thesis, University of Manchester, 2013.

[11]   Desai, J., *Equity in Heritage Conservation, the Case of Ahmedabad*, Routledge: Oxon and New York, 2019.

[12]   Yang, C., Peng, D. & Sun, S., Creating a virtual activity for the intangible culture heritage. *16th International Conference on Artificial Reality and Telexistence-Workshops, ICAT'06*, pp. 636–641, 2006.

[13]   Goodrick, G. & Gillings, M., *Constructs, Simulations and Hyperreal Worlds: The Role of Virtual Reality (VR) in Archaeological Research*, eds G. Lock & K. Brown, On the Theory and Practice of Archaeological Computing: Oxford, pp. 41–58, 2000.

[14]   Abdelmonem, M.G., *Virtual Heritage: Global Perspectives for Creative Modes of Heritage Visualisation*, Nottingham Trent University: Nottingham, 2017.

[15]   Devilat, B., '3D laser scanning built heritage: St. Boniface's Church as a teaching experience. *Proceedings of Digital Cultural Heritage: FUTURE VISIONS London Symposium*, Brisbane, pp. 18–38, 2019.

SECTION 5
SOCIAL, CULTURAL AND ECONOMIC ASPECTS

ROLE HISTORIC ASSETS CAN PLAY IN REVIVING THE RETAIL HIGH STREET: A CASE STUDY OF DERBY'S RETAIL HIGH STREET

SARAH BALL, DAVID HIGGINS & HAZEL ANN NASH
BTS, Birmingham City University, UK

ABSTRACT

The long-term decline in the historic high street has been an important issue for local communities, governments and real estate investors. This has led to significant discussions concerning the triggers of retail decline and consideration for how heritage themed high streets can evolve in the future and the associated resources for this to be achieved. Utilising a case study of Derby's historical Cathedral Quarter, this paper explores (i) the issues involved in reversing the decline of retail in the traditional high street; (ii) the strategies used to sustain and improve the high street as a destination; and (iii) the role of heritage assets in improving business occupancy of high street premises. In order to provide an insight into the processes available to regenerate the high street and attract space occupiers, a series of semi-structured interviews were undertaken with leading real estate consultants and investment professionals. The research findings suggest there are multiple reasons for the decline, including economic, environmental, and functional factors, and that these will continue to impact the evolution of the high street moving forward, unless proactive strategies are put in place. Increased mixed-use development within town centres, including residential and co-working space, is seen by interviewees as significant. Furthermore, the case study provides clear evidence that utilising heritage assets with unique characteristics can positively impact the retail high street. This can include reinstating historical frontages of retail units and so strengthening visiting numbers that creates destination footfall, resulting not only in a decrease in vacancy rates but often improved rental values. A catalyst is to provide an informed strategy to those wishing to undertake projects similar to the funded regeneration works on the historical assets within Derby's Cathedral Quarter Scheme.
Keywords: high street, historic assets, retail operations, repurposing.

1 INTRODUCTION

Over the last few decades, the traditional functions of UK town centres and associated high streets have changed, with the focus shifting towards the expansion of out of town retail parks hosting a variety of outlets, including food, clothing and homeware goods [1]. With the rise of the online retail market and the ease of access to goods from it, people have been drawn away from town centres. Consequently, the character of the high street has continued to change, with retail spending declining in this sector, and this trend set to continue.

As high street retail faces these challenges, they need to consider how to approach changing consumer habits, rising business rates and the online retail market. Whilst consumers enjoy visiting local high streets, they are looking for an improved experience with many people preferring to combine shopping with other leisure activities, such as eating at fine dining restaurants [2]. The rising costs of conducting a business are also affecting the success of high street retail companies. Business rates and taxes can have a significant impact on small independent retailers, but even the larger retailers have been calling for the Government to review business rates in order to safeguard high street retail [3].

The most well-known and widely reported impact on high street retail has been the increase in the availability of online purchasing. Although a number of high street companies have adapted by having an online presence, it is the specialist 'online-only' retailers, such as Amazon and Asos, who are having the greatest impact on physical high street sales [4].

Facing these challenges, the redevelopment of historic high street locations continues to prove challenging in terms of providing a balance between commercial drivers and historical constraints. This research is based on a February 2020 case study of Derby's Cathedral Quarter, a prime example of a historic high street which has become an economic success in a time of widespread retail 'collapse.' The paper will examine how this quarter bucked the general trend of high streets and accomplished success in occupancy of high street premises.

This study adopted a qualitative approach, using carefully conducted, semi-structured interviews with six key professionals from the heritage, funding and retail industries. The interviewees included a funding consultant, a specialist from a funding provider, planning consultants, development consultants and a conservation officer from a local authority. The interviews maintained a clear focus, but with sufficient flexibility to allow for further discussion points. This enabled the participants to draw upon their own expertise and knowledge, in order to explore their opinions about the high street retail market and the redevelopment of the historical high street.

The following section provides a literature review covering the retail sector, key determinants and the high street regeneration with the focus on Derby's Cathedral Quarter. Section 3 will then set out the details concerning the selected methodology and associated details of interviewees. Section 4 provides the empirical findings and implications while the last section gives concluding comments.

2 LITERATURE REVIEW

In this section, the review of literature covers three key areas: the UK's current retail climate and key determinants, High Street regeneration and the Derby Cathedral Quarter.

2.1 The UK's current retail climate and key determinants

The UK retail market has been an increasing source of concern in recent years and, as such, is under constant scrutiny by today's media. Although it is generally recognised that the retail sector has been in a state of decline since the 1960s, it was not until 1995 onwards that solid statistics became available to support this. At this time the British Retail Consortium (BRC) was launched and began conducting research into the retail sector, analysing fast moving markets and long-term structural trends [5]. A further milestone, was the Hughes and Jackson [6] 'Death of the high street: identification, prevention reinvention' [6], research paper which identified the individual factors behind the changes in the retail sector with a conceptual model defining retail obsolescence with four key drivers (see Fig. 1).

Fig. 1 illustrates the multiple factors influencing the decline of the high street and the reasons for the general decline in the physical retail sector, and a real likelihood that the high street may well become obsolete as a retail location in the future. The key retail drivers for the demise of the high street are detailed as (i) Economic, (ii) Environmental, (iii) Functional, and (iv) Locational [6].

2.1.1 (i) Economic

Economic obsolescence arises as a result of changes in population, business innovation and, most importantly, market supply and demand. The biggest impact on supply and demand has been the rise of online shopping which, it is predicted, will account for approximately 20% of consumer expenditure by 2020 [7]. These figures are supported by the Office for National Statistics report for October 2019, which noted that, as a proportion of all retailing, online sales have increased to 19.2% [8], as online retail continues to create an easier retail experience for consumers.

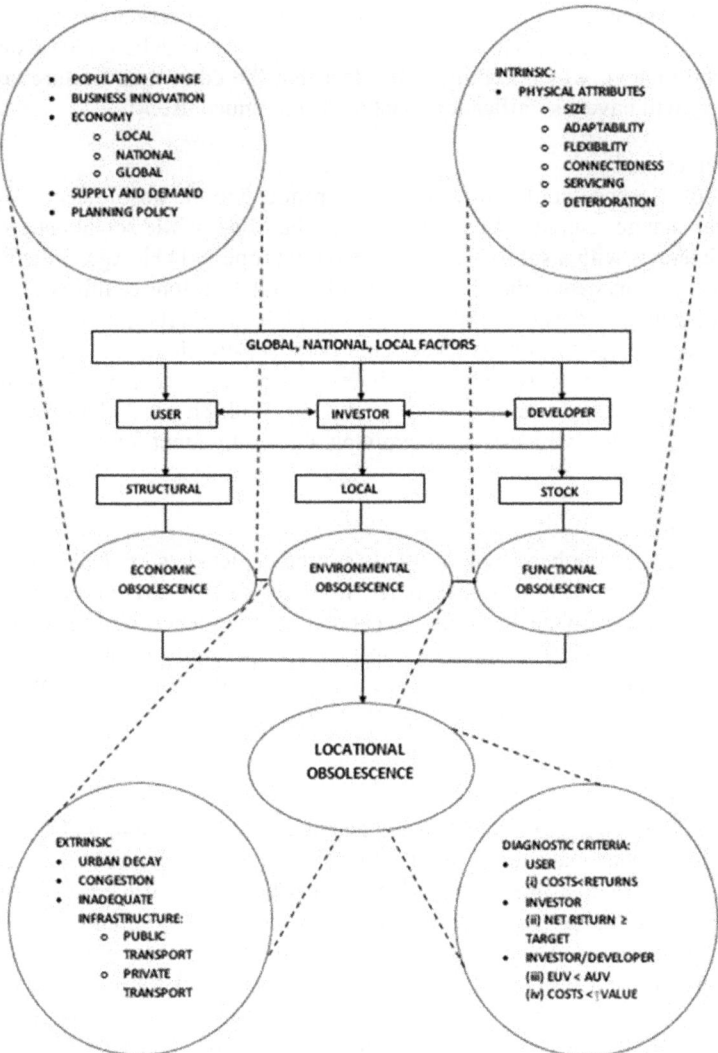

Figure 1: Conceptual model of retail obsolescence.

Part of the economic reasoning behind the changes in the retail market is that competition has arisen from other types of retail space, such as retail parks and supermarkets which offer more floor space than traditional high street shops, and are also able to adapt to changing circumstances, such as the ability to convert to warehousing and distribution uses. The creation of retail buildings with larger floor space, and consequently a wider range of goods, is creating a highly competitive atmosphere for retail outlets in towns and smaller cities [9].

2.1.2 (ii) Environmental

Although not intrinsic to the retail sector, the factors that contribute to environmental obsolescence (types of depreciation) are important if there is an increase in alternative retail destinations. Environmental aspects that impact the footfall of a retail location include urban decay and congestion, as a shopper is more likely to seek out a destination that is aesthetically pleasing and easy to get to [6]. This is supported by the RICS [10] examination of environmental factors, which highlights the fact that the current and future area around a retail property will have a significant impact on its continued use.

2.1.3 (iii) Functional

Functional obsolescence could be seen to arise from economic obsolescence. For example, technological changes can affect the qualities of a building, while social patterns can affect how a user interacts with a subject, in this case retail property [11]. Again, the RICS defines this as the process in which the specification of an asset no longer fulfils the function for which it was originally designed [10]. In terms of high street retail, many stores are able to function for retail purposes, but as those retail purposes are changing, the older high street stores are struggling to meet service and delivery area requirements and this will continue to change with the increased use of internet retail [12]. With physical challenges, operational changes and the speed at which retail is evolving, the requirement for high street retail space is declining.

2.1.4 (iv) Locational

The process of a location becoming obsolete is often depicted as gradual as the above three factors evolve and interact with each other, however certain circumstances may cause a location to become obsolete instantly. Locational obsolescence has been identified as a culmination of the previous three factors. In other words, if all of the above factors apply, then a specific retail property can be identified as 'locationally' obsolete within the retail sector, whether it is on a high street or elsewhere [6].

2.2 High street regeneration

To enable the high street to evolve, it is necessary to look for alternative ways to boost footfall, such as multi stakeholder involvement and funding. This involves changing the focus of the high street by introducing alternative uses such as office space and residential units. Mixed-use units are on the increase, combining the requirement for retail space alongside the suggested alternative uses. Conversion of retail space into other types of uses, particularly office space and residential units, is growing in popularity due to the increasing amount of vacant retail space on the high street. In previous years this focussed mainly on secondary retail space, however the possibility of changing the use of retail sites in areas where there is a concern that retail is no longer a viable use is increasingly under consideration [6].

Portas [13], proved to be a key piece of literature in terms of exploring the use of the high street. She identified the need for town centres to evolve rapidly in response to customer demands in order to remain relevant, however this has not been the case in most situations. Reinvention has tended to be a reactive response once the decline has already happened, rather than a pre-emptive action to head off a fall in the performance of the high street retail sector [14]. One possible solution would be to involve all stakeholders as early as possible, from retail management to the council, and even the public shoppers who ultimately will be supporting and using the environment. Initiatives such as NE1, a Business Improvement District Company focussed on strengthening Newcastle's economy, have had some success from implementing schemes such as free council parking after 5 pm in order to attract people. This is also an example of a way of boosting the high street retail economy without having to physically introduce new development or alter the use of retail spaces [15].

2.3 Derby's cathedral quarter: Historical and current

Derby was originally a town with a small priory, with records dating back to the mid-12th Century. A hospital was added soon after, and it continued to expand in size and trade. Derby continued to grow throughout the 17th Century, with its economy based around brewing and cloth-making. Derby Cathedral has been highlighted as a reference point in Fig. 2. In 1839 the railway network reached Derby, aiding the continued growth of commerce in the area, and contributing to the Market Hall opening in 1866, the site of which is still used for this purpose today. The town's economy received a further boost in 1907 when Rolls Royce moved to Derby [16].

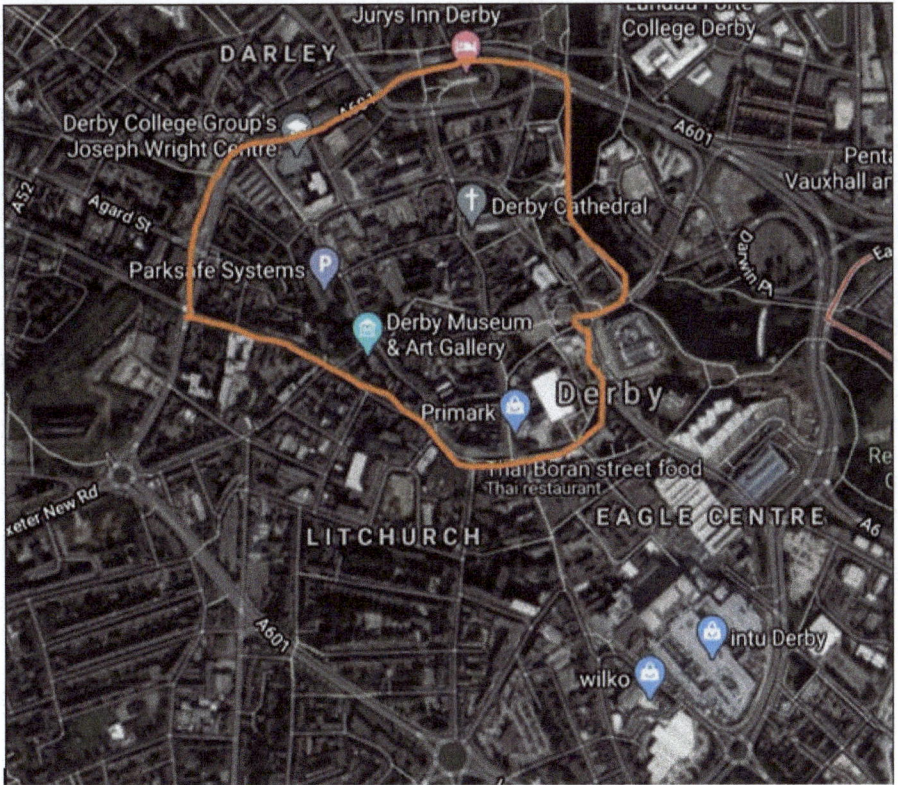

Figure 2: Map of Derby City Centre.

Derby's City Centre was once abundant with thriving high street shops. The Derby County Borough Directory dating back to 1935 holds a selection of advertisements for retail establishments that were flourishing in the area [17], however, since the opening of Westfield Derby in 2007 (the city's indoor shopping centre), many of the city centre's retail outlets have suffered, and consequently 2013 saw an increased number of units that were either boarded up or 'To Let.'

Fig. 2 shows the location of key Derby shopping precincts. The Derby Cathedral Quarter retail presence had initially been hit hard when the large Westfield Centre opened on the other side of the city centre. Colliers 2014 retail study noted that since the 2008 recession the cost of renting shop space had fallen by 22.2% (£40 per square foot) over the five-year period into 2013 due to lack of confidence in the high street. However, at the time the East Midlands was faring better overall than other areas.

The Monitoring and Legacy Report summarising the project in DCQ stated that in the declining economic climate landlords' only response to the reduced demand for retail property was to reduce rents and introduce more flexible lease terms, neither of which is attractive in terms of a building's investment value. With the increase in poor quality and non-heritage frontage signs in a Conservation Area proving difficult for the council to control, it became clear that intervention would be required in order to restore the historic identity of the area. DCC attempted to combat this issue by producing guides for retailers applying for a change in shop front design, however this was not enough to have a significant impact on the damage that had been done.

A Partnership Scheme in Conservation Areas (PSiCA) between Historic England (HE) and Derby City Council (DCC) was set up between 2008 and 2016 following DCC's approach to HE regarding the worsening state of disrepair in Derby's conservation area. Following this, in 2009 HE allocated a large proportion of the area as being 'at risk' of being lost due to decay, neglect or inappropriate development [18]. This wasn't however the first of the funding programmes to be put in place in Derby; in 2001 the Heritage Lottery Fund (HLF) and the Townscape Heritage Initiative (THI) had begun to provide aid through restorations of 16 properties in the area and improvements to the public realm, all of which were completed in 2018. Due to the success of this scheme, HLF has continued to aid the surrounding area.

DCC's approach to Historic England was partly led by HE's proven track record with grant-led regenerations in smaller towns, which demonstrated the potential for this to be applied to larger areas. The partnership was formed, and owners and leaseholders were approached regarding repairs and reinstatement of their shop frontages. At the same time the commercial benefits of a historical shop frontage reinstatement, such as increased footfall and sales were promoted, which led to an oversubscription of owners seeking assistance.

Once the project was completed and analysis had been undertaken, it became clear that the project had been a success. Quantitative footfall data was used to demonstrate the economic and social benefits of historic building regeneration. From 2008 to 2013 national footfall had decreased by 26% yet research conducted by Marketing Derby demonstrated that footfall in DCQ had increased by 12% and specifically, Sadler Gate's footfall had increased by 15%, going against the national trend. This was supported by the fall in vacancy rates between 2010 and 2017, which in 2009 had been at 25%, the highest of any major urban centre in the UK [19]. The above statistics are represented in Fig. 3 and a more detailed map of the Cathedral Quarter can be seen in Fig. 4.

Fig. 3 shows that funding is one way of aiding high street regeneration and has been used for this purpose for over thirty years. Although there are a variety of funds available to premises' owners, in most cases a portion of private funding is also required. However, it is important to note that without the opportunity to draw on funding many listed buildings and heritage assets would be neglected and fall into disrepair, resulting in a detrimental impact on the environmental, functional and locational characteristics of the high street. Ultimately the result would be a spiralling reduction in high street use.

The Derby's Cathedral Quarter case study offers a clear representation of the impact that funding aimed at repair and restoration of heritage assets can have in the successful regeneration not only of a single high street, but of an entire retail area. Furthermore, in physically improving the aesthetic appeal of historic buildings in the area, footfall and retail spending in a market that is otherwise generally declining has increased.

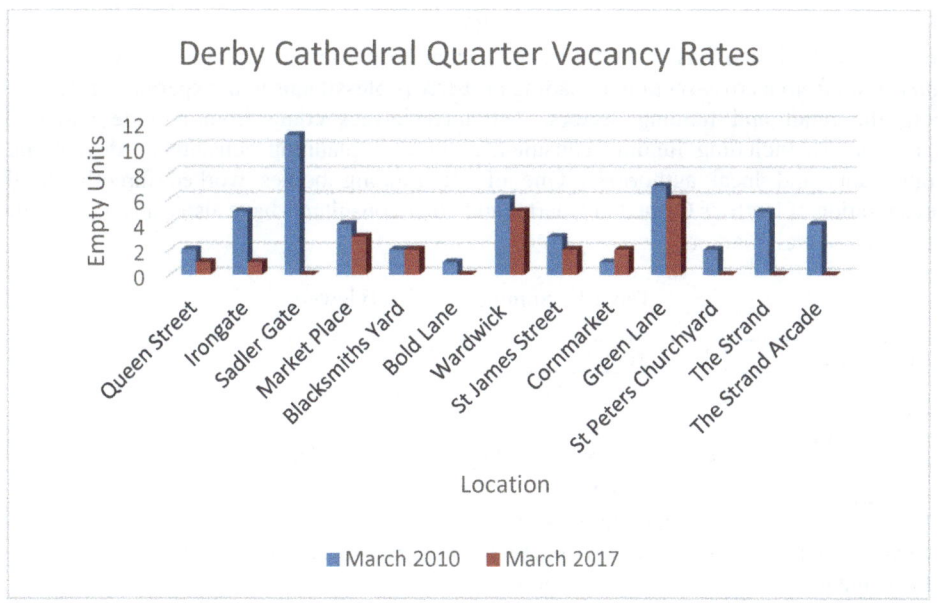

Figure 3: Derby cathedral quarter vacancy rates.

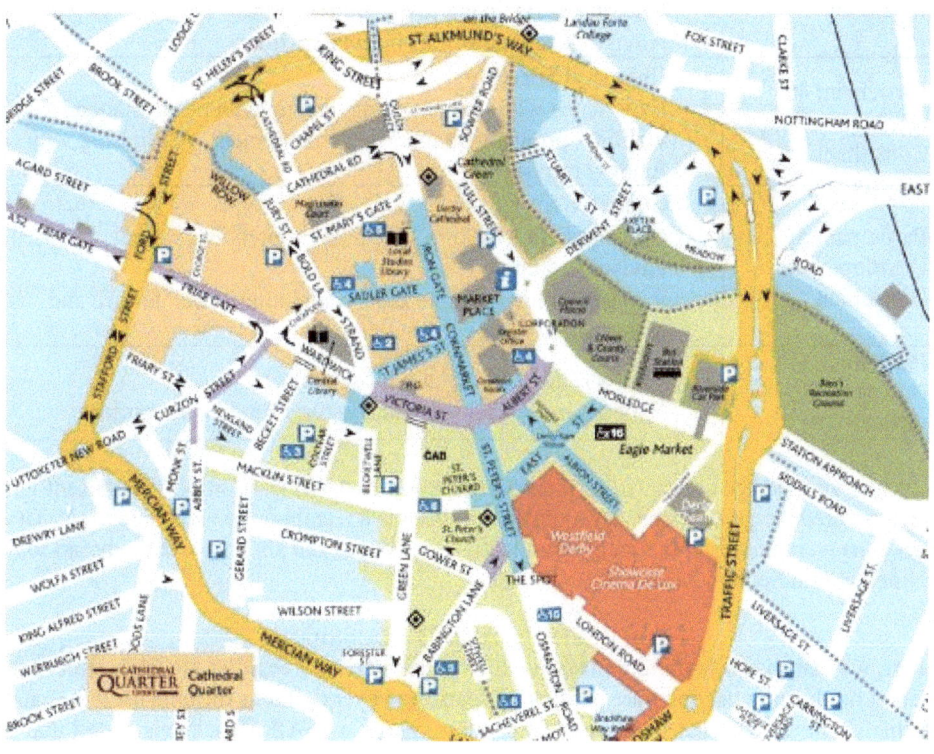

Figure 4: Derby cathedral quarter area map.

3 METHODOLOGY

The research design has been formulated from responses given in six semi-structured interviews from a cross-section of leading property professionals with experience relevant to both the retail and funding sectors. The interviewees come from a wide variety of backgrounds, including funding consultants, funders, planning consultants, development consultants and local authorities. One of the funding bodies worked directly on the regeneration of Derby's Cathedral Quarter and other consultants have also carried out studies on Derby's retail market.

Table 1: Summary of interviewees.

Interviewees	Background
Participant 1: Conservation Specialist	• Principal Advisor and Lead Specialist for a funding provider • Focus on conservation • Focus on struggling towns and cities looking at development and buildings
Participant 2: Regeneration Consultant	• Chartered Surveyor • Advises on development and regeneration projects across a range of sectors including retail, mixed use and heritage
Participant 3: Planning Consultant	• Chartered Town Planner • Experience in planning permission for retail and commercial proposals • Undertakes retail and town centre studies • Advises local authorities on retail matters
Participant 4: Heritage Consultant	• Chartered Surveyor • Experience with grant funding, development and financial viability, project management, procurement, and liaising with funding providers
Participant 5: Development Consultant	• Chartered Surveyor • Company Director of a specialist property development and advisory company • Projects within multiple sectors including retail, leisure and mixed use
Participant 6: Conservation Officer	• Conservation Officer at a local authority • Experience of working with Historic England on projects

Each interviewee consented to their participation by signing an informed consent letter and answered the guided interview questions either via a telephone or video interview. These questions were based on the learning objectives set out in the abstract and the broad topics defined within Section 2.

4 RESULTS AND DISCUSSION

There was an overwhelming consensus among the interviewees that the high street retail market has been in decline for a number of years, however opinions differed as to the causes of this decline, and what the current impacting factors are. All participants stated that the retail market has been impacted more since the financial crisis in 2008, but has been in a state of flux for the last 20 years. In the Regeneration Consultant's opinion, out of town retail has seen a 'cyclical movement'; it was prospering prior to the financial crisis and 'then the financial crisis decimated it, but then it remodelled itself and has succeeded.' This was supported by the Development Consultant, who stated that the introduction of 'discount retail parks, places where you get better value' from 2009, 'kept evolving' and 'the high street stagnated as it were.'

The Regeneration Consultant also stated that 'there is a control shift taking place' in terms of consumers' habits, and this accounts for the current decline. This view backs up the statistics suggesting that department stores were the only high street retailers to see a profit in late 2019 [20]. The Planning Consultant however, looked at the function of department stores in more detail, claiming that stores such as Debenhams and BHS were the 'anchor tenants' and that the increasing trend in the closure of these stores and their move away from the high street, particularly in the last five years, has played a major part in the decline in wider high street retail.

In considering the causes of the decline in the retail market, all the contributing factors mentioned by the interviewees can be categorised into the broad areas theorised by Hughes and Jackson [6].

4.1 (i) Economic

All the participants agreed that the introduction of online retail and increased exposure to the internet have played a role in the decline of high street retail with the Development Consultant stating that 'we discovered internet shopping and shopping habits changed,' and because of this, the Conservation Officer noticed that 'there have been more shops shutting over recent years.' However, both participants also believed that these issues are being overplayed. The Planning Consultant qualified this by stating that 'too much blame is put on the internet' as a sole cause of the decline, and there are other defining factors which should also be considered.

Both the Conservation Specialist and the Heritage Consultant were of the opinion that the closure of other uses within town centres is also contributing to the decline. For example, The Conservation Specialist explained that 'when the banks started shutting down, it meant that people didn't have a destination use to go to which they'd often combine with other shopping visits.' This view was supported by the Heritage Consultant, who raised the point that 'the closure of local civic buildings such as town halls, libraries and leisure centres meant that people had less of a reason to go to the town centres, making shopping trips a less frequent occurrence.' These arguments suggest that usage of the high street is changing, and that requirements for the kind of physical retail presence it offers are also altering, and this has had a significant impact on supply and demand [6].

Another economic factor, which was mentioned by all the participants, was planning policy, but in this regard opinion differed. The Planning Consultant believed that planning policy has not necessarily been a cause of the decline of the high street for retail purposes as a retail centre but stated that it would need to change in order to influence how high street retail would operate in the future. In terms of the decline, the Planning Consultant commented that

> '... clearly it's a market failure and I think planning is an impediment to correct market failure ... I think sometimes it's overplayed as a factor ... yet the planning system needs to evolve at least to a certain degree.' (Planning Consultant)

The Heritage Consultant was of the opinion that previous changes in planning policy had been a factor, as they had paved the way for the creation of larger out of town retail parks with destination retailers such as IKEA and supermarkets, and that this had led people away from the high street.

4.2 (ii) Functional

All the interviewees commented on the increase in vacancy rates, but it was agreed that this was not just down to an increasing number of businesses leaving retail units. The suitability of the space for the functional requirements of the businesses is also a factor. As such, the problem cannot just be attributed to business rates and taxes alone. Indeed, the Heritage Consultant focussed on the physical aspects of many high street retail units, which, when it comes to occupying high street retail units, often have:

> '… difficult storage facilities, some of it might be upstairs, or it might be down a corridor or in a basement and you might not be able to get your lorries around the back very easily to load and unload. So, I think that's something retailers really have to think about.' (Heritage Consultant)

This view concurs with Grimsey et al. who state that the older stores in high street locations are struggling to meet service and delivery area requirements [12]. With larger retailers generally downsizing to smaller units or leasing less space altogether, larger units, which no longer suit the needs of businesses, are left empty. In other words, often the specification of an asset no longer fulfils the function for which it was originally designed [10].

4.3 (iii) Environmental

The Heritage Consultant expressed the importance of the 'actual physical environment of a high street, you know, it's a lot of pretty buildings. People might just go there for a day out, they might not go even to necessarily buy anything,' yet just being able to attract someone to a place will automatically increase the chances of retail sales. It was noted by the interviewees that a town centre's infrastructure will also have an effect on the success of its high street retail units. The Regeneration Consultant stressed the importance of providing good infrastructure and access links to a town centre to encourage people to come. There has been a marked increase in the number of people cycling and therefore better cycle access is required around town and city centres. This highlights the fact that a shopper is more likely to seek out a destination that is aesthetically pleasing and easy to reach [6].

4.4 (iv) Location

The Conservation Specialist focussed on the concept of placemaking as a whole, rather than planning alone, stating that there is

> '… much more of an emphasis on placemaking rather than just development, and retail provisions, land use and planning and … the way in which we think placemaking and provisional retail is having an input both positively and negatively over time depending on the place in question.' (Conservation Specialist)

The concept of placemaking has been under discussion since the 1960s, however it gained more traction in the 1990s [21]. Placemaking is a process by which public spaces and buildings are shaped. It involves multiple factors, including planning and design, and requires input from a diverse range of people from professionals to residents of a local community, in order to bring improvements to a community from a cultural, economic, social and environmental perspective [22]. In terms of addressing the issue of retail in the high street, placemaking may prove to be a more suitable strategy for reviewing how the high street could be maintained by incorporating the wider aspects of community.

The Conservation Specialist expressed the view that even a small investment can have an impact, and

> '… if people see the effect, they see the upward trend… and it attracts investment. When people see a downward trend, it actually inhibits it [investment].' (Conservation Specialist)

It would appear that the Conservation Specialist is correct, as the market report undertaken by Aspinall Verdi confirmed that the Cathedral Quarter in Derby was continuing to grow and benefit from new operators [23]. In the Derby PSiCA report this was described as a 'cluster effect' in which owners and leaseholders were approached and recognised the commercial benefits of reinstating a historical shop frontage, which led to an oversubscription of owners seeking assistance [19].

5 SUMMARY

The traditional retail high street has long been in decline and its future is a matter of contention among property professionals and the local community. More recently, it has become a more pressing issue, and it is widely accepted that urgent action will be required if we are not to lose the high street as a retail location altogether. In the past few years, a multitude of funding projects have been launched in a bid to breathe new life into the historic retail high street.

Given the limited amount of academic research on working solutions to the decline of the retail high street, this paper has drawn upon the findings of a case study of Derby's Cathedral Quarter, to demonstrate the impact funding and utilising heritage assets can have, and the opinions of a cross-section of Real Estate and Funding Professionals within the local area. The analysis was based around the Hughes and Jackson [6] conceptual model with economic, environmental, functional and locational categories.

The interviewees identified several economic factors, including the introduction of online retail which has a knock-on effect on other economic factors, such as supply and demand. The introduction and expansion of online retail has increased the accessibility of retail items and reduced the need for visiting the high street to make purchases. Consequently, this has reduced the stock requirements in stores, but increased the need for warehousing out of town, leading to the repurposing of high street locations for mixed uses and varied functions including office space and residential use. The change in consumer spending habits was also identified as a key economic reason for the reduction in high street spending. The majority of the interviewees commented that consumers are seeking multiple uses and an 'experience' from their local high street, and therefore shopping is no longer the primary reason for visiting town centres. Instead, shopping is seen as a secondary or tertiary reason for visiting them, and is often combined with eating out at restaurants, meeting friends, or going to the cinema.

The statistics from the PSiCA Report demonstrate the enormous impact the funding from Historic England, Derby City Council and private entities have had. All the major streets within the DCQ that underwent restorative works to their buildings experienced a decline in vacancy rates and an increase in footfall and rental values, which went against the national downward trend at the time. It is therefore evident that projects such as the one undertaken for Derby's Cathedral Quarter, which used funding as a resource to improve existing heritage assets, can continue to have a positive impact on the historic high street retail economy.

Overall, it was clear from the literature available and from the views expressed by the interviewees, that retail on the high street will continue to change and shrink and that this will increasingly open up the area to alternative uses such as office space and residential uses. However, retail will not disappear from the high street altogether if the location has an historic appeal and is aesthetically pleasing and offers shopping as a pleasurable activity.

REFERENCES

[1]   Housing, Communities and Local Government, *High Street and Town Centres in 2030*, UK Parliament: London, 2019.

[2]   Carmona, M., London's local high streets: The problems, potential and complexities of mixed street corridors. *Progress in Planning*, **100**, pp. 1–84, 2015.

[3]   Farrell, S., UK's business rates system "broken" says treasury committee. *The Guardian*, 2019.

[4]   Butler, S., Thousands of UK shops left empty as high street crisis deepens. *The Guardian*, 2019.

[5]   British Retail Consortium, *About BRC*, BRE: London, 2019.

[6]   Hughes, C. & Jackson, C., Death of the high street: Identification, prevention, reinvention. *Regional Studies, Regional Science*, **2**(1), pp. 237–256, 2015.

[7]   Colliers, *Midsummer Retail Report 2011*, Colliers: London, 2011.

[8]   ONS, Retail sales, Great Britain: October 2019. Office for National Statistics, The Stationery Office, London, 2019.

[9]   Carmona, M., de Magalhaes, C. & Hammond, J., *The Smaller Towns Report*, BCSC: London, 2004.

[10] RICS, *RICS Valuation – Professional Standards, Global and UK Edition,* 8th edn, RICS: London, 2012.

[11] Williams, A.M., *Obsolescence + Re-Use: A Study of Multi-Storey Industrial Buildings*, School of Land and Building Studies, Leicester Polytechnic: Leicester, 1985.

[12] Grimsey, B. et al., *The Grimsey Review: An Alternative Future for the High Street*, Bill Grimsey, 2013.

[13] Portas, M., *The Portas Review: An Independent Review into the Future of our High Streets*, BIS: London, 2011.

[14] Pal, J., Ntounis, N. & Theodoridis, C., How to reinvent the high street: Evidence from the HS2020. *Journal of Place Management and Development,* **10**(4), pp. 380–381, 2017.

[15] Cape, S., Saving our high streets; With Christmas fast approaching, shops are working flat-out to encourage sales. But retail guru Mary Portas believes high streets need regeneration to save them from failure, find Stephen Cape. *The Journal*, 2011.

[16] Derby City Council, *City Centre Conservation Area: Appraisal and Management Plan*, Derby City Council: Derby, 2012.

[17] Johnson, R., The cost of renting retail space in Derby has fallen by almost a quarter over the past five years because of lack of confidence on the high street. *Derby Telegraph*, 2013.

[18] Historic England, *What is the Heritage at Risk Programme?* Historic England: Swindon, 2020.

WIT Transactions on The Built Environment, Vol 203, © 2021 WIT Press
www.witpress.com, ISSN 1743-3509 (on-line)

[19] Anarchitecture, *Monitoring and Legacy Report: Derby PSiCA*, Anarchitecture: Nottingham, 2017

[20] ONS, Retail sales, Great Britain: October 2019. Office for National Statistics, The Stationery Office, London, 2019.

[21] Project for Public Places, *What is Placemaking?* 2007.

[22] Historic England, *Support for Place-Making and Design*, Historic England: Swindon, 2020.

[23] Nexus Planning, *Derby City Council, Retail and Centres Study*, Nexus Planning: Manchester, 2019.

CAPITAL MODEL PRISON:
A POLITICAL TOOL FOR GOVERNMENT POWER

WEIQI CHU
Department of Architecture, DAAP, University of Cincinnati, USA

ABSTRACT

Beijing Prison, also known as *Capital Model Prison*, was China's first modern prison built after the Republic of China was established in 1912, and functions still as a model of national prison reform. Before its construction, China's prisons were generally regarded as punishment centers. New prison design sought to indoctrinate prisoners and allowed the incarcerated to remain in limited contact with the outside world. The new government introduced modern architectural layouts, incorporating European architectural features while retaining Chinese architectural characteristics. The Qing government originally built this prison to perpetuate class rule, but the Republic of China government refined and adapted it to popularize politics. As a dictatorship tool it served the ruling class, although this did not necessarily rule out its positive contributions to political and historical change. This research explores why the prison was built and describes its intended purpose. It also considers the prison's spatial layout, and its potential to serve as a model for a new generation of Chinese prisons. In particular, it shows how its establishment enabled the government to popularize new ideas during changing historical periods, highlighting the prison's influence on society and culture, and its underlying exemplary symbolic significance.

Keywords: architectural, political, Capital Model Prison, transformation, culture, social.

1 INTRODUCTION

As confinement centers, prisons were primarily punitive institutions that stripped away individuals' liberty. Before the Qing Dynasty, prisons were usually repurposed buildings such as former temples. The general thought was that prisoners should suffer and did not deserve to be treated as valuable human beings; hence, like the prisoners it held, a prison's architectural significance was of little consequence. After the bourgeois revolutions in Western countries, the function of prison buildings evolved in line with social progress. In addition to strengthening the state apparatus, the Qing government improved prison administration. An early agenda item was to improve prison architecture, and that concern subsequently lead to the creation of Capital Model Prison. *Model prison* was a general term applied to improved late Qing Dynasty prisons modeled after Western and Japanese prison designs. A model prison had two characteristics: first, it was a prison system that followed the Western and Japanese administrative structure of separating the judiciary and the prison itself; and secondly, the main prison building's architectural design follows European and American examples [1]. The Qing dynasty's Capital Model Prison contributed to traditional Chinese prison model transformations and responded to modern prison system trends and development.

Many important historical lessons of contemporary value resulted from the establishment of model prisons during the late Qing Dynasty. Under the guidance of historical materialism, the study of model prison architecture in the late Qing Dynasty is of theoretical and practical significance in strengthening modern and civilized Chinese prison construction.

2 PRISON: DEFINITION AND ORIGIN OF REFORM

2.1 Prison defined

In modern jurisprudence, a prison is where criminals are held, and sentences are executed. However, to adequately study the concept of prison, the concept should be treated in both a broad and narrow sense.

Prison in the broad sense refers to institutions and locales where criminals are held or forced to perform hard labor, and the institution is backed by the coercive power of the state. The concept of prison in the narrow sense is "a place for the execution of free sentences": specifically, a place of public creation where people's freedom of movement is restrained by the state's power according to legal provisions. In short, it is a special institution set up by the ruling class to carry out the punishment of convicted criminals according to state law.

In this paper, *prison* is addressed in the narrow sense and refers to a place where the ruling class detains convicted prisoners; it is the penalty enforcement agency established in accordance with national law [2].

2.2 Origin of prison reform

Crime, punishment, and the prison system in modern China have witnessed radical changes during the first half of the 18th century, as evidenced through the lens of the Chinese prison system. This paper explores the profound and lasting repercussions of superimposing both Eastern and Western-derived repentance and rehabilitation models on traditional Chinese crime and punishment methodology, instead of presenting a simple history of prison rules and administration [3]. Prisons reflect a society's notions of law, order, and individual rights, as well as human nature itself and its tractability and capacity to change. During China's tumultuous years from 1895 to 1949, these notions transformed dramatically.

3 PRISON ADMINISTRATION CHANGE DURING CRISES AND DEMONSTRATIONS

After the 18th century when Western capitalist society developed rapidly, legal systems reflected the will of the bourgeoisie and their interests were safeguarded. Their slogans of freedom, equality and fraternity were forceful and powerful, and appeals and movements to implement penitentiary education and improve prisons became global. Capitalist countries represented by Britain, the United States, France, Germany, Japan, Italy, Denmark, Sweden, and Belgium all reformed and improved their prison systems and began to conduct related research. Japan significantly improved its national prisons after the Meiji Restoration. Following the Opium War defeat, China's feudal economic foundation was severely damaged, and around 1901 the imperialist powers intensified their political, economic, and cultural invasion of China while the Qing government became increasingly brutal; it was a time of unprecedented national contradictions. The period between the 1840 Opium War and the founding of the People's Republic of China in 1949 was a period of reflection for modern Chinese prison administrations. During this period, China was reduced from an independent feudal state to a semi-colonial/semi-feudal state, and the nature of society changed drastically, thus marking China's prisons as semi-colonial/semi feudal. Prison reform was imminent and in the ruling classes' interest. Pain and humiliation virtually became the root cause of the contradiction between the purpose of imprisonment and the misdemeanor. Change in the inherent thinking of prison governance was obviously necessary.

4 THEORY

4.1 Political theory and architecture

The relationship between politics and architecture is bi-directional: politics may influence architecture, but architecture may influence politics as well [4]. Model prison architecture in China was influenced by the politics of the Great Powers (primarily Britain, the United States and Japan) and their quest for global hegemony. According to Zhongguang [5], prisons in China were mostly houses that had been converted to accommodate the restrictive functionality of prisons [5]. The beginning of the 20th century marked the introduction of formal prison designs in China when the "Beijing Teacher Model Prison" was built toward the end of the imperial dynastic reign in China [6]. It was the first specifically built prison to accommodate structural and functional requirements. After liberation, the prison was renamed to Beijing Prison whose service lasted for four decades. Since the first Opium War, the jurisdiction of the Chinese over British subjects had been a contested issue, with the British refusing to accept China's jurisdiction over its subjects living in China. This was one of the major triggers of the 1840–1841 military conflict between the two countries [7].

Western countries regarded China's court proceedings and legal punishment as both barbaric and intolerable. Therefore, they insisted that China cede its sovereignty over legal matters pertaining to European subjects in China. Consequently, extraterritoriality was an issue included in the 1840 and 1860 settlements between China and the Great Powers, thus allowing foreign subjects to be tried by their respective foreign consuls. Apart from rendering foreign merchants and missionaries immune to Chinese authorities, this had the effect of perpetuating the influx of foreign jurisprudential thought and practice, leading to the emergence of a substitute legal system. Moreover, using their military leverage, Britain, the US, and Japan pressed for legal reforms to be included in treaties signed with China in 1902 and 1903 [8].

Aligned with this, Qing functionaries and reformers were persuaded that the Chinese state's authority and prestige could only be reasserted by ending extraterritoriality. This required the adoption of western legal practices, and pressure exerted by the Great Powers necessitated the reformation of the country's punishment system. The urgency for this was enhanced by the Chinese state's decline both within and outside its borders toward the end of the 19th century. A new prisons law emphasizing instruction and industrial labor (reformation of the individual) rather than punishment alone was thus passed in 1913 [9], signaling China's eagerness to gain global legitimacy and to be perceived as reformative. The facilitation of transformation that was crucial to establishment of prisons was effected by ministers such as Mr. Xuan Hongci were influential in facilitating prison transformation. Hongci traveled globally, searching for information relevant to the constitutional establishment of prisons. Upon his return, he directed the Ministry of Criminal Justice conversion to the Ministry of Law, and established the Department of Prisons. This enabled governmental authority to oversee the national prison administration [5].

From the Foucaultian perspective in which a societal group is studied in relation to power, this may reflect the need to have a group subjected to an idea via soft coercion and to be productive for the idea to be useful [10]. One major consequence of the pressure imposed by the Great Powers was that the Chinese punishment system was reformed in line with a western template [7]; the Beijing Prison design was based on Great Britain's Pentonville Prison (Fig. 1).

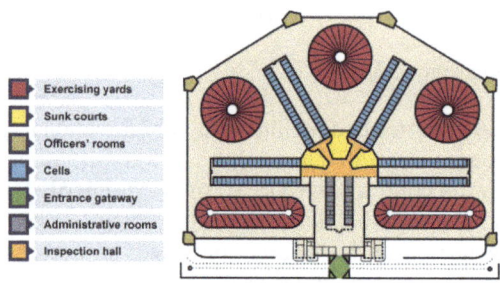

Figure 1: Layout of the Pentonville Prison [11].

A visual depiction of the Beijing Prison in 1917 is presented below (Fig. 2). Note the cellblock radiating from the central tower.

Figure 2: Beijing Prison in 1917 [12].

Political influence on prison architecture was manifested in this manner. Although the Hubei model prison was the first model prison in China, the Beijing Prison whose construction began in 1909 and was completed after the Xinhai revolution, was the most famous of the model prisons in China [7]. As an example of architecture's influence on politics, the Beijing Prison had distinct architectural attributes which helped express the preferred political ideology of the late Qing era, and subsequently, the Republican era.

A layout plan of Beijing Prison is presented below (Fig. 3). According to Kirby et al. [7], London's Pentonville Prison served as the model. Their similar layouts can be seen below.

According to Dikotter [14], the two predominant prison architectural models in China were the radial/fan-shaped (*shanmianxing*) (Fig. 4) and crucifixion (*shizixing*) designs. As shown in the Pentonville Prison plan (Fig. 1), Beijing Prison utilized the fan-shaped plan (which was a variation of the cruciform shape), with blocks of cells radiating from the center (two central points).

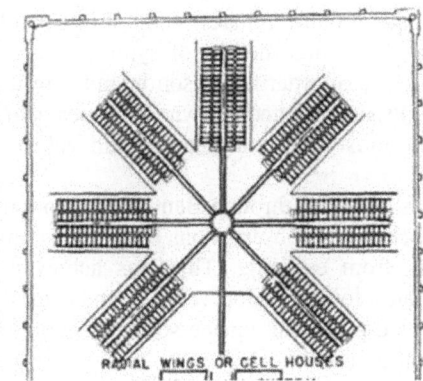

Figure 3: Schematic layout of Beijing Prison formerly known as Jingshi Prison [12].

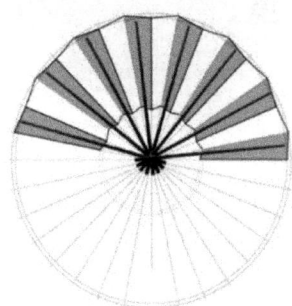

Figure 4: A fan illustrating the radial shape (shanmianxing) [13].

According to Kirby et al. [7], the prison spanned 300 m from east to west, was 330 m in length from north to south, and consisted of three sections. The wings which radiated from the center contained 165 individual, and 40 communal prison cells. The prison was designed in accordance with the Panopticon principle, as elucidated by Bentham [15], an architectural design that assured optimal visibility and surveillance [7].

The basic idea behind such visibility was to attain panoptic control in line with the Panoptic Control Theory which asserts that individuals (inmates) are more likely to adhere to established rules and norms when they are aware of being observed [16]. Given the configuration of the prison architecture, prisoners were transformed into individual objects within the "economy of observation," with no opportunity to hide from observation. From a political perspective, the Panopticon has also been interpreted as a surveillance tool for control and subjugation in a disciplinarian society [14]. This appears to reflect the Foucaultian view of the contemporary penal system's evolution, with the agency of punishment shifting from the corporeal to the spiritual (i.e., from a public, visible, bodily form of punishment to a private, invisible, disciplining of the soul). Bodily incarceration was previously viewed as the state's absolute power over its citizens, but Panopticon control illustrates the use of strict discipline to achieve social control, the overriding aim being to forestall future crimes rather than to achieve revenge [10].

Architectural prison design incorporating a visible and clear center and multiple sub-centers also reflected a prison's hierarchical order, providing unambiguous structures for the ordering, separation, and regulation of inmates within the prison [7]. According to Dikotter [14], the architecture provided a high degree of symmetry and regularity – architectural aspects meant to mirror a sense of order the prison intended to instill in its inmates – along with a sense of character. Prison architecture was also meant to enforce transparency and impermeability, two of the most widely articulated and valued attributes within Chinese society.

While transparency was achieved through Bentham's Panopticon design principles that allowed for around-the-clock inmate observations by wardens, impermeability reflected the need to prevent prisoners from escaping. This was achieved using thick brick walls, reinforced concrete, iron gates, long corridors (Fig. 5), and central towers, creating a bastion that contained evil within and protected outside society from moral decay. It was a factory for the industrial production of morality [14].

Figure 5: Beijing Prison corridor in 1917 [12].

The separation of cells and use of single cells created spaces for solitary confinement viewed as essential to discourage vice and encourage virtue. It was thought that solitary confinement would prompt cathartic reflection and silent meditation, and in so doing would enable an inmate to appreciate the full value of liberty. It could also prevent other inmates' undesirable influences and place the inmate in the therapeutic hands of wardens, hastening the inmate's reformation. It was also believed that solitary confinement could break an inmate's evil instincts and resistance, and thereby transform his/her character [7].

The entire architectural system not only facilitated the achievement of penological goals, it led to the fulfilment of broader political goals. As Dikotter [14] reported, architectural features helped to underpin the new communist regime's penological principles and acted as a tool to pursue Confucian ideals of a cohesive, virtuous and ordered society. From the Foucaultian Theory of Power perspective, prisons provided a venue wherein new social relationships between felons and the state could be forged, the formation of such relationships being mediated within the ubiquitous and ingrained microphysics of power.

Power in the modern age is comprised of much more than punishment and coercion; it also incorporates various cultural techniques to persuade and manipulate subordinates. Accordingly, the dominant discourse is expected to generate narratives, symbols, and representations through which most of the society's cultural and scientific products are manipulated [10]. Within the Chinese context this was expected to spread from a prison which was regarded as a miniature society, to society in general. As such, the prison was also used by the China nation state as a laboratory where "a new subjectivity of the loyal, good, disciplined citizen was developed" and for the Chinese nation-state to arrange a system of representation and formulate new conditions for "individualized self-representation" [14].

5 CHOICES UNDER PRESSURE TO ABOLISH CONSULAR JURISDICTION

Since the 18th century and during the development of the bourgeois revolution, bourgeois legal culture gradually formed with bourgeois humanism as its core. Amid the slogans of freedom, equality, and fraternity, the penal system has undergone major changes. Reform movements using prisons for penalty enforcement quickly emerged on the European continent; penalties characterized by humanity, probation, and education became popular in prison systems. "From humiliation to probation," became the concrete manifestation of humanitarianism in prison governance under the western *rights* concept. This transformation also took place in China during the early 18th century. It represented not only a "learning from the West" mentality of Western prison civilization, but also a chosen mix of Western and Chinese prison systems, the latter carrying the historical identity of an ancient prison governance spirit [17]. Whether it was from a sincere belief in traditional attachments or to make reforms run smoothly, the tradition of benevolent governance provided an effective ideological foundation and psychological support for the acceptance and recognition of probation thought.

Starting in the 1840s, the forced signings of one-sided treaties damaged China's judicial sovereignty, one manifestation being the establishment of other countries' consular jurisdictions in China. Western prisons set up by judicial organs of colonial countries appeared one after another, ultimately influencing the modernization of Chinese prisons. While safeguarding the national sovereignty and interests of the country to some degree, prison administration reforms were strongly political in nature and undoubtedly determined the basic character of future reforms.

6 CAPITAL MODEL PRISON

During the closing days of the Qing Dynasty, the government launched the "New Deal" and "Preparatory Constitutionalism" campaign in an effort to save its regime. A series of judicial reforms were enacted as an important part of this political reform, one of which addressed prison administration. When the Republic of China came into power, a series of new prison reforms were decreed for political and diplomatic reasons and were based on the late Qing Dynasty's prison administration's practices. The new reforms greatly exceeded earlier ones in terms of scale, implementation, and longevity.

Capital Model Prison was selected for the present paper because it has been an exemplary model of national prison reform since the end of the Qing Dynasty [18], and it represents prison reform trends during the Republic of China. Prisons commonly symbolize brutality, darkness, and corruption. They are violent tools set up by a ruling class to maintain dominance, and the institutions can be conveniently separated physically from civilization. As a reflection of China's modernization, prisons become a yardstick to measure the degree of a country's social civilization; the degree of prison civility is an important coordinate for measuring a society's transition to modern civilization. The modern civilization consciousness it realistically embodies reflects China's modernization process. Some of the problems exposed during the transformation also reflect China's other deep-seated problems.

As the first modern prison in China, Capital Model Prison was built in the late Qing Dynasty and was renovated in the early years of the Republic of China [19]. Due to regime changes, its identity has undergone a series of names from Capital Model Prison, Capital No. 1 Prison, Hebei No. 1 Prison, Peiping No. 1 Prison, Beijing Hebei No. 1 Prison, and Beijing No. 1 Prison. As a new-style prison, it differs from those that have come before, especially in terms of management. It has become a prison reform movement model representing the late Qing Dynasty and the Republic of China.

Due to its intended function, multiple factors had to be considered when selecting the prison location. The prison was set outside Beijing to the southwest, a relatively sparsely populated area that is relatively open, low-lying, and conducive to prison placement. During initial construction, large trenches were dug around the perimeter to both prevent escape and to accumulate water. Having been designed by Japanese prison scientist Koizumiro Ogawa (小河滋次郎), the architectural style resembles that of Japanese prisons [20].

Figure 6:   Aerial view of the Capital Model Prison (The Beijing Prison) showing the layout of the entire site [21].

The main gate is oriented toward Yongding Gate Street in the east, creating an east-west main axis, and the entire site is divided into front, middle, and back areas. The gate constitutes the central axis of the front area. South of a rain channel inside the gate is the guard and teaching center, and to the north is the exhibition center and reception room. To the west are two guard dormitories facing each other. A second entrance leads to the middle area and the main central office building. The back area contains the main prison. Additionally, there are two special prison areas: the disease surveillance room to the north, and the female room to the south of the prison room (the original plan for the infant prison) separated by two walls.

A comparative layout of the general prison plan is shown in Fig. 8.

In 1918 a new section was built to the north of the site along with a north gate and an agricultural plantation to the west. Within the prison there are workshops for prisoner use as well as wards, bathrooms, warehouses [23], and both single-living and multi-living cells. Great importance was given to the transformation of Capital Model Prison during the Republic of China, and new workshops, main rooms, and other prison buildings were required additions. The size and dimension of materials used in the construction was highly regulated, as well as certain materials, depths, wall thickness, etc. Prison buildings at that time were mainly made of brick and wood which were relatively simple and practical, but limited by factors such as the national economy, finances, and available construction technology.

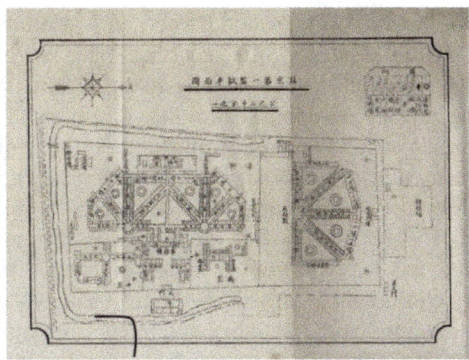

Figure 7: The layout of a Capital Model Prison [21].

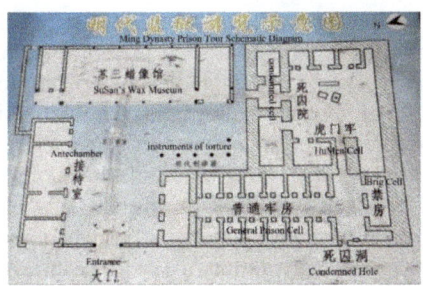

Figure 8: The layout of a general prison in Chinese feudal dynasty [22].

While building new prisons and updating older ones, the Republic of China government mandated specific requirements that found their way into the Measures for Amendment and Improvement of Old Prisons (修正改良旧监所最低限度办法) [24]. The Republic of China's fiscal revenue was extremely limited and was primarily used to maintain a huge military force and support an ongoing civil war, or to repay foreign and domestic debts. Consequently, the national government was unable to invest large amounts to adequately renovate outdated prisons; they could only afford to make modest improvements. In 1912, the Republic of China government formally opened Capital Model Prison and began to accept prisoners and changed its name to Beijing Prison. Yuanzeng Wang, the first warden, had studied penology in Japan. With China's national economic conditions in mind, he presided over the formulation of a series of rules and regulations covering all aspects of prison management including the institutional setting, prison guard service rules, labor system, reward and payment system, commutation and parole system, and the education and teaching system. It is not an exaggeration to call him the founder of the new Chinese prison system.

As noted above, Capital Model Prison was in a low-lying damp area. Drains installed when the prison was built had been clogged for years, inevitably producing many infectious agents living in stagnant reed filled pools that could lead to an epidemic. Moving the prison to a more favorable location was not possible; therefore, Yuanzeng Wang opted to fill the ditch, convert what remained into a gutter, and installed a small floodgate. Yuanzeng Wang objected to soil sleeping platforms traditionally used, believing they contributed to unpleasant smells, and because Beijing is severely cold in the winter and there were many prisoners, he felt the large quantity of firewood burned for warmth was not conducive to the health of the prisoners. He therefore suggested that heating pipes be installed in the prison area. Unfortunately, his suggestions were not fully realized until 1922 because of funding issues. Yuanzeng Wang made prisoner health the forefront of his prison work. His recommendations were based on advanced penology theories that embodied humanitarianism and civilized notions absent in prior Chinese prison management. It was more than a decade later when the Capital Model Prison facility upgrades were completed, a delay Yuanzeng Wang greatly regretted. He did not expect quick success when attempting to educate criminals; he approached the task through a series of steps: first, improve the prison environment and sanitary conditions to protect criminals' basic physical health [25]; second, formulate and strictly implement work, rest, diet, and related systems, and build playgrounds to improve criminals' physical fitness; and third, after achieving the first two goals, instill a sense of ethics in prisoners, and teach them a means of earning a living. The three-step procedure accounts for why this prison has been effective in educating criminals [15].

7   CAPITAL MODEL PRISON AS A GOVERNMENT TOOL

As a state entity and symbol of legal authority, a prison must be tangible and meet basic social requirements. Without a basic understanding of prison facilities, it is impossible to discuss the implementation of execution law and criminal policies, nor probation and humanity issues. Unlike most other building types, a prison itself has no social class structure; its overriding purpose is to serve the ruling class and their objectives [26]. It is a tool of dictatorship. Capital Model Prison was built on the eve of the 1911 Revolution, and its purpose was very clear: a prison for revolutionary parties who opposed the Qing government. It was completed in 1911 when the Revolution of 1911 overthrew Qing Dynasty rule and the prison became the national government's tool – a place to detain progressives. After the People's Republic of China's founding, the nature of the regime fundamentally changed. The prison became a punishment and reform institution for hostile domestic and foreign individuals who endangered national security, and for those who undermined public order or lives, and property. For the Chinese, the prison was not just a building; it was a symbol of national power and the light of revolution. The Qing government's prison reform was based on the construction of prison execution facilities, legislation customization, and the exploitation of talent. Barbaric prison executions were linked with concepts of humanity, probation, and civilization. However, when the old system was eliminated, the prison situation did not fundamentally change after the introduction of Western thought, mainly because old prison management methods and the people who used them were still important components of the system. The desire for reform ran counter to actual results. Through reform at the beginning of the 20th century, various constraints led to tortuous development. Today, as the prison reform continues, basic elements should be understood in order to promote its effective implementation as a government tool.

8   CLAIMS AND REFLECTIONS IN JUDGING THE PROS AND CONS OF CAPITAL MODEL PRISON

8.1   Prison buildings are one of the instruments of a hierarchical dictatorship and are an integral part of the state apparatus

Like any other type of building, prison architecture is not class-specific in itself; it is characterized by the fact that as an instrument of dictatorship, it has always been in the hands of the ruling class [27]. Capital Model Prison built just before the Xinhai Revolution had a clear purpose: to serve the interests of the ruling class. However, when it was completed in 1911 just as the Xinhai Revolution was overthrowing the Qing court, the building became an instrument of dictatorship for the Nationalist government. After the fall of the Beijing government, it was again used to imprison progressives. In this sense, prisons are also a special type of school. In this respect, the role of Capital Model Prison has won international acclaim. Hence, it cannot therefore be said that the prison building has "any attributes of its own"; it serves whoever takes control of it and can be regarded as a neutral and material by-product of the current society.

8.2 Capital Model Prison is a breakthrough in Chinese prison architecture

Following the bourgeois revolution when society progressed and laws were improved, the perceived function of prison buildings also changed [28]. In the Qing court's "New Deal" measures that usurped foreign countries strengths to make up for China's weaknesses, the strengthening of the state apparatus had to be accompanied by prison improvements. Capital Model Prison was therefore a major step forward in Chinese prison architecture and the emancipation of human rights.

8.3 Capital Model Prison is a testament to the development of modern architecture in Beijing and deserves preservation

Built more than 80 years ago, the Capital Model Prison building has served for more than 40 years after the establishment of New China, becoming a rehabilitation center for criminals in China [29]. Mirroring foreign architectural experience, it was the first innovative prison of exemplary quality in modern times.

9 CONCLUSION

The creation of the model prison in the late Qing Dynasty created a new chapter in the modern history of Chinese prison architecture. Some of the new architectural innovations emerging because of the adoption of the Beijing model prison included: the adoption of the radial/fan-shaped design, the separation of cells (which radiated from the center), and the pursuit of the design principles of the Panopticon with overriding focus on visibility.

Other architectural features included: a visible and clear center and multiple sub centers, use of a high degree of symmetry and regularity, thick brick walls, reinforced concrete, iron gates, long corridors, and central towers. The new prison vividly illustrated the relationship between architecture and politics. Its concept of prison architecture and adaptations of architectural rules had exerted a profound impact on the successive reigns and dynasties such as Peking warlord government and KMT government. It also brought an end to the traditional Chinese model of prison architecture and provided the pavement for the beginning of modern prison architecture in China by combining the styles of Western buildings.

The model prisons created in the late Qing dynasty were in line with the world's trend of improving prison architecture and conditions, although the starting point was to maintain and save the interests of the ruling class, which was no doubt a utilitarian one. However, it should be acknowledged that the establishment of model prisons in the late Qing dynasty was a major step forward in breaking with the traditional model of prison architecture. Meanwhile, the capital prison endeavors to educate and illuminate prisoners instead of solely punishment. This breakthrough may not have reflected the basic intentions of the legislators of the time, but it resulted in putting an end to more than two thousand years of jailhouse-style management under the feudal system, thus making a direct transition to modern prison management and ultimately laying the foundations for the modernization of prison management in new China.

REFERENCES

[1] Bie, Z. & Ma, X., Research on the architectural layout and construction techniques of Mixian County Ya prison based on ancient Fengshui architectural culture. *China Cultural Heritage*, (3), pp. 99–104, 2019.

[2] Waller, I. & Chan, J., Prison use: A Canadian and international comparison. *Criminal Law Quarterly*, **17**(1), pp. 47–71, 1974.

[3] Li, X., The construction of the model prison of the capital in the Qing Dynasty. *Research on the History of Qing Dynasty*, (3), pp. 148–156, 2019.

[4] Stankovic, D., Tanic, M., Kostic, A. & Nikolic, V., Balcan cities, political influences and architecture: The case of Serbia. *ACE Journal*, **2**(10), pp. 3–11, 2017. DOI: 10.18503/2309-7434-2017-2(10)-3-11.

[5] Zhongguang, J., Review of the first new type prison building in China: The Capital Model Prison. *Journal of Beijing Institute of Civil Engineering and Architecture*, **1**, 1998.

[6] Zhou, J., The Chinese correctional system and its development. *International Journal of Comparative and Applied Criminal Justice*, **15**(1–2), pp. 15–32, 1991.

[7] Kirby, W.C., Leutner, M. & Mühlhahn, K., *Global Conjectures: China in Transnational Perspective*, LIT Verlag: Münster, 2006.

[8] Gelber, H., *Battle for Beijing, 1858–1860: Franco-British Conflict in China*, Springer: New York, 2016. DOI: 10.1007/978-3-319-30584-4.

[9] Spence, J.D., *The Search of Modern China*, W.W. Norton & Co. Inc.: New York, 1990.

[10] Driver, F., Power, space, and the body: A critical assessment of Foucault's discipline and punish. *Environment and Planning D: Society and Space*, **3**(4), pp. 425–446, 1985. DOI: 10.1068/d030425.

[11] BBC, Methods of punishment, 2021. https://www.bbc.co.uk/bitesize/guides/z938v9q/revision/6. Accessed on: 8 Feb. 2021.

[12] 17 QQ, Beijing prison, 2021. https://line.17qq.com/article/qcewhwtax_p7.html. Accessed on: 8 Feb. 2021.

[13] Patrick, L., Create an illustrated, Japanese style hand fan in photoshop, 2010. https://design.tutsplus.com/tutorials/create-an-illustrated-japanese-style-hand-fan-in-photoshop--psd-9052. Accessed on: 9 Jul. 2010.

[14] Dikotter, F., Crime and punishment in early Republican China: Beijing's first model prison, 1912–1922. *Late Imperial China*, **21**(2), pp. 140–162, 2000. DOI: 10.1353/late.2000.0008.

[15] Miller, J.A. & Miller, R., Jeremy Bentham's panoptic device. *JSTOR*, **41**, pp. 3–29, 1987. DOI: 10.2307/778327.

[16] Strub, H., The theory of panoptical control: Bentham's panopticon and Orwell's nineteen eighty-four. *Journal of the History of the Behavioural Sciences*, **25**(1), pp. 40–59, 1989. DOI: 10.1002/1520-6696(198901)25:1<40::AID-JHBS2300250104>3.0.CO;2-W.

[17] Wang, X., The architectural features of model prison in late Qing Dynasty. *Journal of Henan Judicial Police Vocational College*, **10**(3), pp. 15–19, 2012. DOI: 10.3969/j.issn.1672-2663.2012.03.003.

[18] Mahoney, J., Path dependence in historical sociology. *Theory and Society*, **29**(4), pp. 507–548, 2000. DOI: 10.1023/A:1007113830879.

[19] Holbig, H. & Gilley, B., Reclaiming legitimacy in China. *Politics & Policy*, **38**(3), pp. 395–422, 2010. DOI: 10.1111/j.1747-1346.2010.00241.x.

[20] Yang, B., Research on Koizumiro Ogawa and the transformation of prison system in the late Qing Dynasty. *Journal of China Three Gorges University (Humanities & Social Sciences)*, 35(4), pp. 92–95, 2013.

[21] 17 QQ, Beijing prison, 2021. https://line.17qq.com/article/qcewhwtax.html. Accessed on: 8 Feb. 2021.

[22] Yueguan, History of "Guan" language can you distinguish the "three, six, nine, etc." of Qingming prison? 2018. https://www.sohu.com/a/243741442_488248. Accessed on: 20 Mar. 2021.

[23] Melossi, D. & Pavarini, M., *The Prison and the Factory: Origins of the Penitentiary System*, Macmillan: London, 2018.

[24] Kiely, J., *The Compelling Ideal: Thought Reform and the Prison in China, 1901–1956*, Yale University Press: New Haven, 2014. DOI: 10.12987/yale/9780300185942.001.0001.

[25] Jiang, J., Liu, B. & Wang, Y., Exploration and practice of low alert prison. *Research on Crime and Reform*, (4), pp. 19–30, 2018.

[26] Li, F. & Qiu, J., The baseline of leniency: Rational rethought on Chinese juvenile criminal justice. *Juvenile Delinquency Prevention Research*, (4), pp. 29–35, 2013.

[27] Xu, M., *On Special Procedures of Juvenile Criminal Justice*, Law Press: Beijing, 2007.

[28] Zhang, J., Approaches to juvenile procuratorial works. *People's Procuratorate*, (23), pp. 165–167, 2011.

[29] Chen, W., Prosecutorial reform in the context of judicial reform. *Procuratorial Daily*, 23 Jul. 2013. https://www.spp.gov.cn/llyj/201307/t20130723_60693.shtml.

REMAINS OF THE BEAUTY:
A DEFINITION OF BEAUTY IN THE BUILT SPACE

SILVANA KÜHTZ
Department of European and Mediterranean Cultures: Architecture, Environment, Cultural Heritages (DiCEM),
University of Basilicata, Italy

ABSTRACT
The study Cultural Heritage Counts for Europe (2015), demonstrated how cultural heritage is central to the processes of value production also in terms of its contribution to continuous social innovation. The focus is not only on the tangible and physical assets, but also on the intangible ones. Thoughts and beliefs stemming from very different historical and geographical contexts are confronting each other with increasing frequency. This research asks whether the cultural values and ideals we relied on in the last century are also somehow changing. In the light of modernity and of the present uncertainties (i.e. migrations, pandemic, etc.), we probably have to redefine and reaffirm what we took for granted during the second half of the 20th century, i.e. the idea of beauty. A series of interviews with experts to deepen the perception and definition of the concept of beauty are at the core of this paper.
Keywords: beauty and heritage, interviews, definition of beauty.

1 INTRODUCTION
The 2005 *Faro Convention* asserted that cultural heritage lies at the centre of local community life and the 2013 HangZhou UNESCO Declaration identified it as the pillar of economic, social and environmental sustainability. The study *Cultural Heritage Counts for Europe*, promoted by the European community and presented in Brussels in 2015, demonstrated how cultural heritage is also central to the processes of value production. Not only in terms of the revenues generated by the streams of public or cultural tourism, but also in terms of cultural heritage's contribution to continuous social innovation. The focus is not only on the tangible and physical assets, but also on the intangible ones. Richard Rogers affirmed: "culture gives meaning and pleasure to life. It encourages us to understand our place in history and helps us challenge established social convention. Urban regeneration, economic growth and cultural development are all inter-related" [1].
2018 was the European Year of Cultural Heritage (EYCH).

Thoughts and convictions stemming from very different historical and geographical contexts are confronting each other with increasing frequency, and therefore we asked ourselves whether the cultural values and ideals we relied on in the last century were also somehow changing. In the light of modernity and uncertainty, we probably have to redefine and reaffirm what we took for granted during the second half of the 20th century [2]. Let's take for example one of our Western cities – would we be able to truly and honestly say what we find ugly? Does the concept of beauty conform to a fashion or does it belong to something deeper? What is beauty? What remains of the beauty, then?

One of the aims of this research project is to investigate our current conditions through the concept of beauty, referring to what it means on an individual and global level. This investigation started in 2015 by questioning the rationale behind the habit of keeping rusty, ugly and inhospitable structures in place, instead of demolishing them.

WIT Transactions on The Built Environment, Vol 203, © 2021 WIT Press
www.witpress.com, ISSN 1743-3509 (on-line)
doi:10.2495/STR210191

The concept of beauty is a long-standing philosophical question that does not have a definitive answer, but it can give incentives for personal and social improvement and reflection. This research's intention is to trigger discussions and queries rather than provide ultimate answers/solutions. The questions posed refer both to beauty and to the tactic of demolishing ugliness, and they refer in particular to the built space.

2 EXPLORATIVE QUESTIONS

The interviews (with experts, opinion leaders, citizens) open up a wide field of perception and concur in widening the definition of beauty and of an environment worth living. The narration of our perceptions, tastes and experiences of beauty creates awareness and knowledge, allowing for interactions among designers, architects and citizens.

The questions posed were always the same, i.e.: What is beauty for you? Where do you find beauty? Is beauty subjective or objective? What does the word demolition mean for you?

Beauty is in the eye of the beholder is a sentence that has existed for nearly 2300 years. To restate it in the words of those anonymous medieval scholars: *de gustibus non disputandum est.*

Is the concept of beauty really subjective then?

Recent works in the fields of neuroscience and behavioural psychology have begun to confirm that there are indeed means for measuring beauty by studying human responses and preferences. Not only do we make individual judgements about what we find beautiful, but we also hold within us predetermined criteria that lead us to make decisions that are not as personal as we might suppose [3]. To a large extent, researchers tell us, our judgements can be predicted. Preferences for certain shapes, colours, textures, materials and various other criteria can be tabulated and even ascribed to human qualities such as age, gender, cultural background and experience [4].

It is interesting to note how Matera, once considered shame of the world, was appointed European capital of Culture 2019. What was not demolished and cancelled out in the past, simply because it was too expensive, is now a place of precious beauty and tourism.

What is beauty then? Has heritage always to do with beauty?

"What is beauty and why do we need it? All thinking people recognise that this is a real question, and one of especial relevance in the disordered times in which we live. Indeed, in the case of the built environment, there is no question more urgent" wrote Sir Roger Scruton in 2019 [5].

3 THE INTERVIEWS ON BEAUTY

The interviews that compose this research are numerous, at the moment more than 60 people have given their views on the topic. Hereafter the answers of the experts who specifically deal with architecture and space, for their competencies and for the nature of their answers.

3.1 Alfredo Brillembourg, architect, film maker, writer

There is a nostalgic view of beauty based on proportions that comes from our western education. A lot of people do not like New York, Le Corbusier hated it, it is grey and messy. Others do not like Caracas, Lagos, Mumbai, but I love them. I would reflect on what Slavoj Zizek said. Human beings as an extension of nature should probably like what humans have produced the most, that is trash. So standing in front of a hill of trash we should reflect that this is what we have created in the name of progress.

I believe that beauty somehow is an attitude. Beauty is the same with people and personality. I like imperfections, roughness, unfinished buildings. The early films of Wim Wenders give an idea of this concept of beauty in the built space. We should not emphasise on the nostalgic beauty, we should find it in all the leftovers spaces of the cities, the favelas, the slums, the villages. They have modularity.

Since Kant we have pushed an agenda of rationalization and science, thinking that this could bring order and harmony, we have lost our sense of understanding that science, poetry and art must be together. Architecture is one of the greatest arts as it encompasses everything and cities are the greatest examples of civilization. The beauty of cities is in the diversity. What makes a city interesting is the chiaroscuro. Without it we cannot manage cities.

In 1980s people were revaluating the classical language of architecture almost in nostalgic ways; in 1990s with globalization we were given a pattern of industrial architecture. After the twin towers tragedy, where the object of the attack was architecture, there is a break in societies. I find that the work of the architect should go more in the direction of roughness. Steel and concrete last and have character. We do not need decorations, we need what can withstand time.

I find beauty in Gaza where people have to share a very small space. Beauty is an expression of human endeavour and what it has produced. Beauty is something that impacts me, where human beings are taking over the space in a unique way, or are utilizing the space in an unusual way.

In the past beauty could be identified with form, today with process, with an expression of complexity.

3.2 Mario Cucinella, architect

Beauty is naturally associated to Italy, where it is very democratic: wherever you go there is beauty. Beauty is difficult to define because is intertwined with the codes of one's time. There is something that remains recognizable, however, over time, and is linked to the ability to also give profound contents beyond trends. I'm talking about solid content connected to the invisible, not just to the aesthetic aspects. But beauty is also something that is not seen, such as the care of space, or the secular Casentino beech-woods. There must be profound reasons for doing things in architecture that go beyond the form. As architects we must build with a vision of solidity, be aware of the responsibility we have in doing things and always consider their consequences. Aesthetics is not to be denied but has to embed the founding principles, beauty should be understood and realized, it should not be chased.

Demolishing is a beautiful word, I am for demolishing a lot. Too easy to criticize what has been done forty years ago. To build in excess for speculative reasons is useless, also, we don't need to reuse everything. Let's find the necessary courage to demolish and give value to the empty landscape.

3.3 Dan Pitera, Dean of the School of Architecture, Detroit Mercy School

For me beauty is how I feel, feeling joy and being passionate about what I do is beautiful, as well as it is beautiful being calm and at peace, relaxed. Beauty is all around me, external but also internal. It is not necessarily in what we see but in what we do. Working together is an example of beauty. It is not one thing or the other: it is the thing and the process, internal and external, something that brings a person joy, it is a noun and a verb, the system and the things, subjective and objective.

If an activity brings many people together and engages their passions, it is beautiful. To a child I would say that beauty is not fixed, it can change with knowledge, perspectives, complexity. Understanding the beauty of the complexity of life is the process of growing up. Beautiful things are complex, this is something that in architecture should be emphasised.

Demolition brings to mind a very strong attitude that people have in Detroit nowadays: they think that everything abandoned should be demolished or renovated to its original conditions. I think that there is a space in between. We could evaluate things differently: if there is something that is empty and abandoned and it is a symbol of negativity I can agree for its demolition. If a building brings great heritage we should keep it and renovate it to move forward.

As for myself, I do not demolish, I de-construct, i.e. I reuse fragments, I start off with something people think should be demolished and instead of demolishing I keep pieces and materials that can be reused and remove the rest. "Architectorial de-spoilure" is how we call this process. A building that should be demolished is basically blight in the landscape. But also an empty space after demolition can be blight in the space, because it can affect the citizens also on a psychological level. In thinking about demolition we should therefore think about all the consequences, not only physical ones, that's why we have to work in all the directions and in time.

3.4 Jean Philip Vassal, architect

Beauty is what we love. We start being interested in something, and by looking at it then beauty appears. To bring beauty in the world, you have to take care, to concentrate your attention and go beyond. You can see the most beautiful things in the worst situations. It is necessary to search for optimism, for closeness. When you feel with your body, with the sensibility to search and to generate the possibility to solve a problem.

I do not like the word Demolition. Poetry is a sort of demolition of words, it is a subtraction, a reduction, you keep only what is very important. It is a long poem that progressively is made essential. Many buildings are demolished because of their façade and not because of what they really are.

3.5 Francesco Erspamer, Professor of Romance Languages and Literatures, Harvard

The reason that truth and good are essential to a society is obvious. But for beauty it is not so obvious. Nevertheless has always been one of the three fundamental concepts for many civilizations. The fundamental question is not what beauty is, but why beauty. Kant, in the critique of judgment, introduces the idea of the universal subjective. Beauty is what allows you to have a subjective experience but having that experience entails that all others, as human beings like you, share, in the presence of this cathedral with this light at this moment, the same experience of beauty. Beauty is not something that you can look for, like an object, an experience, a truth, beauty is something that happens.

Recently I happened to discuss the beauty of some sporting gesture. The most beautiful things that I remember are those that have no purpose. A totally free gesture. There is a very significant episode of the Odyssey in which Ulysses tells of when he was about to leave for the war, with the fleet waiting to go to the Trojan war. But he stops because he sees a palm, he loses time in front of this palm. Observing this tree, meaningless in context, means that you are able to open up to other possibilities, not just those predictable. It is a metaphor for the fact that through beauty the perception of unpredictable, unusual solutions is exercised.

To Demolition I prefer the word change, evolution, but I don't think everything should be preserved. However, I don't like conceptually that replacing something with something new happens through a demolition process.

I mean, it's already so easy to demolish, time demolishes, everything breaks down. So I am against demolition, aware that anyway it happens.

3.6 Pietro Guida, sculptor

Beauty is the base element of life, it is a feeling that invades you completely. We are all able to notice beauty, it is an inner feeling; as for me, I feel it in my hands, I immediately want to reproduce it. Beauty is an absolute value, pleasure-ness instead is something that indulges in itself, shallow, and must be avoided. Demolishing is a reaction that cancels everything out, sometimes it is an exaggerated reaction, but it is useful to demolish oneself, one's ego.

3.7 Michele Mari writer

Beauty for me is the correspondence to one's own mental scheme, and it brings to me emotions like the desire for possession, I find beauty in the trees, in the music and if I have to think of an urban environment, of the city, I find beautiful everything that is not sparkling, dynamic, contaminated by advertising and specifically I find beauty in old houses. In fact, I don't recognize my city, Milan, as beautiful. Beauty is objective, it is in the canons that we have within us even though we have not created them. The canons evolve slowly and collectively, but we have them inside. I would demolish everything that is abusive and everything that rapes the landscape.

3.8 Moni Ovadia, actor, singer, artist

Beauty is the attention to the fragility of others, of life itself. Beauty in a word is interiority and is always amazing, it is care. Beautiful is a look, a glimpse of nature, in my daily life I find it in the mutual respect. All the places of justice are beautiful, beauty is both subjective and objective.

3.9 Andrea Semplici, photographer

Beauty for me is a deep emotion, adrenaline, heart beating, butterflies in the stomach. Beautiful is also a moment of silence, it is meeting people who surprise me, people who bring their own light. The adventure of travelling for me is beauty, a kinetic balance that I compare to running on the rocks, the balance is given by speed, if you stop you fall down. For me, beauty is also this continuous movement and is both objective and subjective. A child who touches an old tree could appreciate the beauty, understand it by touch. I once saw a poet who recited poetry aloud on his own, believing he was not seen, I find this a beautiful thing.

I like the idea of demolishing walls, houses, and getting rid of ourselves, our smallness, so we have something to demolish every night.

4 CONSIDERATIONS

After analyzing the answers there are perhaps two important aspects we found out: there is taste, which is personal, subjective, and which concerns the strictly aesthetic aspect of something, its pleasantness, and then there is beauty, which functions as the trigger of an emotion shared, in a certain sense, by everyone. So, we can say that it is the experience of what we like, the sentiment of beauty, that unites us all (as the research of Zeki et al. [6] confirms) and is triggered by what we like. This sentiment is common to us all. I may like flowers and a sunset, you may like mountains, heavy metal music and a dinosaur, but we feel the same thing. On the other hand, our interviews underline how the word Demolition of ugliness is at times associated with negative thoughts but also with the idea of making space for something else, new and refreshing.

Why the concept of beauty is important in the city? Beauty in architecture brings us joy and happiness. Merriam-Webster defines beauty as "qualities in a person or thing that gives pleasure to the senses or pleasurably exalts the mind or spirit." Or more to the point, Stendhal, the 19th century French writer, wrote, "Beauty is the promise of happiness." And happiness is one of our fundamental human needs. Beauty is then more robust than it seems.

Padelford in [7], "I think that we are not at all aware of the immense social asset that uniformly good architecture would be. Fancy a city in which all of the buildings are beautiful, and trace the influence on the lives of the inhabitants. In the first place, it would add greatly to the happiness of people, for, as has been observed, it is the normal function of beauty to make us happy. Unless we have allowed ourselves to become diseased, happiness will attend beauty as naturally as flowers turn to the sun."

De Botton in *The architecture of Happiness* [8] writes: "The buildings we admire are ultimately those which, in a variety of ways, extol the values we think worthwhile – which refer, that is, whether through materials, shapes or colours, to such legendary positive qualities as friendliness, kindness, subtlety, strength and intelligence. Our sense of beauty and our understanding of the nature of a good life are intertwined."

Beauty is not the exclusive property of the landed and wealthy. It belongs to us all or it should do. We should strive to ensure that every citizen, however deprived or disadvantaged, has a proper share of it. At present this is not happening. Beauty is unequally distributed. "Beauty comprehends all that feeds into the sense of being at home in a shared world. People make sacrifices for beauty as they do for love. (…) Beauty includes everything that promotes a healthy and happy life, everything that makes a collection of buildings into a place, everything that turns anywhere into somewhere, and nowhere into home. (…) Whereas ugliness means buildings that are unadaptable, unhealthy and unsightly, and which violate the context in which they are placed" [9].

Stemming from what written so far, this is a tentative definition of beauty in the built space. The beauty of cities, in cities, is in their diversity, it's a process that explores complexity, that needs the responsibility of the architect, the builder, the citizen. It is in the ability to give content beyond any trend, it is in the invisible, in the care and attention devoted to the space, the community, mankind. It is in the joy of realizing one's desires, it is in the passion for knowledge and sharing space and life. Beauty is what we love, what we end up loving after close search, with generosity and attention. Beauty happens, but we have to be prepared to experience it, to feel it. It is in a space of justice and joy. Beauty is, despite everything else, even when not seen. In the built space it is the ability to build for the community, with the best available materials, as an inviolable refuse of ugliness.

REFERENCES

[1] Bianchini, F. & Parkinson, M. (eds), *Cultural Policy and Urban Regeneration: The West European Experience*, Manchester University Press, 1994.

[2] Kühtz, S. & Rizzi, C., *The Demolishers Manifesto – What Remains of the Beauty the Choice of the Architect*, ed. Spagine, Lecce, 2019.

[3] Ishizu, T. & Zeki, S., Toward a brain-based theory of beauty. *PLoS ONE*, **6**(7), p. e21852, 2011.

[4] Grice, G.S., Why are some buildings so ugly? *OAA Perspectives*, p. 8, Oct. 2012.

[5] Scruton, R., The need for beauty. *Building Beautiful – A Collection of Essays on the Design, Style and Economics of the Built Environment*, ed. J. Airey, Policy Exchange, 2019.

[6] Zeki, S., Romaya, J.P., Benincasa, D.M.T. & Atiyah, M.F., The experience of mathematical beauty and its neural correlates. *Frontiers of Human Neuroscience*, **8**, p. 68, 2014.

[7] Padelford, F.M., The civic control of architecture. *American Journal of Sociology*, pp. 45–46, Jul. 1908.

[8] De Botton, A., *The Architecture of Happiness*, Knopf Doubleday Pub, 2008.

[9] Living with beauty: Report of the Building Better, Building Beautiful Commission, Jan. 2020.

ADAPTIVE REUSE HERITAGE BUILDINGS ADDRESSING SUSTAINABILITY POTENTIALS: ANALYTICAL CASE STUDIES IN SHARJAH, UNITED ARAB EMIRATES

IMAN IBRAHIM & FATMA ELTARABISHI
University of Sharjah, UAE

ABSTRACT

Due to the rapid growth and development worldwide, nations tend to preserve their historic sites that represent their heritage. The adaptive reuse strategy is used to help communities maintain their own local identity and culture. The adaptive reuse terminology is used interchangeably with terms such as renovation, refurbishment and rehabilitation. However, this study will distinguish each term based on a framework discussed in the literature. The research aims to determine the additional value added when the potentials of three pillars of sustainability are considered in the adaptive reuse heritage buildings. Two case studies, namely The Chedi Al Bait hotel and the Sharjah Art Foundation (SAF) in Sharjah, United Arab Emirates (UAE), are selected on the basis of their historical significance and unfitness. The sustainability potentials inherited in each case will be assessed and analyzed. Based on the analytical studies, it is concluded that the local government in the UAE as decision makers have successfully achieved the added value of sustainability enhancing the adaptive reuse of the two case studies discussed.

Keywords: adaptive reuse, sustainability value, Al-Bait Hotel, Sharjah Art Foundation (SAF).

1 INTRODUCTION

Heritage buildings are considered as an integral part of a nation's identity, tradition and culture. It is crucial to preserve such buildings in order to maintain local identity and culture. Unfortunately, governments tend to be absent-minded and fail to invest in them to have a fruitful source of outcome for the country. For example, the Antoniadis Palace in Alexandria, Egypt was deteriorated due to lack of maintenance and repair [1]. In fact, the biggest threat governments pose to heritage and ancient buildings is in deciding to demolish them. The obsolescence of heritage buildings is economically unsustainable [2] and contrary to achieving sustainability goals generally [3]. One of the most implemented ways to ensure the maintaining of historic buildings is to adapt them to modern use.

Adaptive reuse refers to the process of reconstructing an existing building for a different purpose other than the original purpose that was built for [4]. This process is applied in different facilities such as museums, mosques, residential buildings and offices [5]. Adaption of existing old buildings results in many tangible and intangible benefits to the public and government and also in terms of environmental sustainability, through the reuse of existing buildings and increasing the building's life expectancy. Adaptive reuse is an eco-friendly solution to built environment needs through reducing material, transport and energy consumption and lowering pollution associated with new building projects [3]. Developers can save cost and time when investing in the reconstruction of old buildings rather than new construction projects [4]. Accordingly, reconstruction sites expose neighbours to less noise pollution compared to typical construction sites. Zulkifli advocated that demolition cost is as high as to reach 10% of the total cost of new project construction [4]. One important intangible benefit is that younger generations will be able to visualize and respect their history and connect with their past, heritage and locality [6]. Bringing life back to dumbed buildings

WIT Transactions on The Built Environment, Vol 203, © 2021 WIT Press
www.witpress.com, ISSN 1743-3509 (on-line)
doi:10.2495/STR210201

create jobs for locals where rehabilitation and restoration skills are required. Although reconstruction prevents depopulation of urban areas in a country [2], eventually it may result in a high urban density [7].

Adaptive reuse is a complex process which similar materials and construction method should be used. In Australia, architects, developers and building managers were interviewed by Bullen and Love [8] where results revealed about 85% of participants stated that the viability of recycling existing materials is one of the critical barriers to implement the adaptive reuse. Because buildings are old and neglected, workers may have to strictly follow safety and health requirements. The assessment of an old building's plan and its operation is a time-consuming exercise [7].

Adaptive reuse projects can be described by a wide range of terminologies such as retrofitting, refurbishment, rehabilitation. The wide diversity of terminologies is due to the type and scale of building, existing conditions and the construction activities conducted during these projects. In other words, authors in literature used the terminologies based on the status of the building and what is necessarily needed for reconstruction. No standards to select a terminology for each project, each case was dealt with differently. Authors have justified their terminology selection based on their own perceptions and reviewed literature. For example, Ebbert [9] used the term refurbishment to indicate the replacements of office facades and Ishak et al. [10] used refurbishment to accommodate modern building services and energy saving techniques. However, Bhuiyan et al. [11] have used refurbishment and rehabilitation interchangeably as the reviewed literature showed. Therefore, in order to set minds out, this study used the proposed framework by Shahi et al. [2] which illustrates the clear definition of each terminology based on the literature reviewed. The study has eliminated the terminologies "revitalization" and "modernization" due to their less common use in the literature. Fig. 1 summarizes the most common terminologies based on Shahi's literature review.

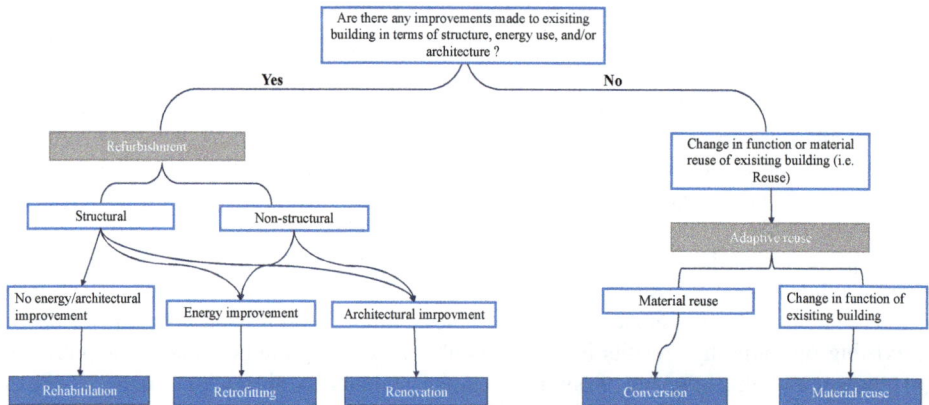

Figure 1: Refurbishment and adaptive reuse definitions framework. *(Source: Authors.)*

The authors categorized the adaptation of buildings to two categories: Refurbishment and Adaptive use. The term refurbishment is considered as an umbrella to retrofitting, renovation and rehabilitation. Simply, applying any additional structural or non-structural improvements to an existing building falls under the term "Refurbishment." Besides, any changes to the structural plan of a building falls under the adaptive reuse category. The adaptive reuse term also includes removal or change in a building's materials. Using the discussed framework, this study will identify which terms best describe the following two case studies in the United Arab Emirate: The Chedi Al Bait hotel and the Sharjah Art Foundation (SAF) in Sharjah.

2 ADAPTIVE REUSE VALUE

In this modern world, the adaptation of old buildings has increased in trend. Mısırlısoy and Günçe [6] believe that adaptive reuse calls for architectural conservation which ensures the significance of economic, social and cultural sustainability to the urban communities. Similarly, Elsorady [12] states that heritage buildings are heritage symbols that should be preserved to retain a country's identity.

Adaptive reuse can transform heritage buildings to accessible and useful places as well as to benefit the area in a sustainable manner. The adaptive reuse of buildings in the Old City of Bethlehem in Palestine and Visby old city in Sweden are two cases that presented their positive contribution to build a sustainable environment [3]. In terms of social sustainability, a stronger connection between past and future generations have been created. For economic sustainability, more jobs are offered, and tourists are attracted to the cities. Regarding the environmental sustainability, embodied energy (i.e., energy consumed from the acquisition of natural resources to product delivery) was reduced.

Adaptive reuse is a complex process where decision makers have to consider many aspects in order to obtain a well-developed heritage building that follows the modern community. This is discussed in the study conducted by Mısırlısoy and Günçe [6]. They designed a holistic model for developing adaptive reuse strategies for heritage buildings. The model illustrates five steps that should be holistically considered to reach a decision of the new use for the heritage buildings. One of the important steps in the model is to analyse the existing condition of a building such as its physical character and heritage values. Following strategies and analysing the heritage building is crucial in order to make successful decisions. Several adaptive reuse projects reported various problems for example cracks in masonry walls and concrete slabs, roof erosion and roof leakage and effective solutions were given [13]. Sandbhor and Botre [14] highlight the importance of evaluating the building's structure as it provides a valuable source of knowledge regarding their weaknesses and the ways they respond to extreme conditions.

Correspondingly, Reiner [15] set out 13 elements that should be considered in heritage buildings for an outstanding adaptive reuse. Bin Zulkifli [4] used these elements to evaluate three case studies in three different countries namely the UK, Sarawak and Denmark. Elsorady [12] has identified her own indicators including physical characteristics, building function, public perception which were produced from the literature to evaluate the Alexandria National Museum in Egypt. The results showed that most of the indicators were achieved and sustained. However, Günçe and Mısırlısoy [16] assessed the success of adaptive reuse of buildings based on one factor: users experience. This shows that no standards are set to measure or determine the success of the adaptive reuse practices. The aim of this paper is to determine the sustainability potentials added value for heritage buildings adaptive reuse in Sharjah showing two different case studies with unique successful practice.

3 CASE STUDIES IN SHARJAH, UNITED ARAB EMIRATES

Sharjah is one of the seven emirates of the United Arab Emirates (UAE) and is considered the culture capital of the UAE. Sharjah enjoys a very rich heritage culture to date and it is well known for its strong commitment to art, culture and history [17] as Fig. 2 shows the old urban fabric of Sharjah. Most tourist places in Sharjah to visit are heritage and cultural. The emirate is also well-known for its dry climate. Constant efforts are made by the UAE government to preserve the historical identity of the UAE. The modern architecture in the UAE seems to hold the same values as the traditional architecture. This can be visualized through several project in the UAE such as the Abu Dhabi central market, Sharjah Central Souq, Al Bahr Towers. However, this paper will focus on the Al bait hotel and SAF in Sharjah briefing the history, design and building structure and nature of materials used.

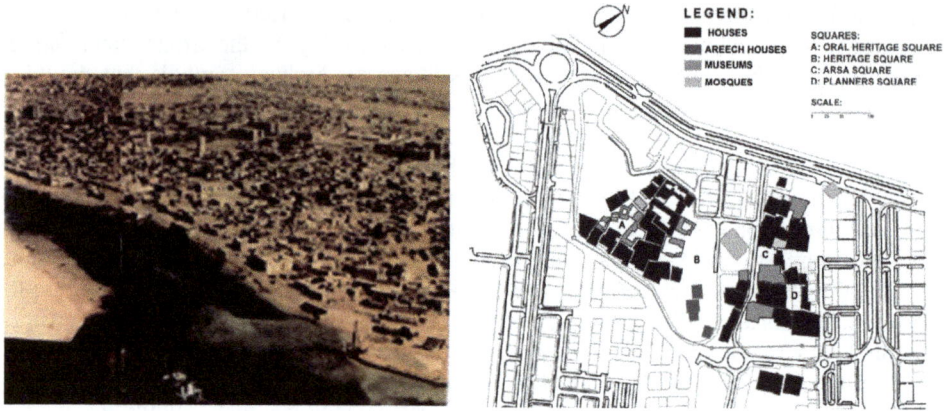

Figure 2: Heart of Sharjah Urban Fabric, with focus on Al Mirija historic area [18].

Figure 3: Sharjah urban fabric heritage area historic changes [17].

Sharjah's local architecture is a mix between the past and present. It is the place where old mid-rise and new high-rise constructions are observed in the same area [17] as shown in Fig. 3. Most of the old buildings are 50 years old and were built of coral stone walls and palm tree roofs [19]. The Marija area is a unique zone that accommodates all the old surviving constructions; located in the center of Sharjah city. In fact, it is the most historically and architecturally significant area in Sharjah city. In the late 19th, the Marija area construction situation suffered badly and as a result, the conservation process in Sharjah started with that area. Sharjah government recommended to preserve three buildings in the Marija area: Al-Midfa House, Al-Naboodah House and the Al-Mugsuba Tower due to their uniqueness and architectural significance. Indeed, the buildings were restored and conserved starting with Al-Naboodah House [18]. The success of these projects has motivated owners to handle other reconstruction and replication projects. Therefore, the area become an attraction spot for heritage tourism. In 1998, Sharjah's rulers decided not only to restore old buildings, but to replicate and rebuild as the Sharjah Art Foundation case. All the buildings were given a contemporary touch while maintaining the traditional architecture with the aim of increasing the cultural and social activities in the region. This paper will mainly focus on the Al bait hotel and SAF cases due to their high economic and social values. The aim is to determine the sustainability values of the heritage adaptive reuse for the buildings from different aspects; social, environmental and economic.

3.1 Sharjah Art Foundation (SAF)

The Sharjah Art foundation (SAF) is the producer and leader of contemporary art within the Emirate of Sharjah and the Middle East region. The foundation consists of a series of existing old historic buildings and five new buildings to match the old series. It has accessible areas and various facilities such as cafes, gallery spaces, and courtyards for concerts. This project took three years and in 2013, doors were open for visitors. It is close to the intersection between Al Merija Street and Corniche Street as well as near to Al Zahra mosque (Fig. 4). The SAF is a legal public body funded by the government and is a non-profit international organization [20]. The foundation's commitment is to sustain Sharjah's heritage through exhibitions, performances and programs in Sharjah and across the UAE that are mainly hosted in historical buildings that are now reused as educational centres. One of the core events is the Sharjah Biennial; a global event where contemporary artists attend and participate from around the world.

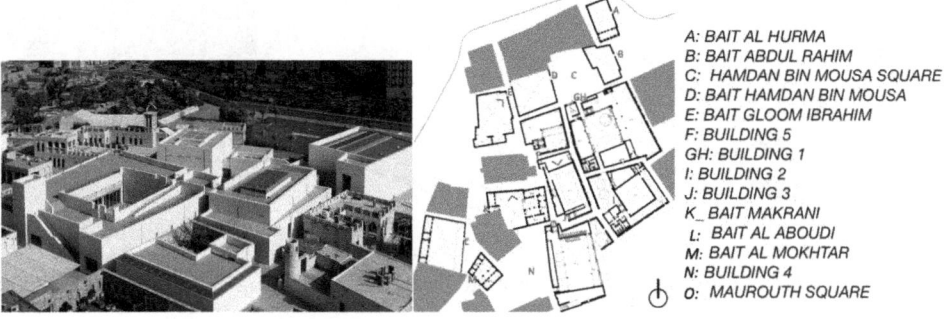

A: BAIT AL HURMA
B: BAIT ABDUL RAHIM
C: HAMDAN BIN MOUSA SQUARE
D: BAIT HAMDAN BIN MOUSA
E: BAIT GLOOM IBRAHIM
F: BUILDING 5
GH: BUILDING 1
I: BUILDING 2
J: BUILDING 3
K_ BAIT MAKRANI
L: BAIT AL ABOUDI
M: BAIT AL MOKHTAR
N: BUILDING 4
O: MAUROUTH SQUARE

Figure 4: Sharjah Art Foundation (SAF) ground floor plan and space functions [17].

As the Sharjah Biennial event started to attract attention, more spaces were needed to facilitate the installations and display of contemporary art. Therefore, five new buildings were needed to fit in the heritage area and to match the old heritage buildings. The white-cube-like buildings (Fig. 4) followed the street or circulation pattern that must have existed back in time. Each building's scale varies where larger buildings are built at the center and shorter ones are at the edges of the foundation. They are designed simply with no curves or sloping planes. All building used concrete structure except one that used steel structure. Concrete pavements, white plastered/painted walls and clear insulated glass are the materials used throughout the project.

3.1.1 Environmental sustainability value
The value to the environment is a crucial part of heritage as it can significantly affect the visitors experience and it improves attractiveness to the heritage-reused place. The environmental value is closely associated with the natural landscape and green spaces. In the

year 2020, Tu [21] study stated the environmental value as the most important factor when assessing heritage buildings for adaptive reuse feasibility.

Back in the old days, courtyards were an important feature in all houses. The SAF have designed the new five buildings to be surrounded by courtyards, as shown in Fig. 5. Such outdoor spaces create natural ventilation and as a result, visitors can smell a clean air breeze. Similarly, the narrow circulations between the modern and heritage buildings are designed to create shaded areas. In fact, the shaded areas act as a cooling technique for the pedestrians due to the hot and dry weather in Sharjah.

Plants, greenspaces, parks and gardens are important natural elements to give life to the heritage site. However, most of the SAF is paved, with little landscape. Thus, a low chance of plants could survive due to the dry-humid weather in the UAE. Close to the entrance of the foundation, a parking lot has been transformed to a landscaped pedestrian space to create a welcoming space for visitors. One of the courtyards have relatively denser trees than other courtyards since it may be exposed to a better weather condition. Lighting is another factor that has been taken care of as windows were adjusted using blinds and applying films to achieve the desired levels of natural light.

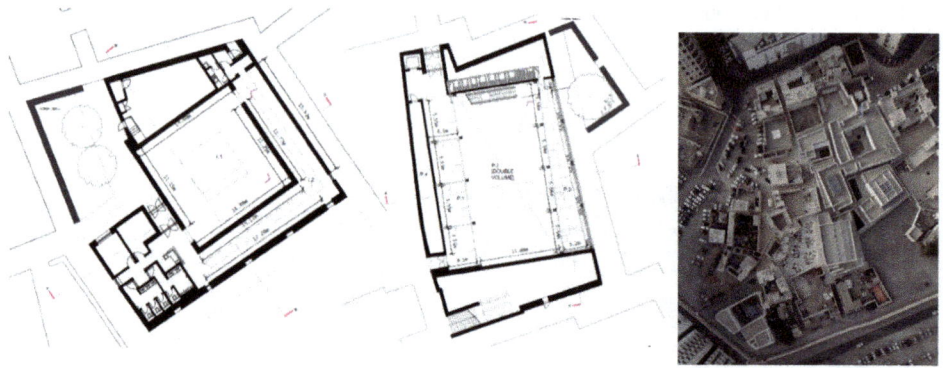

Figure 5: SAF exhibition buildings ground floor plans showing the harmony with the heritage urban fabric [17].

The traditional compact urban fabric is used in the very narrow spaces to provide shaded aisles for pedestrians. Also, the irregular geometry of streets allows wind circulation bringing cool breezes (Fig. 6).

Figure 6: SAF narrow streets and irregular spaces allowing wind circulation [17].

3.1.2 Economic sustainability value

Several adaptive reuse projects have failed since they do not meet the requirements and needs of the neighbourhood and consequently, no income was generated. A high economic efficiency is only achieved when economic and non-economic aspects (including environmental and social aspects) are considered [21]. Indeed, SAF has attracted visitors locally and globally which is economically beneficial, positively affecting the city's economic development. The SAF project won the Aga Khan award for architecture in 2019. This has definitely served the local government economic situation. Accordingly, investors and developers were interested to be involved in adaptive reuse projects.

The architects and developers tried to maximize the efficiency of the material used, preserving 40% of the urban fabric. The standard construction methods were used for all the new buildings. The project relied on local labour and no specialized labour. To make the foundation accessible, air conditioning system, mechanical, electrical and plumbing equipment were installed. Also, the UAE's first nitrogen-based fire-suppression system was installed. Despite the fact that such systems may be expensive, but they serve the visitor's safety and enjoyment.

Figure 7: Sharjah Art Biennial, the International hub for artists [17].

3.1.3 Social sustainability value

The adaptive heritage reuse raises the social values the most. A harmonic relationship occurs between visitors and the environmental features discussed earlier. Courtyards allow social gatherings and meetings as well as to provide some stress relief and tranquillity for visitors. The natural landscape and green spaces evoke tranquillity and an interesting experience to talk about to friends. Bull and Al-Thani [22] concluded that leisure time for gulf countries nationals involves the family gatherings and rarely going to cultural institutions. Therefore, courtyards, cafes and green spaces are facilities that will bring in positive social responses to the project. In fact, the combination of displaying the modern art with the restoration of the cultural heritage as well as the different social activities afforded has built a relationship with national, resident and non-resident visitors from different ages. This relationship helps SAF increases visitors' knowledge about art and artists within the community, thus attracting more visitors to exhibitions and their other programs.

The cultural value is not to be overlooked. Visitors are exposed to an interesting experience as well widening their knowledge. The heritage site shapes the cultural value. For example, stone coral, a traditional building material for walls is used in the circulation areas and appears like masonry walls. The combination of the modern and traditional material will bring the attention of the occupants and give them the opportunity to visualize the history of the city. For international visitors, they will have the chance to know the history of the community and feel the sense of belonging.

Figure 8:   Art exhibitions encouraging local and international community to be introduced to local heritage [17].

3.2 The Chedi Al Bait hotel

The Chedi Al Bait Sharjah hotel is a five-star hotel that is located in the heart of Sharjah. The term 'Al Bait' is referred to as 'the home' which is considered as a unique place that brings the UAE history back to life. The luxury hotel is 10,000 m^2 with 53 room resorts that is managed by General Hotel Management (GHM) in partnership with the Sharjah Investment and Development Authority (Shurooq). It is located in Sharjah (20 minutes ride from the city of Dubai) adjacent to Corniche Road running along Sharjah Creek and Al Hisn Street. The hotel features events and business facilities where resorts or private rooms are booked for events [23].

 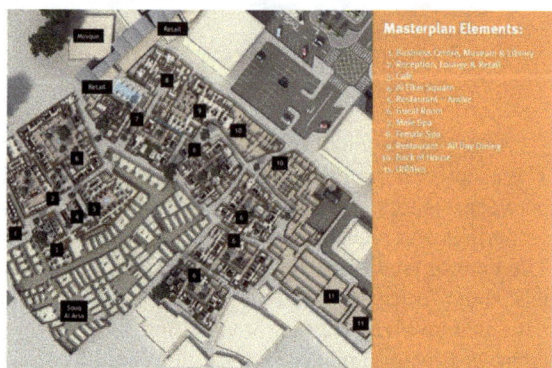

Figure 9: Al Bait Hotel master plan complex of old heritage buildings [24].

The preservation of the hotel is one of the region's biggest restoration projects that has not ended yet and is expected to end by 2025. However, the hotel was opened and available for booking in 2018. Al Bait Sharjah guests experience the traditional souks in the area, including Souk Al Shanasiyah, one of the most ancient markets in the Gulf region [25]. Guests are tasting the history of UAE mixed with the modern touch.

Figure 10: Unique architectural elements showing the original building before and after the adaptive reuse [18], [23].

Previously, Al bait was named as "Al-Midfa house" based on Al-Midfa family who are considered as one of the major merchant families in Sharjah. Back then in the early 1920s, it was designed and used as the main 'majlis' or meeting room for the men of the Al-Midfa family and as a rest house [18]. Mangrove poles supported the roof and mats of palm leaves were used as insulators. In the summer, Al-Midfa family would sleep in the courtyards under trees for a better quality sleep [18]. Unfortunately, it was neglected for many years and as a result saline erosion has occurred except in two elements: the wind tower and a fortified residential structure, as shown in Fig. 12. The rounded barjeel, or wind tower was a unique symbol of UAE's culture that was used as a natural ventilation during summer seasons. In 2012, Dr. Sheikh Sultan Bin Mohammad Al Qasimi announced the initiative of reviving the UAE's heritage and in 2013, the restoration of the Al-Bait hotel was started [26].

Figure 11: The original traditional house "Al Midfa house" before the adaptive reuse [18].

3.2.1 Environmental sustainability value

To significantly impact the visitors experience, Al bait hotel developers have restored several features. A significant architectural development to the courtyards is observed in Fig. 13. Trees are also planted to decorate the courtyard and allow visitors to enjoy the experience. Also, the round wind tower, a traditional method of ventilation, is conserved and guests visit the café to discover the wind tower. Natural lightning is ensured to occur in the rooms of the hotel during the day.

Figure 12: Cross section drawing showing the central courtyard space [27].

3.2.2 Economic sustainability value

The main aim of Al bait hotel is to have a unique tourism icon and to draw the attention of tourists and residents. Indeed, this is the case of Al Bait hotel. Economically, the developers of the project have used local materials such as traditional clay flooring, masculine leather and timber ceilings, shutters and doors. They have also selected traditional Emirati fabrics, patterns and ornaments [28]. This contribution increases the local market economy and local employment.

Figure 13: Interior views for Al Bait hotel showing the traditional treatment in luxury hotel [23].

Based on the visitors reviews, a travel agency for booking hotels, tourists come from everywhere in the world, including France, Germany, Saudi Arabia, United Kingdom, etc. The variety of nationalities and the good reviews for the hotel allow visitors to convey the cultural heritage to others, which improves the economic development of the city.

3.2.3 Social sustainability values

Based on many touristic booking sites, it is noticed Al Bait hotel is exceptionally rated 9.6 out of 10. All reviews reported by visitors were delightful, satisfied and enjoyed a comfortable tranquil stay in a busy city. The library facility available in the hotel has given a chance to the guests to induce knowledge about the history of the heritage site and know more about the history of the UAE. Gathering places in Al Bait hotel were important for guests to communicate with the heritage traditional touch.

Al Bait hotel experience is not limited for tourists from different nationalities, but also it gave locals the curiosity to go through a unique experience of their ancient's traditional architecture and the local social life, habits and traditions that the owner represented in the UAE government were very keen to present along with the luxurious services.

4 CONCLUSION

One of the aims of the paper is to identify the terminologies that best describes the two case studies discussed earlier. Fig. 1 is used to identify the SAF and Al Bait hotel terminologies. Regarding Al Bait hotel, it includes material reuse, change in function of existing building, architectural and structural improvements. Therefore, several terminologies can be used to describe Al bait hotel experience, such as adaptive reuse: conversion or material reuse, or refurbishment: retrofitting or renovation. For the SAF case, it includes material reuse, no change in function of existing building, architectural and structural improvements. It can be described as adaptive reuse: material reuse, or refurbishment: retrofitting or renovation.

The discussion of the two case studies within assessing sustainability aspects potentials proved that adaptive reuse for the heritage building has clear sustainability potentials. When explored the environmental sustainability values going through the architecture heritage elements and treatments adopting passive design strategies and vernacular architecture elements to deal with the hot dry climate, using the courtyards and wind towers for natural ventilation and natural lighting, and the compact urban fabric that provide shaded spaces and narrow streets allow wind circulation. Along with the economic aspects, that encouraged UAE government to support such adaptive reuse projects, where SAF become one of the most international artists' hubs in the world, and the social sustainability potential was clear when spreading the Emirate culture worldwide through the international Sharjah Biennale or Al Bait hotel experience, that provide the community with a local experience.

REFERENCES

[1] Elsorady, D.A., Adaptive reuse decision making of a heritage building antoniadis palace, *Egypt. International Journal of Architectural Heritage*, pp. 658–677, 2018.

[2] Shahi, S., Esnaashary Esfahani, M., Bachmann, C. & Haas, C., A definition framework for building adaptation projects. *Sustainable Cities and Society*, **63**, 2020.

[3] Ijla, A. & Broström, T., The sustainable viability of adaptive reuse of historic buildings: The experiences of two world heritage old cities; Bethlehem in Palestine and Visby in Sweden. *International Invention Journal of Arts and Social Science*, **2**(4), pp. 52–66, 2015.

[4] Bin Zulkifli, M.A.A., Design Principles of Adaptive Reuse: Case Studies on Dockyard, University Teknologi Malaysia, 2019.

[5] Alpler, Z.B., Şahin, N.P. & Dağlı, U.U., A critical discussion of industrial heritage buildings adaptive re-use as film spaces, case study: Industrial heritage buildings at Istanbul. *Journal of Architectural Conservation*, **26**(3), pp. 215–234, 2020.

[6] Mısırlısoy, D. & Günçe, K., Adaptive reuse strategies for heritage buildings: A holistic approach. *Sustainable Cities and Society*, **26**, pp. 91–98, 2016.

[7] Bullen, P. & Love P., Factors influencing the adaptive re-use of buildings. *Journal of Engineering, Design and Technology*, **9**(1), pp. 32–46, 2011.

[8] Bullen, P.A. & Love, P.E.D., Adaptive reuse of heritage buildings: Sustaining an icon or eyesore. *RICS Construction and Property Conference*, pp. 1652–1662, 2011.

[9]   Ebbert, T., Refurbishment Strategies for the Technical Improvement of Office Facades, RWTH Aachen University of Technology, 2010.
[10]   Ishak, N., Ibrahim, F.A. & Azizan, M.A., Analysis of factors influencing building refurbishment project performance. *E3S Web of Conferences*, **34**, 2018.
[11]   Bhuiyan, N.W., Islam S. & Jones K., An approach to sustainable refurbishment of existing building. *31st Annual ARCOM Conference*, pp. 1093–1102, 2015.
[12]   Elsorady, D.A., Assessment of the compatibility of new uses for heritage buildings: The example of Alexandria National Museum, Alexandria, Egypt. *Journal of Cultural Heritage*, **15**(5), pp. 511–521, 2014. DOI: 10.1016/j.culher.2013.10.011.
[13]   Subramaniam, S.R., A review on repair and rehabilitation of heritage buildings. *International Research Journal of Engineering and Technology*, **3**(4), pp. 1330–1336, 2016.
[14]   Sandbhor, S. & Botre, R., A systematic approach towards restoration of heritage buildings – A case study. *International Journal of Research in Engineering and Technology*, **2**(3), pp. 229–238, 2013.
[15]   Reiner, L.E., *How to Recycle Buildings*, 1979.
[16]   Günçe, K. & Mısırlısoy, D., Assessment of adaptive reuse practices through user experiences: Traditional houses in the Walled City of Nicosia. *Sustainability*, **11**, p. 540, 2019.
[17]   Dada, R.A., Aga Khan for architecture report: Al-Mureijah art spaces. Aga Khan, 2019. https://www.akdn.org/architecture/project/al-mureijah-art-spaces. Accessed on: 15 Feb. 2021.
[18]   Anderson, G., *The Urban Conservative Dilemma*, Durham University: Sharjah, UAE, 1991.
[19]   Hadjri, K. & Boussaa, D., Architectural and urban heritage conservation in the United Arab Emirates. *Open House International*, **32**(3), pp. 12–26, 2007.
[20]   Universes in Universe, Al Mureijah Art Spaces. https://universes.art/en/art-destinations/sharjah/art-spaces/saf-art-spaces#c65424. Accessed on: 25 Feb. 2021.
[21]   Tu, H., The attractiveness of adaptive heritage reuse : A theoretical framework. *Sustainability*, **12**, 2020.
[22]   Bull, J. & Al-Thani, S.H., Six things we didn't know: Researching the needs of family audiences in Qatar, 2013.
[23]   UAE, Ghmhotels, The Chedi Al Bait, Sharjah, 2020. https://www.ghmhotels.com/en/the-chedi-al-bait-sharjah/.
[24]   ProTenders, Al Bait Hotel Renovation, 2019. https://www.protenders.com/projects/al-bait-hotel-renovation. Accessed on: 24 Feb. 2021.
[25]   Kumar, K., Al Bait Sharjah: Combining authentic Emirati heritage with modernity. *The Arab Weekly*, 2018. https://thearabweekly.com/al-bait-sharjah-combining-authentic-emirati-heritage-modernity. Accessed on: 25 Feb. 2021.
[26]   Waqas, M., Traditional 5-star hotel to open in the heart of Sharjah. *Arabian Gazette*, 2013. https://arabiangazette.com/traditional-5-star-hotel-to-open-in-the-heart-of-sharjah-20130316/. Accessed on: 25 Feb. 2021.
[27]   Shurooq, The Chedi Al Bait, 2021. https://shurooq.gov.ae/project/the_chedi_albait/. Accessed on: 24 Feb. 2021.
[28]   Abdel-Razzaq, J., Sharjah's Al Bait Hotel creates an old village experience. *Architectural Digest*, 2019. https://www.admiddleeast.com/sharjahs-al-bait-hotel-offers-the-experience-of-an-old-village-retreat. Accessed on: 24 Feb. 2020.

HERITAGE SITES: THE PROBLEM OF ECONOMIC, SOCIAL AND CULTURAL VALUATION

RONI DEMIRBAG[1], VIVIENNE SAVERIMUTTU[1], YIYANG LIU[1] & SAJAN CYRIL[2]
[1]School of Business, Western Sydney University, Australia
[2]Australian Institute of Higher Education, Australia

ABSTRACT

The flooding of Hasankeyf raises an important problem of economic valuation of heritage sites. Hasankeyf, located on the Tigris river in South-eastern Turkey, prior to its flooding by the Ilisu dam project, was considered to be one of the oldest continuously inhabited human settlements that goes back 12,000 years. The Turkish government justified the construction of the dam by emphasising its economic benefits despite strong objections from other stakeholders who were concerned about the impending loss of cultural and social heritage. The paper highlights the historical and cultural significance of Hasankeyf, and the impact of the dam from the perspective of the Government and other stakeholders. The economic benefits of potential infrastructure developments often outweigh the benefits of the cultural and social value of heritage. Valuing heritage sites is inherently subjective and difficult to quantify. The focus of this paper is to determine a rigorous method of quantifying the economic value of cultural and social heritage. Treating cultural heritage sites as public goods the application of three methods of valuation is assessed with reference to Hasankeyf: The Coase theorem, the travel cost method and the contingent valuation method. These methods are examined in terms of practical application, socio-economic level of population and political will. Unlike the Coase theorem, the travel cost and the contingent valuation methods could have applied in Hasankeyf to assess the economic value of its cultural loss with support from the international community. However, the low socio-economic level of the local population and lack of political will would have proved counterproductive. Nevertheless, where the political will exists, these methods are practical and proven to be rigorous in their application. Thereby transforming cultural and social heritage into economic value that could endure in the future.

Keywords: Hasankeyf, loss of heritage sites, Ilisu dam, valuation of social and cultural heritage.

1 INTRODUCTION

In 1972, at the UNESCO World Heritage "Convention Concerning the Protection of the World Cultural and Natural Heritage", works of cultural heritage were categorised as monuments, groups of buildings and sites that are works of man or combined works of man and nature 'that are outstanding and of universal value' from various points of view. Natural heritage was considered as natural features consisting of physical and biological formations, geological and physiographical formations and precisely delineated areas which constitute the habitat of threatened species, natural sites or precisely delineated natural areas of outstanding universal value from the point of view of science, conservation or natural beauty [1]. The destruction of natural and cultural heritage sites can occur due to decay and natural disasters but also due to the actions of Governments facing development pressures as a result of population growth or armed conflicts.

In the face of such destruction, Organisations such as UNESCO, the International Union for Conservation of Nature (IUCN) and many other non-governmental organisations (NGOs) 'seek to preserve a balance in this interaction between man and nature' by actively sponsoring the preservation of these natural and cultural heritage sites for future generations by explicitly recognising their importance as 'legacies from the past' and 'irreplaceable sources of life inspiration' [1]. In promoting a universal attitude of 'stewardship' [2] these organisations are implicitly fostering the building of 'social capital', widely recognised as the underlying

WIT Transactions on The Built Environment, Vol 203, © 2021 WIT Press
www.witpress.com, ISSN 1743-3509 (on-line)
doi:10.2495/STR210211

source of differences in levels of socio-economic development leading to a higher quality of life [3] and democratic governance that strive to achieve justice and stability [4].

Hasankeyf (Fig. 1), was a unique historical site of cultural and natural heritage in the Republic of Turkey. The construction of the Ilisu dam, due to its capacity for energy production, was promoted by the Government and led to the inundation of Hasankeyf and surrounding archaeological sites. While the local population and international communities bemoaned the resulting loss of the cultural and social heritage, its economic benefits were advocated by the Government. The Government reasoned that the building of the dam would result in significant benefits to the country including the elevation of the socio-economic status of the population of Hasankeyf.

Figure 1: Hasankeyf, south east Turkey.

The focus of this paper is to assess methods that are able to quantify the future benefits that can ensue from cultural and social heritage such that their economic value can be taken into account when threatened by large development projects. The Methodology involves a review of the applicability of three different methods of evaluation of cultural and social heritage sites with respect to Hasankeyf. The assumptions and conditions under which each method is applicable, and whether or not they may be effectively and practically employed for heritage sites such as Hasankeyf will be critically examined. Identifying what could have worked for Hasankeyf will enable us to apply it more effectively in assessing the economic value of cultural heritage which are likely to be lost due to developmental projects in the future.

Section 2 highlights the uniqueness and importance of Hasankeyf as a historical site of cultural and natural heritage. Section 3 details in brief the origins of the Ilisu dam project, its impact during the life of the project, the Government's position on the economic and environmental benefits of the project, and the perspectives of other stakeholders who wish to prevent the loss of heritage due to the project. Section 4, is a pictorial abstract of the flooding of Hasankeyf with highlights of its impact on its people and culture. Section 5 while examining counter measures in an attempt to save Hasankeyf, explores the feasibility of different methods of valuing loss of cultural heritage and the conditions under which they are applicable. The paper concludes by highlighting important insights for future application in cases where development projects threaten significant cultural and social heritage sites.

WIT Transactions on The Built Environment, Vol 203, © 2021 WIT Press
www.witpress.com, ISSN 1743-3509 (on-line)

2 HASANKEYF

Hasankeyf is located in the Southeast region of Turkey near the Batman Province. It is one of the oldest continually occupied settlements in the world with an estimated history dating back approximately 12,000 years [5]. A medieval settlement, located on approximately 660 ha, Hasankeyf lies along the Tigris River, which extends all the way down to the Arab Gulf [6]. The ruins of the 4-arch bridge (old Hasankeyf bridge) built by the Artuqids stood (until recently) on the River Tigris. Man-made cave dwellings, archaeological masterpieces from Hasankeyf, date back to the Neolithic period [5].

There are three main reasons why Hasankeyf is a globally important heritage site. Having such a long history provides an invaluable research opportunity for archaeological excavation and study in order to further our collective understanding of the early stages of human civilisation. Various studies have been conducted that shed important light on the early transition from the Neolithic period [7]–[10].

Secondly, being located on major trade routes had bestowed Hasankeyf with geostrategic importance for various Empires and States during its very long history. Being part of the Silk road, artefacts from various stages belonging to different cultures and civilisations made Hasankeyf a very important heritage site that could not be assigned solely to one single ethnic and religious group or culture [11]. The Roman, Byzantine, Seljuk and Ottoman empires have also influenced Hasankeyf's unique historical and cultural atmosphere, which is impossible to replicate [5].

Hasankeyf, as part of the broader Tigris basin, also performed a very important ecological role in maintaining biodiversity. For over 10,000 years, the Tigris River has been the mainstay to humans, plant, fish and reptile species and their habitats as a source of fresh water [6]. In Hasankeyf and its surroundings, 472 taxa belonging to 279 genera and 64 families were identified during a study. Many of the species of wild plants in the area are of significant cultural importance not only as edible food but also for economic use [12].

For economic growth to keep pace with population growth, Governments are often faced with controversial decisions, as they need to engage in large infrastructure projects to expand their production capacity to meet the growing needs of its citizens [13]. Such infrastructure projects pose a serious threat to archaeological sites, that could not only result in loss of cultural and natural heritage, but also have an impact on biodiversity and the environment through loss of habitat [12], [14]. The residents of Hasankeyf, have lost much of their cultural heritage as a result of the Ilisu dam project in South-eastern Turkey.

3 THE ILISU DAM PROJECT, TURKEY

The South-eastern Anatolia Project known as the GAP, is a mega project consisting of 22 dam projects and 19 hydropower schemes. This GAP project includes, the Ilisu dam on the Tigris river, the largest and last of its dam projects the Turkish Government embarked on to address its infrastructural needs [15]. Located in the South-eastern part of Turkey, the population of that area being mainly Kurdish, Ilisu is a small village along the Tigris. The initial attempt to build the dam in 1996 [16] suffered an early setback, during the environmental assessment stage. As a result of an environmental controversy, some members of the international consortium, and the Bank coordinating the financing withdrew from the Project [17]. After nearly 10 years of delay in getting the project off the ground [16] the construction of the Ilisu dam officially began in 2007 [18] with an estimated cost of $2 billion and expected to generate as much electricity as a small nuclear reactor [19]. The dam, with a storage capacity of 10.4 billion cubic meters (bcm), is for the purpose of hydroelectric power generation [20]. The dam commenced operation of its first turbine in 19 May 2020 and the second on 30 June 2020 [21].

From the perspective of the Turkish Government, the Ilisu dam project was important to address the country's dependence on energy imports and provide economic opportunities to the impoverished South-eastern region. With Turkey's demand for electricity and energy, second only to China, the Government argued that the country's dependence on imported coal and gas would decrease as the dam contributes to the economy, and irrigates local agricultural land. Further, in the context of Turkey's proximity to the large reserves of natural gas and oil reserves, it allows Turkey to strengthen its energy supply security in developing into a regional trade centre in energy [22].

Once completed, the dam would only have a functional life span of 30 to 50 years. The dam has generated much controversy due to its environmental impact and more pertinently the loss of natural and cultural heritage. Local residents were unhappy due to the loss of centuries old human settlements with its unique culture that was impossible to replicate [5]. The government's determination to build the dam was met with strong resistance from human rights and heritage activists as well [11]. To criticisms of environmental damage, the Government's response is that the Ilisu dam would result in major environmental benefits, as thermal power plants would reduce emissions of greenhouse gases from using coal and gas [22]. The Ilisu dam has also drawn protests from downstream co-riparians, Iraq and Syria, from time to time on the grounds that the Ilısu dam reduces the Tigris's water flow downstream [11]. Iraq's Mosul dam and the Ilisu dam share the same watershed, and the worst-case scenario assessment reveals that Ilisu's impact on Mosul could be as high as a 78% reduction in inflow [20]. From the perspective of Iraq and Syria, the Ilisu dam has given Turkey absolute control on the water flowing down from the river [11].

4 HASANKEYF LOST

Once the hydro-electric power plant began operations, Hasankeyf gradually submerged under water. Even though the Turkish Government was persistent in eventually building the dam, knowing that the historical and cultural heritage of Hasankeyf would be lost forever, they did take action in order to preserve some of the artefacts. Fig. 2 depicts the 500-year-old Zeynel Bey tomb in its original location, and Fig. 3 after its relocation.

Figure 2: 500-year-old Zeynel Bey tomb in its original location.

Figure 3: Zeynel Bey tomb after relocation.

Figure 4: Iconic Minaret of Hasankeyf in its original location.

Figure 5: Iconic Minaret of Hasankeyf after relocation.

Fig. 4 depicts the iconic Minaret of Hasankeyf in its original location and Fig. 5 after its relocation. When the Ilisu Dam was first approved in 1996 [11], archaeologists had been invited to start excavations in and around Hasankeyf, in order to assist in the preservation of artefacts before the flooding of the dam [23].

The flooding of Hasankeyf and surrounding areas resulted in the evacuation of approximately 200 villages and the resettlement of 78,000 people [18]. Pictured in Fig. 6 is the town of Hasankeyf when it was inhabited. The town has since been fully submerged.

Figure 6: The township of Hasankeyf before evacuation.

As part of resettlement program the Turkish government established a new Hasankeyf town, located 3 km from the original town. Zeynel Bey tomb, Eyyubi Mosque and Artuklu Hamam were relocated to the newly established town. Furthermore, the Government established a museum dedicated to the artefacts salvaged from various excavations in and around Hasankeyf [5]. Fig. 7 is the town of Hasankeyf, after evacuation for resettlement, while it was being submerged. In the background the new Hasankeyf town is visible just below the mountains.

Figure 7: Hasankeyf town while it was being submerged.

Many experts have pointed out that the museum did not actually salvage sufficient amounts of artefacts [24]. In fact, some artefacts could not be salvaged because they were too fragile to survive the relocation process [25]. With the exception of a few high positioned caves and artefacts the whole town, its ancient monuments, cave dwellings (Fig. 8) and churches remaining are completely submerged [5].

Figure 8: Cave dwellings before inundation.

5 VALUING CULTURAL AND SOCIAL HERITAGE

For the Kurdish people, who predominantly inhabited the reservoir area in Hasankeyf, prior to its inundation [16], this regional centre was their inheritance from the past, with emotional and spiritual links endowed by their cultural heritage [26]. Aykan argues, that the construction of the Ilisu Dam, which denied this group of people the right to participate in their cultural heritage, can be considered as a violation of their human rights [5]. However, as Hodder [27] states, the human rights argument seldom works as it tends to pit, humans against humans. In this case the argument would be pitting the Turkish Government, arguing for economic and social development for many groups, against the cultural heritage rights of one group.

In recognition of its historical relevance and unexplored archaeological sites, in 1978, the Ministry of Culture of the Republic of Turkey, had declared Hasankeyf a first-degree Conservation Area [28]. Some argued that Hasankeyf met at least nine of the ten requirements for World Heritage site status. Thus, being declared a UNESCO World Heritage site was another avenue for saving Hasankeyf. However, only a National Government could apply to UNESCO for World heritage status to be awarded to a site [5]. In 2013, based on scientific opinion a campaign was launched to collect signatures, which were presented to the Prime Minister of Turkey to support an application to UNESCO for world heritage status, but that was declined by the Prime Minister. By 2013, the Turkish Government had the necessary funding, after the initial delays, to continue work on the dam [29] to its completion. The arguments put forth by stakeholders, intent on preventing the loss of Hasankeyf, emphasising the aesthetic value of cultural and social heritage, the educational potential to future generations, though of immense value, are all subjective in nature and therefore proved ineffective against the tangible economic benefits of a development project.

WIT Transactions on The Built Environment, Vol 203, © 2021 WIT Press
www.witpress.com, ISSN 1743-3509 (on-line)

The construction of large infrastructure projects by governments, to meet the developmental needs of a country, are often controversial due to the loss of cultural and social heritage sites that are of value to its population. To counter this irretrievable loss, it is necessary to convince governments, using an economic argument that reflects the invaluable loss in future benefits. A cost benefit analysis based on the economic, and social benefits of the Ilisu dam, as proposed by the Turkish Government, could be performed easily from the perspective of the Government. But how does one represent cultural heritage in monetary terms? Cultural and social heritage, in some sites, is often intangible, and therefore valuation is subjective and difficult to ascertain in precise economic terms. To value the benefits of cultural and social heritage in economic terms' three models are explored in the context of Hasankeyf, and examined with respect to its ease of applicability, the socioeconomic conditions of the population in the threatened site and the prevailing 'political will' of the governing regime. According to Bedate et al. [30] Cultural heritage sites, share many aspects of public goods, although resources to maintain them are scarce. Similar to public goods cultural heritage sites provide benefits and externalities. Thus, all three models explore valuing cultural heritage as a public good [26], [30]. A public good, has no attached property rights assigned. Further, it has two key characteristics of non-excludability and non-rivalry and is also freely available. In other words, due to its key characteristics there will be no private provision of a public good, and will instead be provided by the Government.

Throsby [26] applied the Coase theorem to Australian cultural heritage in order to mimic a market solution and reach a voluntary negotiated outcome. The Coase theorem assumes that the stakeholders can be identified, property rights are well defined, transaction cost are negligible or zero and contracts are monitored and enforceable. As cultural heritage is analogous to a public good, it was concluded in the Australian case, that the assumptions of the Coase theorem could not be met, hence the solution was considered inapplicable. However, based on an extension of the Coase theorem being applied to Taiwan it is argued that when Governments are involved, they can set the market rules without creating a State monopoly and also the rules perform the function of property rights.

This particular case study is based on Taiwan which has an excellent record for post-colonial heritage conservation [31]. As in the Australian case, applying the Coase theorem to Hasankeyf is not possible due to lack of well-defined property rights. In addition, it cannot be assumed that there is political will for the Government to step in for conservation purposes as has been proven. Thus, it follows this model is inapplicable for Hasankeyf and other cultural sites in similar situations.

The travel cost method (TCM) focuses on the economic valuation of non-marketed resources and links travel costs faced by visitors to a cultural heritage site [32]. Two approaches may be applied. TCM is the longest established revealed preference approach [33]. The revealed preference method is based on observed behaviours and indirectly obtained data. In the stated preference method, individuals are asked to value their willingness to pay for a resource [34]. In this instance the methodology applied is contingent valuation, which is our third method of valuation, namely, the contingent valuation method (CVM). The TCM is based on costs incurred to visit a site including flights, accommodation and meals, and an estimation of the value of their time spent travelling from their homes to their accommodation while visiting and the time spent travelling from their accommodation to tour sites.

The two methods identified, to value Travel cost and cultural heritage using willingness to pay, are based on demand that has been created, and, in most cases with the assistance of the Government, such that people are willing to pay, to travel to the destination in the case of TCM, and, for non-priced benefits in the case of CVM. Both methods, in order to be applied, require already established demand for the cultural product or experience such that it is possible to conduct surveys to estimate the economic benefit of consuming these non – market goods [35].

The international community and NGOs were very supportive in lobbying and looking for the means of saving Hasankeyf. In applying these methods to Hasankeyf, data to be collected via questionnaires and surveys could have been administered with the assistance of the international community to estimate the demand for travel and the willingness to pay for a non-market resource such as the cave dwellings in Hasankeyf or Zeynel Bey's tomb. If sufficient interest in Hasankeyf could have been built up for tourism purposes, convincing the Government of its economic value as a future source of wealth generation may have been possible. Of course, where Governments recognise the value of cultural goods for its economic and social benefits, especially in terms of the economic benefit that could be garnered from tourism, preservation will be considered as a viable option. In this instance, methods such as the Travel Cost Method TCM [32] or the Contingent Valuation Method CVM [34] could be applied to assess the economic viability of preserving cultural heritage, provided there are no conflicting development priorities. However, the low socioeconomic status of the population would have meant the necessary investment would have been lacking unless it came from the Government. The question remains whether the political will of the government could have been influenced in its favour.

In conclusion, despite the fact that cultural heritage is widely recognised by Governments, academics and public authorities as essential in building social capital [3], it is often sidestepped when faced with the need for infrastructure development to meet the demands of growing populations. Among the many infrastructure projects, dams in particular are controversial. Across the Middle East and North Africa, Marchetti et al. [36] have identified approximately 1,300 km of ancient rivers and 2,500 archaeological sites that have been submerged in the process of dam construction.

Hasankeyf, a unique site of immense cultural and social value has been inundated as a result of the Ilisu dam project. The Government justified its actions based on the demand for energy to meet the needs of a growing population. Despite calls from the international community and the local population the dam is now in operation. Hasankeyf, when it was declared a first-degree Conservation Area, in 1978, was not being recognised for its potential for tourism. Rather the emphasis was more on its cultural worth for posterity, its research potential due to its past, its biodiversity, and its unique history. The undertaking in 2013, to seek UNESCO World heritage site recognition, was also a desperate attempt to save Hasankeyf for the same reasons. These are all valid reasons for preservation of the cultural and natural heritage of Hasankeyf. However, only an economic argument may be considered if at all when it comes to cultural heritage.

In the case of Hasankeyf, where the socio-economic status of the population is low attempting to quantify its cultural worth in economic terms would have been near impossible or possibly subjective. Under these circumstances trying to convince a national Government bent on economic development, building a dam of political significance to change its plans is impossible as has been proven. Despite the loss, some lessons have been learned. The importance of focussing on economic arguments to prevent the loss of heritage sites as a result of development projects is undeniable. There are proven rigorous methods such as the TCM and the application of CVM to cultural heritage that are easily applicable in quantifying the value of non-marketable resources. It is important to focus on the need for such valuations to prevent the destruction of cultural heritage sites.

ACKNOWLEDGEMENT
All photographs are courtesy of Seyda Goyan. We greatly appreciate his assistance.

REFERENCES

[1]   UNESCO World Heritage Centre, Convention concerning the protection of the world cultural and natural heritage. 1972. https://whc.unesco.org/archive/convention-en.pdf.

[2]   Liburd, J. & Becken, S., Values in nature conservation, tourism and UNESCO world heritage site stewardship. *Journal of Sustainable Tourism*, **25**(12), pp. 1719–1735, 2017.

[3]   Murzyn-Kupisz, M. & Działek, J., Cultural heritage in building and enhancing social capital. *Journal of Cultural Heritage Management and Sustainable Development*, 2013.

[4]   Putnam, R.D., *Bowling Alone: The Collapse and Revival of American Community*, Simon and Schuster, 2000.

[5]   Aykan, B., Saving Hasankeyf: Limits and possibilities of international human rights law. *International Journal of Cultural Property*, **25**(1), 2018.

[6]   Röhr, C., *Archaeology in 21 ST c: Case of Hasankef*, Turkey.

[7]   Carter, T., Moir, R., Wong, T., Campeau, K., Miyake, Y. & Maeda, O., Hunter-fisher-gatherer river transportation: Insights from sourcing the obsidian of Hasankeyf Höyük, a Pre-Pottery Neolithic A village on the Upper Tigris (SE Turkey). *Quaternary International*, **574**, pp. 27–42, 2021.

[8]   Maeda, O., Lithic analysis and the transition to the Neolithic in the Upper Tigris Valley: Recent excavations at Hasankeyf Höyük. *Antiquity*, **92**(361), pp. 56–73, 2018.

[9]   Tatsumi, Y., A Neolithic sedentary hunter–gatherer settlement with densely arranged buildings: Results of geophysical prospection at Hasankeyf Höyük in south-eastern Anatolia. *Archaeological Prospection*, **27**(4), pp. 329–342, 2020.

[10]   Karul, N., Gusir Höyük. *The Neolithic in Turkey*, **1**, pp. 1–7, 2011.

[11]   Warner, J., The struggle over Turkey's Ilısu Dam: Domestic and international security linkages. *International Environmental Agreements: Politics, Law and Economics*, **12**(3), pp. 231–250, 2012.

[12]   Yeşil, Y. & İnal, İ., Traditional knowledge of wild edible plants in Hasankeyf (Batman Province, Turkey). *Acta Societatis Botanicorum Poloniae*, **88**(3), p. 3633, 2019.

[13]   Ministry of Foreign Affairs, Turkey's energy strategy. https://www.mfa.gov.tr/turkeys-energy-strategy.en.mfa. Accessed on: 15 Apr. 2021.

[14]   Machat, C., Ziesemer, J. & Petzet, M., *Heritage at Risk: ICOMOS World Report 2006–2007 on Monuments and Sites in Danger*, Hendrik Bäßler Verlag, 2008.

[15]   Balat, M., Southeastern Anatolia Project (GAP) of Turkey and regional development applications. *Energy Exploration & Exploitation*, **21**(5), pp. 391–404, 2003.

[16]   Eberlein, C., Drillisch, H., Ayboga, E. & Wenidoppler, T., The Ilisu dam in Turkey and the role of the export credit agencies and NGO networks. *Water Alternatives*, **3**(2), p. 291, 2010.

[17]   Égré, D. & Senécal, P., Social impact assessments of large dams throughout the world: Lessons learned over two decades. *Impact Assessment and Project Appraisal*, **21**(3), pp. 215–224, 2003.

[18]   Hommes, L., Boelens, R. & Maat, H., Contested hydrosocial territories and disputed water governance: Struggles and competing claims over the Ilisu Dam development in southeastern Turkey. *Geoforum*, **71**, pp. 9–20, 2016.

[19] Hockenos, P., Turkey's dam-building spree continues, at steep ecological cost, 2019. https://e360.yale.edu/features/turkeys-dam-building-spree-continues-at-steep-ecological-cost. Accessed on: 5 Apr. 2021.

[20] Al-Madhhachi, A.S., Rahi, K.A. & Leabi, W.K., Hydrological impact of Ilisu dam on Mosul dam; the river Tigris. *Geosciences*, **10**(4), p. 120, 2020.

[21] Bimay, M., Spatial, social and environmental effects of forced displacement due to dam construction: The case of Hasankeyf. *Econharran*, **5**(7), pp. 49–83.

[22] Ministry of Foreign Affairs, Turkey's energy strategy. https://www.mfa.gov.tr/turkeys-energy-strategy.en.mfa. Accessed on: 15 Apr. 2021.

[23] Kucukgocmen, A., Turkey starts filling huge Tigris river dam, activists say. https://www.reuters.com/article/us-turkey-dam-idUSKCN1US194. Accessed on: 20 Apr. 2021.

[24] Drazewska, B., Hasankeyf, the Ilisu Dam, and the existence of "Common European Standards" on cultural heritage protection. *Santander Art and Culture Law Review*, **4**(2), pp. 89–120, 2018.

[25] Topal, T. & Kaya, Y., Assessment of deterioration and collapse mechanisms of dolomitic limestone at Hasankeyf Antique City before and after reservoir impounding (Turkey). *Environmental Earth Sciences*, **75**(2), p. 131, 2016.

[26] Throsby, C.D., Paying for the past: Economics, cultural heritage, and public policy. *Australia's Economy in Its International Context*, p. 527, 2006.

[27] Hodder, I., Cultural heritage rights: From ownership and descent to justice and well-being. *Anthropological Quarterly*, **83**(4), pp. 861–882, 2010.

[28] Kocabas, H.M., Planning in fragile sites in Turkey: In case of Hasankeyf. PLANNING TIMES – You better keep planning or you get in deep water, for the cities they are a-changin. *Proceedings of 18th International Conference on Urban Planning, Regional Development and Information Society*, CORP – Competence Center of Urban and Regional Planning, pp. 691–700, 2013.

[29] Gourlay, W., Turkey's Hasankeyf: The plight of archaeological and architectural treasures in Southeast Anatolia. *TAASA Review*, **14**, 2013.

[30] Bedate, A., Herrero, L.C. & Sanz, J.Á., Economic valuation of the cultural heritage: Application to four case studies in Spain. *Journal of Cultural Heritage*, **5**(1), pp. 101–111, 2004.

[31] Go, V.K. & Lai, L.W., Learning from Taiwan's post-colonial heritage conservation. *Land Use Policy*, **84**, pp. 79–86, 2019.

[32] Torres-Ortega, S., Pérez-Álvarez, R., Díaz-Simal, P., Luis-Ruiz, D., Manuel, J. & Piña-García, F., Economic valuation of cultural heritage: Application of travel cost method to the National Museum and Research Center of Altamira. *Sustainability*, **10**(7), p. 2550, 2018.

[33] Blakemore, F. & Williams, A., British tourists' valuation of a Turkish beach using contingent valuation and travel cost methods. *Journal of Coastal Research*, **24**(6(246)), pp. 1469–1480, 2008.

[34] Majumdar, S., Deng, J., Zhang, Y. & Pierskalla, C., Using contingent valuation to estimate the willingness of tourists to pay for urban forests: A study in Savannah, Georgia. *Urban Forestry & Urban Greening*, **10**(4), pp. 275–280, 2011.

[35] Venkatachalam, L., The contingent valuation method: A review. *Environmental Impact Assessment Review*, **24**(1), pp. 89–124, 2004.

[36] Marchetti, N., Curci, A., Gatto, M.C., Nicolini, S., Mühl, S. & Zaina, F., A multi-scalar approach for assessing the impact of dams on the cultural heritage in the Middle East and North Africa. *Journal of Cultural Heritage*, **37**, pp. 17–28, 2019.

SECTION 6
INDUSTRIAL HERITAGE

URBAN REGENERATION OF INDUSTRIAL SITES: BETWEEN HERITAGE PRESERVATION AND GENTRIFICATION

RAFAELA SIMONATO CITRON
University of São Paulo, Brazil

ABSTRACT

This work is part of PhD research that addresses two recurrent problems in large Brazilian cities: the risk of demolition of important industrial buildings – due to several factors, such as the advanced state of degradation given the lack of use, the lack of recognition of this heritage in the country and the pressure of the real estate market, increasingly interested in the land these sites occupy – and the great demand for housing in central areas. The two themes – the preservation of industrial heritage and social housing in central areas – are rarely addressed together. Internationally, especially in the UK, the reuse of industrial heritage for residential use is quite common and has been going on since the first factories were closed with deindustrialisation and consequent industrial deconcentration, leading to the abandonment of several industrial sites in areas with complete urban infrastructure. Although successful in terms of preserving industrial heritage, since they enabled this heritage to be kept in the urban landscape, the adaptive reuse projects and the site's urban regeneration usually result in the gentrification of the regenerated area through projects carried out via a partnership between the public and the private sector that, even by offering a portion of onsite affordable housing, fail to serve the local community, let alone solve the country's housing problem. This article will show as a case study the Royal Arsenal district, in the south-east of the docks in London, with the aim of demonstrating how the urban regeneration, while preserving industrial heritage, divided the neighbourhood and contributes to its gentrification.

Keywords: urban regeneration, industrial heritage, gentrification, conservation, heritage.

1 INTRODUCTION

The United Kingdom, as well as being a pioneer in the industrialization process, was one of the first nations to suffer from deindustrialisation and the consequent industrial de-concentration. As Stratton [1] explains, the industrial sites demanded the proximity to the city centre, where the main services were, to the river, through which much of the freight transport took place and with the working-class dwellings, since most of the worker used to walk to work. The evolution of the means of transportation and, more recently, environmental concern has caused industries to be removed from the central areas.

In the United Kingdom, the reuse of industrial heritage is a well-established practice, in particular its adaptive reuse for housing purposes. With a growing demand for new housing, the City of London encourages the practice and, since 2017, industrial buildings (small and unlisted) can be converted to residential use without the need for planning permission, in a scheme called Permitted Development Rights. However, while the practice of reuse for residential purposes has enabled the conservation and valorisation of this heritage, on the other hand urban regeneration projects have resulted in elitist spaces, which end up gentrifying areas previously predominantly occupied by the working class. With several urban regeneration projects taking place at the same time, as in Battersea and Elephant and Castle, for example, London in going through a widespread gentrification. According to the government's official list, there were 246,575 families on the housing waiting list in the City of London in 2020.

WIT Transactions on The Built Environment, Vol 203, © 2021 WIT Press
www.witpress.com, ISSN 1743-3509 (on-line)
doi:10.2495/STR210221

In addition to social problems, the heritage issue itself is also questioned. Industrial heritage is seen by many authors as a different kind of heritage, or at least with some characteristics that differentiate them from other heritage buildings. One of the reasons is the proximity that this heritage still has to the communities that created them. As deindustrialization is still a recent event, many worker's families still live in the vicinity of the former industrial sites. The feeling of belonging of the community with these sites bring to the industrial heritage an intangible dimension that has been little explored and preserved in the urban regeneration projects of industrial sites. Cossons [2] talks about these values attributed to industrial heritage, which go far beyond the documentary and material value of the buildings themselves and argues that the preservation of this heritage from the arguments and values attributed to other heritage buildings is often not well received by the community. According to the author, industrial heritage brings challenges that are not found in other sectors of heritage, such as the need to recover socially and economically these areas that are affected by unemployment and economic collapse [2]. It is on top of the argument of economic and social recovery that urban regeneration projects arise. In Architectural Design's special edition on urban regeneration, Littlefield [3] explains:

> "The term regeneration is most usefully invoked as part of an attempt to reverse economic decline and social deprivation, a set of actions to restart the process of generating wealth and providing inhabitants with viable social and economic choices. Property development, or 'urban renewal', might well be the catalyst for these things, but there is always the danger that it becomes a placebo" [3]

When mentioning market-led urban regenerations, the author talks about the various transformations that took place in east London driven by the Olympics and quotes Susan Betty, who in 2003 said that the danger was not that these projects would cause the removals of former residents, but that they would simply ignore the communities [3]. This is the case of Woolwich and the regeneration of the Royal Arsenal, a military industrial site who's regeneration started in the 2000s and comprised of the reuse of the old warehouses for residential purposes, the construction of new residential buildings as well as new cultural uses, shops and a new public transport link, the Crossrail, which will connect Royal Arsenal to Canary Wharf and the City, London's two main financial districts. Because of this, the district has attracted and continues to attract middle-class young workers from other areas of London. Thus, today the District of Woolwich is divided into two very distinct parts: the regenerated part and the rest of Woolwich.

2   URBAN REGENERATION AND INDUSTRIAL HERITAGE IN THE UK

As already mentioned, the United Kingdom was one of the first nations to industrialise, to suffer from the consequences of deindustrialisation in the post-War period and, consequently, also pioneers in initiating the process of economic reconversion of abandoned industrial sites. The first interventions in these sites, in the post-war period, comprised of the demolition of various sets of work class housing, under the influence of the Garden City movement. Couch et al. [4] explains that these interventions became known in the United States and the United Kingdom as Urban Renewal, which can be understood as physical interventions for the re-development of the affected area, also known as slum clearance. Urban renewal occurred as the same time as post-war reconstruction until the 1960s, when it became clear that it was needed much more than just interventions in the physical environment, but instead it was necessary to socially and economically recover the old industrial areas [4].

The example came from the United States, which in the mid-1960s regenerated several former ports, such as the Port of Baltimore, which began its regeneration process in 1962 with a public investment of $180 million. Although the reuse and conservation of industrial heritage were also present, in the United States the construction of new buildings was predominant over adaptive reuse, as well as new uses focused on tourism, such as shopping malls and marinas. In other American ports, mixed-use projects – offices, housing for the middle class and leisure – became a formula that would later be exported to Europe, Africa and Asia [2].

With regards to industrial heritage, until the 1960s, even though some buildings had already been included in the list of buildings to be preserved, there was not yet a national policy for preserving industrial heritage. The destruction of the arch on the main façade of Euston station in London in 1963, is known as a symbol of the Industrial Revolution, causing great pressure from the public and led the country to conduct a national survey of sites to be protected – Industrial Monuments Survey (IMS) [5].

Between 1963 and 1981, more than 4,000 sites were inventoried and 2,325 were listed as sites worthy of preservation. With the creation of English Heritage in 1984, the Monument Protection Program was created, which evaluated 5,000 more sites and buildings, recommended more than 1,000 for the designation of Ancient Monument and 350 candidates were listed by the end of the program in 2004 [5]. Today, there are more than 30,000 industrial buildings of interest for preservation in England, 4% of which are listed as Grade I, the highest level of listing in the country.

Until the 1970s, most investments in urban regeneration projects came from the government [6]. The Urban Programme, created in 1968, distributed funds to municipalities to invest in social programmes, focusing on solving poverty problems in economically decaying areas. Over time, it became clear that the complexity and challenges of regenerating an area in social and economic decline required the participation of more actors in both the design and the delivery of demarcated areas. In the 1980s, with the arrival of Margaret Thatcher in the power in the United Kingdom and President Reagan in the United States, the state's welfare policies were heavily criticized, and state intervention came to be seen as a hindrance to economic development. Thus, in the 1980s there was a change in planning policies and, from then, urban regeneration projects were mainly led by the private sector [6]. This model remains present to this day and urban regeneration projects led by the housing market are called property-led regeneration in the UK.

These projects were made possible through the creation of Urban Development Corporations (UDC), entities chosen by the government that had the task of creating attractions to the private sector to bring investments to specific areas. From 1981 to 1993, the UDC regenerated 13 central areas in the United Kingdom, including the well-known London Docklands. The UDC received more government funding than the previous program, the Urban Programme, where there was no private sector participation [7]. The idea that market-led regeneration was the solution to the problems of urban degradation was also popular among municipalities in the late 1980s. Property-led regeneration became the formula to regenerate central areas without much justification or explanation and the relationship between interventions and the improvement of economic and social conditions of the communities where there was intervention were very little studied [7].

As pointed out by the author [7], urban regeneration projects led by the real estate market were not addressing the urban and social problems facing the country. The expected trickle-down did not happen and, by the power given to the UDC, communities and municipalities were not being heard, just as poverty and social problems were not on the agenda of urban regeneration programs [6].

Thus, from 1991 a new program is implemented in government, where municipalities could apply to receive government funds to direct the regeneration of degraded urban areas: The City Challenge. Specific municipalities were invited by the government to submit applications for the program, many of them involving former industrial areas in decline [8]. The regenerations that occurred as part of this program continued to take place from partnerships with the private sector but led by local government and with popular consultation. This programme can be considered the first British experience where there was integration between the interests of the private sector and local communities [6]. However, the success of the program is also questionable from the social point of view.

In 2000, urban regeneration projects focused again on the real estate market, when the government published the Planning Guidance Note 3, where it disclosed urban planning and housing strategies. In this note, the government set a target of 60% of new housing being in brownfields, through the adaptive reuse of existing buildings [9]. This measure further encouraged urban regenerations led by the real estate market, through the reuse of old buildings for residential purposes.

Questions are being made whether communities have been heard and actually contribute to decision-making or whether popular consultation is just a box to be ticked, with communities being manipulated to legitimize urban regenerations, with little or no decision-making power in relation to what will be delivered by public-private partnerships in their neighbourhoods [6].

3 ROYAL ARSENAL, WOOLWICH

3.1 Brief contextualization

Woolwich is a district in south-east London and is part of the Borough of Greenwich. With a long history, with archaeological records indicating occupation since the Iron Age era, Woolwich housed for 300 years the largest royal artillery factory – Royal Arsenal. Since 1565, the factory has produced and ammunition and had its heyday during the first world war, being the largest source of British ammunition used in the conflict. During this period, Royal Arsenal employed 80,000 people [10].

Masters [10] – who was born in the neighbourhood and saw Royal Arsenal at its peak – when he was 6 years old and lived with his father, Superintendent of the Factories of Arms and Transport, in one of the buildings of the Royal Arsenal – points in his book to the fact that few people, even local residents, know the history of Royal Arsenal and that the neighbourhood ends up being better known due to the football team Arsenal F.C., which emerged on site at the end of the 19th century, formed by factory workers. The author reports that after its peak between 1915 and 1920, factories began to be dispersed in the 1930s, with the realization that future conflicts would happen through the air, making the concentration of ammunition factories in the same place unsafe. With this, until the beginning of the Second World War, the number of workers went from 80,000 to 30,000. In fact, the site was bombed a few times during the second war, leaving 103 workers dead and another 770 wounded [10].

The site began to decline in the 1950s, when ammunition production was transferred to private industries. The Ministry of Royal Defence occupied for some years some of the buildings until, in 1967, the Royal Ordnance Factory was officially decommissioned. Of the 1,300 hectares that belonged to the artillery complex during its years of operation, there were only 76 hectares left, which were still occupied by the Ministry of Defence (MoD) until 1994.

At the time of the publication of Roy Masters' book in 1995, the area was unoccupied and "awaiting redevelopment" [10]. According to the author, with the announcement that MOD would leave the site, some residents came together and created the Royal Arsenal Museum Advisory Group (RAMAG), with the aim of creating a Heritage Centre with several museums in some of the most important buildings in the neighbourhoods that the history of the site would not be lost. The author also comments, at the end of his book, that the group had already presented the proposal to the Greenwich Council, which encouraged them by saying that "the proposal was very aligned with the municipality's strategy, which aims to raise public and private money to promote jobs, housing and leisure on the 7 miles of land along the river" [10].

In 1992, Greenwich entered the competition to participate in the City Challenge, a government program that, unlike the urban regenerations of the 1980s, was led by the municipalities and allowed greater community participation and greater control under interventions by the private sector. In a video available online, submitted to the government as part of the material needed to join the competition, it is clear the willingness of the community to integrate Royal Arsenal with the rest of Woolwich.

Unfortunately, Greenwich did not win the City Challenge but its urban regeneration began a few years later. The SHARP (Sustainable Historic Arsenal Regeneration Partnership) project, funded by the European Union and led by English Heritage, published an analysis of intervention projects in former military areas, with case studies from England, Malta, Spain and Estonia with the aim of devise sustainable urban regeneration strategies based on the appreciation of industrial heritage. The English case analysed in the publication is the Royal Arsenal, which is therefore the largest source of information on the case.

3.2 The urban regeneration of Royal Arsenal

In 1997 31 hectares equivalent to the part where the listed buildings were, were sold to English Partnerships (EP), at the time the urban regeneration agency of England, for £1 (one pound) [11]. At the time of the sale, 22 buildings had already been listed and the area had been demarcated as a Conservation Area in 1981. According to the author, in the mid-1990s there were more than 60 buildings on the site, with years of construction between 1690 and 1970, most of them in good condition. This number decreased considerably by the end of the 1990s (Fig. 1).

Figure 1: Royal Arsenal, compared between 1914 and 1994.

As already seen, in 1993, the Greenwich Waterfront Development Partnership had already included Royal Arsenal in its masterplan, with the strategy of regenerating the neighbourhood through the idea of diversified economy, given that the neighbourhood, because of the closure of factories, had an unemployment of 17% [11]. The urban regeneration strategy of the neighbourhood was based mainly on its heritage value and economic potential, with the objective of creating a living neighbourhood, mixing new and old buildings and uses such as housing, culture, work and leisure.

The year after the Royal Arsenal's land purchase in 1998, the English Partnership began to draw up a plan for urban regeneration, where the main objective was to promote private investment. One of the first strategies was to trace a zoning dividing the neighbourhood into 4 zones (Fig. 2): housing, along the River Thames; shops and services, in the eastern portion, near the highway and the rest of the Woolwich district; cultural uses near the pier on the main street of the neighbourhood (street number 1) and, finally, leisure activities in the east.

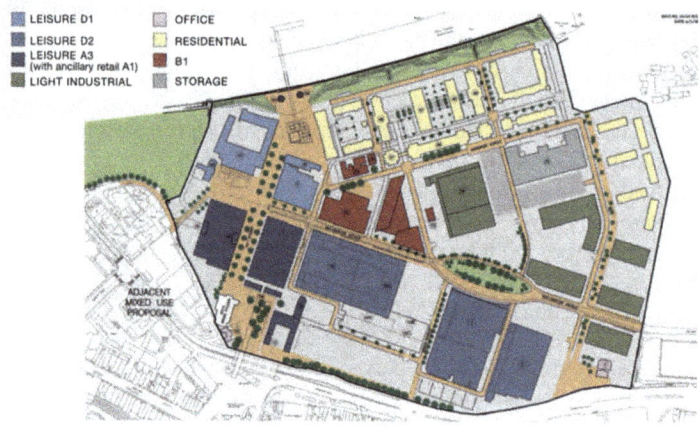

Figure 2: Masterplan 1998.

The masterplan developed by Llewelyn Davis Yeang was approved by the municipality of Greenwich in the form of an Outline Application, which means that the project delivered for approval was not detailed but rather a plan that made clear the intentions and especially respect for architectural heritage to be preserved but allowed flexibility for changes to occur over the years [11]. Stevenson [11] judges this decision as correct, since it was not possible at that time, in 1998, to establish with certainty what use each of the buildings would have, since there was not yet at stake the partner of the private sector, which entered two years later. The most important buildings, such as the Grand Store, a complex of warehouses built between 1803 and 1813 were preserved externally almost in its entirety, while its interior underwent further modifications. Most of the buildings had free plants and had to be partitioned to receive residential use. In the initial approved plan, building 22 – former Administrative Centre had been demarcated as possible demolition and would give way to a new office building (use B1), since it was not listed. However, later in the regeneration process it was eventually preserved and reused thanks to the insertion of three other floors.

In the year 2000, Berkeley Homes Ltd. was selected as the housing developer, who would be the private sector actor to develop the adaptive reuse projects of the first warehouses, transforming them into both private and affordable housing.

Since the publication of the Town and Country Planning Act in 1990, municipalities have defined a percentage of dwellings that must be delivered as affordable housing. This policy, defined by Section 106 of the legislation, is the biggest guarantee yet of affordable housing, as the UK no longer produces Council Housing, built in large numbers under the Welfare State, especially in the post-war period until the late 1970s. In Greenwich, the requirement is that new developments, whether in new construction or in reuse of old buildings, deliver a minimum of 35% in affordable mode. Of these, one third refers to social rent, one third as shared ownership and the remaining third is for keyworkers. These are usually run by non-profit housing cooperatives.

According to Berkeley Homes [12], the regeneration programme received an initial £43 m investment from the Ministry of Defence for the decontamination of the former industrial site. Berkeley Homes initially won the competition with a £150 million project for the adaptive reuse of four listed buildings and to turn them into 700 homes, 35% of them affordable and the rest at market price. In 2003, two years before the scheduled completion of the works, the company acquired some other buildings and obtained approval to turn them into an additional 550 apartments and, just a year later, bought the rest of the EP site (western portion of the neighbourhood, which would initially be destined for an urban park) for the installation of bars and shops, in addition to 2,500 more apartments and the creation of the new Cross Rail station [12]. Thus, thanks to the flexibility of the masterplan initially approved, in 2005 a revised masterplan, this time prepared by the office Allies and Morrison.

The plan was submitted at least twice for approval, once in 2009 (Fig. 3) and another in 2013. At this time, much of the old buildings had already been converted into housing and these masterplan revisions refer to the western portion of the neighbourhood, acquired by Berkeley Homes in 2004. As already mentioned, this area had been demarcated in the 1998 masterplan as a leisure area and was planned the construction of a park, whose area was drastically reduced in the submissions of 2009 and 2013.

Figure 3: Additional site included in the 2006 masterplan.

In addition to the loss of area initially destined for leisure, another important change in the plan was related to the parcel destined to affordable housing. The first masterplan had promised that 35% of the total apartments would be destined to affordable housing divided into three modalities, as previously mentioned: social renting, shared ownership and key worker housing. In reviewing the plan, this percentage was reduced to 25%, with the justification that Berkeley Homes was investing in the construction of the Cross Rail station.

The municipality accepted the justification, adding that the number of dwellings had increased in absolute values, despite the decrease in percentage.

It is noteworthy that the values of apartments rented or sold at affordable prices are based on market values. According to the Berkeley Homes website, apartments in the new developments range from £470,000 (apartments with connecting bedroom and living room) to £1,050,000.00 (2 or 3 bedroom apartments). Housing cooperatives sell the apartments at 25% of the market value in the Shared Ownership scheme, where the buyer owns this percentage and pays rent for the remainder of the amount they did not purchase, around £600 per month. Families with incomes of up to £90,000 a year can buy in this mobility. The social housing modality is the only one that is really affordable for the low-income population, with rents around £106 (if rented directly from the council) or £122 (if rented from housing cooperatives).

With regard to cultural uses, which Roy Masters hoped would be included in the plan in 1995, in fact in 2001, in the first year of regeneration, the Firepower Museum was opened and operated in the neighbourhood until 2016, when it was closed with the promise of reopening it in Salisbury Plain, near Stonehenge, 150 km from the Royal Arsenal district, where the sniper training used to take place. On the site of the museum and in four other listed buildings will be held the last stage of the regeneration project, which will be inaugurated in spring 2021: Woolwich Works, a cultural centre with space for exhibitions, shows and cultural activities in general, as well as cafes and bars.

Figure 4: Updated zoning of the neighbourhood, building 10 dissemination material.

3.3 Woolwich and gentrification

In September 2014, Berkeley Homes interviewed 226 residents of the neighbourhood, which at that time corresponded to 11% of the inhabitants. Among these, 67% were residents in the private housing modality (private owners or tenants) and 32% in the affordable modality, with the intention of representing the real percentage of the different ways of living in the neighbourhood. The result of this survey indicates a community happy with their neighbourhood, where 93% of residents consider themselves happy with the place where they live and 75% think that the place where they live contribute so much to the sense of identity, the latter increasing considerably among the inhabitants of affordable units (90%). The research highlights that people in Royal Arsenal feel safe in the neighbourhood compared to the rest of Woolwich and that regeneration has changed the reputation of the neighbourhood [12].

The survey would have shown very different data if it had involved residents of the Woolwich neighbourhood as a whole, not just the residents of the regenerated part who are mostly people from other parts of London, attracted by the new developments and the good connection of the neighbourhood with the rest of London. Still, research with residents of Royal Arsenal raised the issue of visible separation between the two parties. In fact, what you see today are two distinct neighbourhoods. The wall and the highway are two physical barriers that symbolize a social separation between the new residents of the regenerated neighbourhood and the residents of Woolwich from the process of urban regeneration presented.

In March 2020, a group of students from University City London (UCL) conducted interviews with residents of the Royal Arsenal and with the rest of Woolwich. The results, not yet published, show this disconnect between the regenerated part and the rest of the neighbourhood. Royal Arsenal residents generally say they don't like the rest of the neighbourhood and avoid going there because they don't feel safe, while Woolwich Arsenal residents say they don't visit Royal Arsenal because they don't feel welcome, as it's a place for wealthier people.

Twenty years have passed since the start of the Royal Arsenal's urban regeneration process and now Woolwich residents fear that the neighbourhood's next stages of urban regeneration will cause even more gentrification as well as displacement, something that had not happened in the regeneration of the Royal Arsenal, as there were no residences within the industrial site. Among the new projects planned for the neighbourhood, two stand out:

1. Spray Street Quarter: the project, carried out from the partnership between the Royal Borough of Greenwich (municipality) and St. Modwen Companies, Notting Hill Housing, signed in 2016, provided for the demolition of a block, south of the highway that divides Woolwich and Royal Arsenal (Beresford Street) and east of the old portal. The project was barred in 2018, when the old market located on the block was listed. The project is under review.

2. Woolwich Estates Regeneration: Second site of the Royal Borough of Greenwich, the project consists of the demolition and re-development of 3 housing estates of the neighbourhood. In all, 1,064 social housing apartments will be demolished to make way for new 1,500 units, of which 35% will be affordable, among which only a third will be of similar values to social rent, which previously represented the totality of the apartments.

The Speak Out Woolwich group, made up of neighbourhood residents and shopkeepers, is a resistance group against the gentrification of Woolwich. The main demand of the group is for the fulfilment of the minimum of social housing and the maintenance of the identity of the neighbourhood, nowadays mainly formed by local commerce, which runs the risk of being gradually expelled to make way for the stores of networks, attending to a public of higher standard, as the one that currently lives in the Royal Arsenal.

In September 2002, Woolwich received a new government fund to regenerate the main avenue in the centre of the neighbourhood, Powis Street (Fig. 5), again focusing on historical heritage and with the same goal of promoting the economic and social improvement of the neighbourhood.

Figure 5: Powis Street.

4 FINAL CONSIDERATIONS

It is considered that the example presented, Royal Arsenal, is a good example with regard to the preservation of industrial heritage. Since the site was acquired by the EP to start the process of urban regeneration, most of the buildings have been preserved. The demolitions had taken place years earlier, after the closure of the factories. The listed buildings underwent few external changes and even unlisted buildings were saved, but with major interventions, such as building 22, where three new floors were added at the top. The new buildings within the conservation area respect the importance of historic buildings, both for the height and design and design used and contribute to the valorisation of industrial heritage.

However, especially since the 1980s, it is recognized that the preservation of heritage does not only concern the physical preservation and materials of buildings and as already seen, when it comes to industrial heritage are not always the most important values to be preserved. The relationship of these sites with their communities, either due to the number of years in which the factories were in operation or because the impact that their closure caused is very strong. In the case of the Royal Arsenal, one can see in the speech of people interested in regenerating the area how much there was an expectation that the site would be opened and connected with the rest of the neighbourhood, so that all people would use that space so privileged by the proximity to the river and its history, which is the history of Woolwich. Royal Arsenal is now the noble part of Woolwich and there is no longer a sense of belonging between the community of Woolwich and its former industrial site. Residents of Woolwich today are battling the gentrification of the neighbourhood that has been going on since the urban regeneration of the Royal Arsenal.

From this case study, there is the possibility of thinking about alternatives so that old industrial areas can be regenerated, preserving their assets, generating employment and housing, but that can also meet the needs of the community that is generally strongly related to industrial sites. It is understood the need for private sector involvement to bring investment that the state often lacks, but the participation of the government is critical to ensure that the project delivered meets the needs of local communities. With the example presented, it is clear the need for legislation that requires counterparts from the private sector, but this cannot be the only way to guarantee housing with affordable values for the population, since very low percentage of social housing ends up not allowing a true mix of different social classes.

The next phases of this research intend to study the policies of housing, planning and protection of historical heritage in Brazil and especially in the city of São Paulo to verify the barriers and opportunities in the viability of projects to reuse industrial heritage for social housing.

REFERENCES

[1]   Stratton, M. (ed.), *Industrial Buildings: Conservation and Regeneration*, Taylor & Francis, 2003.

[2]   Cossons, N., Why preserve the industrial heritage. *Industrial Heritage Re-tooled: The TICCIH Guide to Industrial Heritage Conservation*, pp. 6–16, 2012.

[3]   Littlefield, D., (Re) generation: Place, memory, identity. *Architectural Design*, **82**(1), pp. 8–13, 2012.

[4]   Couch, C., Sykes O. & Börstinghaus, W., Thirty years of urban regeneration in Britain, Germany and France: The importance of context and path dependency. *Progress in Planning*, **75**(1), pp. 1–52, 2011.

[5]   Falconer, K., Legal protection. *Industrial Heritage Re-tooled: The TICCIH Guide to Industrial Heritage Conservation. Part III: Realising the Potential*, p. 94, 2012.

[6]   Carpenter, J., Decision-making in regeneration practice. *Architectural Regeneration*, pp. 47–60, 2020.

[7]   Turok, I., Property-led urban regeneration: Panacea or placebo? *Environment and Planning A*, **24**(3), pp. 361–379, 1992.

[8]   Bonshek, J., The 1991 city challenge bids: A review article. *The Town Planning Review*, **63**(4), pp. 435–445, 1992.

[9]   Jones, P. & James, E., *Urban Regeneration in the UK: Boom, Bust and Recovery*, Sage, 2013.

[10] Masters, R., *The Royal Arsenal*, Alan Turner: Woolwich, p. 1005.

[11] Stevenson, M., *Regeneration through Heritage: Understanding the Development Potential of Historic European Arsenals*, English Library: Swindon, 2016.

[12] Berkeley Homes, Royal Arsenal Riverside: A case study about transport and placemaking. http://www.berkeleygroup.co.uk/media/pdf/n/b/Berkeley-royal-arsenal-riverside-case-study.pdf.

POSTCOLONIAL INDUSTRIAL HERITAGE IN NORTH AFRICA: INVESTIGATIONS AND INSIGHTS INTO THE CITY OF CASABLANCA, MOROCCO

CHAIMA SEDDIKI, JEREMY CENCI & ISABELLE DE SMET
Faculty of Architecture and Urban Planning, University of Mons, Belgium

ABSTRACT

Casablanca, which was the laboratory for Neo-Classical, Neo-Moorish, Art-Deco and Bauhaus architecture, is plundering its past and disposing of its colonial industrial heritage. The lack of inventories, tools and expert structures to evaluate this heritage works in favor of the real estate speculation that is taking over strategic post-industrial territories in a generally carefree manner. The notion of "industrial heritage" in Moroccan culture is not very well known, and the status of the heritage of "the other" is rather complex. But through our inductive methodology and the practice of a Down-Up process within the territory of action, a sizeable dimension has emerged from the field, and has been used as a measure. This is the human and social dimension of the postcolonial industrial heritage. The approach of the citizens, with their observations and verbalizations being placed at the heart of the industrial heritage, reveals, for the first time, the identity, historical, memory and cognitive values of the postcolonial industrial heritage. A significant change is taking place. The values attributed to this industrial heritage by the citizens neutralizes the colonial fact, legitimizes the work of the colonists and even raises it to the rank of heritage. Reconciliation has been made and the status of the heritage of the 'other' clarified. Today, the human and social dimension constitutes a real turning point in the field of heritage in Morocco. The post-colonial industrial heritage is recognized by one of the main actors of heritage, by the citizens. This article is based on research carried out over the past five years combining Qualitative Methodology and Grounded Theory on the industrial zones of East Casablanca and Mohammedia, the cradle of Moroccan industry.
Keywords: heritage, Morocco, Casablanca, industrial wasteland.

1 INTRODUCTION

Morocco has experienced several upheavals since its independence in 1956. The major domestic changes began under King Hassan II and were later extended by King Mohammed VI who, since his coronation in 1999, has made social and development projects a priority. These projects have resulted in the opening of huge social, economic, touristic, and industrial projects in the twelve regions of Morocco. It is within this framework that the four imperial cities, Fez, Meknes, Rabat and Marrakech, have been able to benefit from development projects focusing on the reinforcement of the preservation, safeguarding and rehabilitation of their respective tangible and intangible heritage. But what about the post-colonial industrial heritage of Casablanca, the country's business capital?

Interest in heritage in Morocco has a long history. At the institutional level, the starting point is the Cherifian Dahir (Law in Arabic) of 1912, relating to the conservation of historical monuments and listings. The law has undergone several modifications since then (the Dahir 1945 and the Dahir 1980).

The brief analysis of the content of the legislation explains the interest in heritage shown by the imperial cities, but also demonstrates the inadequacy of the texts concerning the architectural heritage of the new cities and the industrial heritage, works of the French protectorate. The vestiges of the post-Protectorate architecture, apart from some public buildings in the city center, do not meet any criteria and do not rise to the level of national heritage. Several cities, and mainly Casablanca, are thus becoming the scenes of the decay of this architecture, sometimes for speculative reasons, sometimes out of simple ignorance.

Figure 1:   Former installations of the national water and electricity office (O.N.E.E), called the Southern Power Station of Roches Noires, 1952, © unknown.

In 2013, a new legal project reinforces and replaces the provisions stipulated by the Dahir of 1980. This project aims "to harmonize the national legal system relating to the protection, enhancement and transmission of the national cultural heritage with the international criteria to which Morocco has adhered and to integrate the new internationally recognized concepts of heritage ..." (© http://www.sgg.gov.ma/portals/0/AvantProjet/47/Avp_Loi_52.13_Fr. pdf.) Industrial sites "wastelands, mines, factories or other period installations ...," and new towns "presenting features of historical, architectural and artistic interest" are included for the first time. In fact, the consideration of industrial heritage was supported by King Mohammed VI in 2014 as part of the Greater Casablanca development project 2015–2020. Nevertheless, despite this promising impetus, safeguarding projects are almost nonexistent.

In fact, in the absence of an inventory, tools, and expertise structures charged with evaluating the potential heritage, the real estate frenzy continues and is taking over the strategic post-industrial territories. Casablanca, which was the laboratory of Neo-classical, Neo-Moorish, Art-Deco and Bauhaus architecture, is ransacking its past and dispossessing itself of its colonial industrial heritage in a careless manner. Faced with this observation, we will try to answer several questions in this article: Is there a colonial industrial heritage in Casablanca? If so, how is it perceived by the population? Are they attached to it in the same way as they are to the existing ancestral heritage, such as the medinas, a UNESCO world heritage site?

2 FIELD EXPLORATION: METHODOLOGIES AND SURVEYS

To answer these questions, we borrowed two inductive methodologies from the social sciences in order to have the latitude to go back and forth between the theory and the practice of the territory, and thus obtain a global and strategic vision. Therefore, we chose the qualitative survey method on the one hand and the grounded theory method on the other. Before applying them to our case study, we would like to briefly explain these principles.

2.1 Qualitative survey method

There are several techniques of qualitative investigation: "mapping and documentary surveys, observation procedures, interviews, group interviews, participant observation, etc." Here, we opt for the techniques of observation and interview. Observation precedes the actual survey and aims to understand the situation(s) experienced by the population, the associations, the administration, etc. Observation allows the optimization of the interview and helps to prepare for the recruitment of the different actors of the industrial heritage of Casablanca and to finely detect the interactions between them. It also allows the identification of possible variations between the speeches collected during the interview and the real practices. During this phase, an alternation between "observation sessions" *in situ* and "reflection and writing sessions" is carried out. The interview allows the issues in the field to be clarified on the basis of the real-life experience of the targeted actors. Each interview is to be considered as expertise in its own right. The specificity of the interview, especially the semi-directive one, is to frame the questioning while giving those interviewed the opportunity to speak freely.

2.2 Grounded theory methodology

Grounded Theory (GT) methodology has been around for over fifty years [1]. It is an inductive approach whose objective is to generate theories while proposing a set of procedures to achieve this [2]–[5]. We chose to add this method to the survey for two reasons: on the one hand, it is more flexible than the quantitative method which is based on 'logico-deductive' ("*The logico-deductive model of science is to seek truth by testing hypotheses using a pre-existing framework and statistical analysis. The goal of research is to arrive at conclusions that can be generalized to other populations.*" [5]) reasoning and, on the other hand, it allows the generation and development of theories 'rooted' in the field data. The GT method is also based on the premise that there is not one truth but several truths, depending on one's definition of a given phenomenon; a definition that varies according to time, place, quality of the observer, cotext and context.

The GT approach is inductive. It allows the identification of relevant concepts and to generate a series of hypotheses during the exploration process. There is no question of imposing quantitative research procedures, standards and evaluation criteria. The criteria are derived from the methodology itself during the process. According to Corbin [5], the purpose of a GT exploration does not lie in the recognition of a 'single truth,' or in the testing of hypotheses, but rather in the interpretation of several truths that coexist and that allow us to understand a phenomenon in its context [5]. GT does not exclude quantitative analyses, however, and resorts to rather deductive methods when it comes to confronting theories with data, as its founders confirm [1]. The specificity of the GT method lies in the reversal of the traditional order of the scientific approach. First, we give the advantage to the data, to the field, and then we resort to the scientific literature. Three principles will guide us in this approach:

- The principle of emergence in the work consists of bringing the person who is at the core of the phenomenon to give themselves up and express their lived experience: Let the meaning emerge while confronting it with empirical data to verify its coherence [1]–[4].
- Theoretical sensitivity is the ability to go beyond intuitive interpretation and to look for what is hidden in common sense [2].

- Sensitizing concepts are the set of theoretical, experiential or cultural tools and knowledge that will allow us to analyze the empirical data in a relevant way [2].

3 APPLICATIONS IN THE FIELD: THE INDUSTRIAL HERITAGE OF CASABLANCA – A DIVISIVE CONCEPT

In order to be able to answer the initial questions and obtain convincing results, we carried out two successive surveys. The first one had a broad vision; the second one was more focused and followed from our first results.

During the observation phase, our mission consisted of observing the field of investigation and rubbing shoulders with its inhabitants, without actually approaching them. During the interview (The survey of the population was mainly conducted in Arabic, the country's first language. Some citizens expressed a certain ease with French, the country's second language. I adapted the interview to the intellectual level of each citizen interviewed. The wording of the questions was adapted and translated simultaneously to facilitate comprehension and extract a maximum amount of information on the subject.) phase, the interviewee was guided in a semi-directive way and had the possibility of leaving the framework if they wished.

3.1 Survey 1: interpretation of the data

Several post-industrial districts within Casablanca were chosen at random. A sample of 26 people from all active age groups was studied. This first survey questioned 4 districts in East Casablanca: Mers Sultan – La Gironde, Belvédère – Roches Noires, La Villette, Hay Mohammadi.

Starting from the premise that each interview is unique, each singularity describes a process "at the origin of an action" to be analyzed. This allowed the identification of individual practices and brought out the singular coherences. We can summarize these as follows:

3.1.1 Denunciation of the structuring and structural problems of the city of Casablanca

The citizens of the industrial districts in the east of the city broke their silence and dare to denounce the structural problems of the city of Casablanca, which have a direct impact on their territory. The citizens of the industrial districts were primarily concerned about the failure of the health, education, waste management and public transport systems. The economic crisis of the 2008 trapped them in unemployment and over-indebtedness. The abandonment and degradation of certain industrial sites led to a climate of insecurity. Young people are increasingly dropping out of school and a phenomenon of delinquency is gradually taking hold in these mythical districts. The citizens of the industrial districts expressed that they are aware of the recent changes taking place in their district. They understand the speculative stakes and even speak of the "real estate mafia," but until now they have been unaware of the heritage issues. They suffer from densification and have a very difficult time with the gentrification that is taking place on their territory.

3.1.2 Secondary interest in safeguarding the civil heritage of the city of Casablanca

After highlighting the dysfunctions, and after having oriented the citizens towards the heritage issue, some of them showed their interest in the Art Deco, Modernist and Neo-Moorish civil heritage of the city of Casablanca. They even defended the interest of safeguarding and enhancing it, the media coverage and the multiple actions around it being the main factor without any doubt.

3.1.3 The division of the citizens regarding the symbolism of the industrial past of the city

In a large majority of the interviews, the disused economic activity sites were accepted to be the result of the colonial period. They are therefore the factories 'of the French,' in opposition to the urban planning and architecture of this same period. A real symbolic separation was present in the mentalities.

Beyond that, when we oriented the discussion and approached the notion of Casablanca's industrial heritage (CIH), the reaction was surprising, to say the least. Although the average citizen was not familiar with this notion and was even surprised by this composition "heritage + industry," they wished and persisted in referring to this heritage as "colonial." By this position, the citizens associated the CIH with the colonial achievement. This is probably an important clarification. The demarcation of the CIH explains, to some extent, why it is not on the list of primary concerns for the citizens of the eastern districts. The colonial CIH is, by its connotation, the work of the other and, in fact, the concern of the other. Although all the citizens agreed on this colonial connotation of the CIH, they were divided as to its evaluation:

- The industrial wastelands and old disused factories are the essence of the city according to the population over 30 years of age.

Citizens over 30 years old were more vocal and explained the symbolism of the witnesses of this past industry. For them, it is, above all, a place of memory, a working-class identity, a working-class culture, a symbol of working-class struggle, an industrial identity, a Casablanca identity, a symbol of progress and glory, a symbol of greatness and prosperity, a symbol of resistance, an architectural landmark, a landscape force, a brand, a sign, a know-how, an apprenticeship, and unique architecture.

Citizens over 30 were sometimes nostalgic when they talked about sites that have disappeared, sometimes helpless when they talked about sites that are being demolished, and sometimes optimistic when they talked about sites that are still standing. Some citizens over the age of 30 expressed their desolations and regrets: the abandonment of certain sites with great potential, the progressive deindustrialization of industrial districts has caused chaos and social disaster, the insensitivity of young people to the symbolism of their district's industrial past, the progressive change of the territory and the lack of information and participatory consultation with the local inhabitants. In addition, some citizens took advantage of this opportunity to describe and share their expectations: The improvement of the image of the industrial districts by cleaning them up, the demolition and cleaning of abandoned and uninteresting industrial sites, the rehabilitation of sites still standing (in very good condition) and the maintenance of the industrial use, the rehabilitation of resistant sites (in the beginning of degradation) and their reconversion into socio-economic and socio-cultural projects.

This category includes the industrial past, and it remains in all its projections and conditions the improvement of its territory by improving it. It even suggests the 'rebranding' of the city, the 'making the city shine,' and changing the negative image through this industrial past.

- The industrial wastelands and old disused factories are a nuisance and symbolize misery and chaos according to those under 30.

For this sample of the population, the vision of disused economic activity sites is quite different. At first sight, they are degraded elements with no future. Objects of the past that symbolize economic and social failures. They are in no way a heritage to be developed. The urban priority is elsewhere.

3.2 Survey 2: interpretation of the data

The systematic industrial past denial of the under-30 age group during Survey 1 led to a second survey. The area observed is at the heart of the industrial heritage of East Casablanca. It is located in the "Hay Mohammadi" district. It is the first industrial center of the Greater Casablanca region. This study area was chosen for two major reasons: the abundance of disused industrial sites and ageing working-class housing estates that are still inhabited, and the rejuvenation of the local population, 44.5% of the inhabitants are under 30 years old.

We therefore have a recognized post-colonial industrial center, a large disused industrial land base and a population from the adjoining working-class housing estates. Moreover, the population that uses these places today, and that is one of the actors of the future of this territory, is increasingly young. All reasons that reinforce the legitimacy of our choice of target.

This study area is close to the slaughterhouses, the disused tanneries, the poultry market, the former shanty town "Douar Ouled Ahmed," which has since been razed to the ground, and the working-class housing estates. The area is in permanent tension and is one of the 264 districts targeted by the INDH ("On 18 May 2005, His Majesty King Mohammed VI gave a historic speech in which he announced the launch of the National Initiative for Human Development (INDH). A large-scale initiative to fight against poverty and exclusion," http://www.maroc.ma/fr/content/indh), which fights against poverty and social exclusion. It represents several deficits: insufficient basic social infrastructure, aging and unhealthy housing, aging and abandonment of several industrial sites, increasing unemployment, impoverishment of the population, increasing school dropout, lack of training and integration opportunities for young people, and lack of quality green spaces.

Given that the challenge of a qualitative survey does not lie in the number of people interviewed but in the way they are interviewed, and their comments are analyzed, 16 people from the Hay Mohammadi district were interviewed (for a listening time of 30 to 60 minutes). Each survey was based on a semi-directive questionnaire and was carried out individually and anonymously if the interviewee so wished.

As a result, the young people did not immediately express themselves on the industrial heritage aspect but more on what concerns them most: the deficiency of the infrastructure, the deficiency of the public education system, unemployment, their uneasiness, etc. The question of industrial heritage only came up halfway through, or even at the end of, each interview. As for the question of heritage, which was of such concern to the over-30s, it was secondary, if not non-existent for the youth. The young people express more interested in the possible reconversion of the heritage than in its preservation or classification. The conversion of industrial wastelands into facilities was stated as more than a wish for these young people, it was an ambition. The ambition to improve their environment and make it sustainable, the ambition to counter the installation of new high-class apartment buildings, the ambition to stop the gradual gentrification of the district and to strengthen the bonds of solidarity between the original inhabitants.

4 CASABLANCA'S INDUSTRIAL HERITAGE: THE HERITAGE OF THE OTHER

In the process of identification and attachment to the territory, heritage appears to be one of the resources allowing this process. It forms a common heritage for a group of individuals and allows a lasting cohesion in time (between the past and the future) and in space (by structuring the territory). From this relationship between heritage and territory emerges a "conceptual kinship" [6]. Heritage, through its production and representations, is a way of "producing fathers" (P. Legendre) [7]. "Producing fathers" was expressed by Pierre Legendre during a debate on the metamorphoses of the notion of heritage and taken up by Henri Pierre Jeudy in the introduction to his book "Patrimoines en folie." Therefore, the process of heritage development expresses a territorial identity and refers to a form of appropriation by human societies [6].

The construction of heritage requires a certain sacredness and reappropriation of a set of achievements, of production objects recognized as witnesses of a civilization and know-how, around which a society or a group recognizes itself and makes itself recognized. When a society appropriates its heritage and exposes it to the public gaze, it is because it wishes to be recognized in this specificity and singularity.

The question then arises: how to appropriate the 'other's' heritage, the 'colonial' heritage? This is the paradox of the Maghreb countries and specifically that of our case study: Morocco.

4.1 Exploring the colonial past in Morocco

Although it symbolizes confrontation and violent occupation, the protectorate, strengthened by its Algerian and Tunisian experience, was keen to preserve the local memory and to safeguard a certain identity in the Moroccan Medinas. This is what Arrif [8] describes as "taking charge of the memory of the other." General Lyautey said in this regard: "We arrived in a country which, with Arabia and certain regions of Central Asia, contained the only cities in the world where exoticism had retained its purity. The character of these cities had to be saved …." It is for this reason that the 'Lyauthean (In reference to General Lyautey)' urbanism practiced in Morocco has been described as the 'urbanism of the protector' versus the 'urbanism of the victor' practiced in Algeria. There are many interpretations, but the majority of Moroccan researchers and historians agree that the colonial construction of heritage and the preservation of Moroccan Medinas representing the "memory of the other," the indigenous, supports the rupture and seals the difference with the latter. The "Old Medina," an ancient territory that shelters and closes in on the natives to isolate them and better control them, versus the "European City," a new territory that meets all the needs of the colonists [8], [9]. What is certain, leaving aside the main motivations of General Lyautey, is that the city of Casablanca developed outside the limits of the Old Medina with the aim of preserving it behind its ramparts and raising it to the level of a sacred heritage.

In 1912, Lyautey created the Department of Antiquities, Fine Arts and Historic Monuments, intended to preserve the heritage of the past and preserve the local memory. From 1924 onwards, the object of heritage evolved, and the classification was extended to the public buildings of the "European City." The law on the preservation and enhancement of the Old Medina was relaxed in 1945 for obvious cultural and hygiene reasons. We retain the dualism of preservation and separation from the colonial construction of heritage in Morocco. Today's Moroccan has made choices and appropriated the colonial heritage, but in what way? Since independence, Morocco has been confronted with the reality of the field

Figure 2:   Former installations of the national water and electricity office (O.N.E.E), called the southern power station of Roches Noires, 2018, © Seddiki.

and the prevailing dualism. It has had to deal with two references imposed as such by General Lyautey: the indigenous heritage (that of its Muslim ancestors) and the colonial heritage (manufactured by the colonists and inspired by their culture). In fact, the French protectorate transposed the policy of safeguarding applied in metropolitan France in 1912, in order to safeguard a different memory on a different territory. Today, it is Morocco's turn to deal with the memory of the other (the colonist) on its own territory.

4.2 Industrial heritage in East Casablanca

At first sight, we could identify that the people of Casablanca are clearly divided on the subject. There is no unanimity on this issue. This new heritage object is in the making and its future is of concern exclusively to the generation over 30.

As the survey shows, the under-30 generation attaches negative connotations to the world of the factory and refuses to identify with or appropriate this industrial legacy. It is also a generation that considers itself globalist and refuses to take root or to be attached to its territory. It is clearly a generation without an identity, or that is at least in search of one.

The generation over 30 accepts the industrial legacy, the relics passed down from generation to generation, and considers them worthy of protection in various respects. Firstly for their historical, technical and architectural value, expressed by the intellectual stratum. Secondly for their social, memorial and symbolic value, all three of which are advocated by the working-class mass that lived and worked there. The dismantled factory or wasteland takes on a symbolic value that acts as a link between the present and the past from which it emerged. Palmer [10] calls this symbolic link, this value of the history and social memory of the place, 'lived experience' [10]. The factories become symbols and icons for this working-class mass, destined for collective appropriation and identity enhancement. The notion of industrial heritage has the power to transform these places into "sanctuaries of memory" [11].

This quest for identity values can be observed in Western societies from the 1960s and 1970s, which saw the emergence of alternative movements: feminist, anti-nuclear and anti-capitalist movements, etc. According to Di Méo [12], this was a time when the younger generation, unlike Moroccan youth, wanted to return to their roots and to what they identified as authenticity. All these currents were characterized by attitudes of opposition to totalitarian values and ideologies. Their obvious infatuation with heritage is what the sociologist Yvon Lamy calls 'return investment' [12].

5 CONCLUSION

The role of industrial architecture is intriguing, and its value is still being debated: the factory as a place of experimentation with modernity and architectural innovation, or the factory as a purely rational envelope for increasingly complex machines. Industrial architecture has always evolved between fashion and economic constraints. Hamon and Cartier [13] remind us of the words of the historian of industrialization Jones [14]: "…industrial buildings are, almost as much as the buildings of power (social, royal, etc.) subject to the effects of fashion…." Sometimes it borrows neutral, functional or fashionable forms and sometimes it invents its own complex and innovative forms. It is, moreover, innovation that marks the Moroccan specificity and crystallizes the singularity of colonial industrial architecture in Morocco.

The generation over 30, the most representative, has gone beyond the stage of heritage awareness and its heritage impulse seems to indicate change and to announce the commitment towards a concrete process of heritage. This process cannot exist without actors and a political context which is favorable to the conservation, valorization and exhibition of Moroccan industrial heritage.

The perception of the Casablanca population of the history of their city has changed fundamentally since the publication of the book and the exhibition "Casablanca, birth of a modern city on African ground," at the Villa of Arts in Casablanca in February 2000.

The impact of the book and the exhibition was immediate. The people of Casablanca were finally able to rid themselves of their inferiority complex facing the "legitimate" Moroccan from the imperial cities. For a long time, Casablanca was considered by the bourgeoisie of the imperial cities of Fez, Rabat or Marrakech as a new colonial city with no historical value, no cultural value and no interest.

Through the knowledge of the heritage, the Casablanca citizen finally found the material means to speak about the history of their city and to defend their identification with it.

This improved knowledge of a heritage that has long been despised does not go beyond the circle of academics and does nothing to stop the cycle of destruction. The demagogic division of Casablanca into 29 communes at the time created a state of incoherence conducive to speculators.

Today, the destruction continues and fills the lack of plots in a city that is renewed daily.

The long-sought identity of Casablanca's citizens is being decimated by real estate fever and speculative interests [3], [15]. In fact, it is more necessary than ever to accompany the various heritage actors in the heritage process and to equip them in an efficient manner.

REFERENCES

[1]   Glaser, B. & Strauss, A., *The Discovery of Grounded Theory: Strategies for Qualitative Research*, Aldine: Chigago, 1967.

[2]   Corbin, J. & Strauss, A., *Basics of Qualitative Research*, Thousand Oaks, 2008.

[3]   Glaser, B., *The Grounded Theory Perspective: Conceptualization Contrasted with Description*, Sociology Press: Mill Valley, 2001.

[4]   Laperrière, A., La Théorisation Ancrée (Grounded Theory): Démarche analytique et comparaison avec d'autres approches apparentées. *La recherche qualitative: enjeux épistémologiques et méthodologiques*, G. Morin: Boucherville, pp. 309–340, 1997.

[5]   Corbin, J., Préface: Méthodologie de la théorisation enracinée. *Fondements, Procédures et Usages*, Presses de l'Université du Québec: Montréal, 2012.

[6]   Di Meo, G., Patrimoine et territoire, une parenté conceptuelle. *Espaces et Sociétés*, **78**(4), pp. 15–35, 1994.

[7]   Jeudy, H.P., *Introduction: Patrimoines en Folie*, Editions de la Maison des sciences de l'homme: Paris, 1990.

[8]   Arrif A., Le paradoxe de la construction du fait patrimonial en situation coloniale. Le cas du Maroc. *Figures de l'orientalisme en Architecture*, **73–74**, 1996.

[9]   Cattedra, R., Casablanca: La réconciliation patrimoniale comme enjeu de l'identité urbaine. *Rives Nord-Méditéranéennes*, 2003.

[10] Palmer, M., Understanding the workplace: A research framework for industrial archaeology. *Industrial Archaeology Review*, 2005.

[11] Lucas, P., *La religion de la vie Quotidienne*, PUF: Paris, 1981.

[12] Di Méo, G., Processus de patrimonialisation et construction des territoires. *Patrimoine et industries en Poitou-Charentes: Connaitre, Poitiers-Châtellerault*, 2007.

[13] Hamon, F. & Cartier, C., L'architecture industrielle, travaux et publications, un bilan international. *Revue de l'Art*, 79, pp. 52–62, 1988.

[14] Jones, E., *Industrial Architecture in Britain, 1750–1939*, Bastford: London, 1985.

[15] Toulier, B. & Pabois, M., *Architecture Coloniale et Patrimoine: L'expérience Française*, Institut National du Patrimoine: Paris, 2003.

RESEARCH OF THE INDUSTRIAL HERITAGE CATEGORY AND SPATIAL DENSITY DISTRIBUTION IN THE WALLOON REGION, BELGIUM, AND NORTHEAST CHINA

JIAZHEN ZHANG[1], JEREMY CENCI[1], VINCENT BECUE[1] & SESIL KOUTRA[1,2]
[1]Faculty of Architecture and Urban Planning, University of Mons, Belgium
[2]Faculty of Engineering, Erasmus Mundus Joint Master SMACCs, University of Mons, Belgium

ABSTRACT

Industrial heritage reflects the industrial civilization and industrial development process of the city in different periods, and its research and protection are conducive to the sustainable development of the city context. The Walloon Region in Belgium is one of the leading regions of industrial development in Western Europe, and Northeast China is the leading region of industrial development in China. The horizontal comparison of industrial heritage protection between the two regions is helpful to provide a reference for industrial heritage protection and urban regeneration of the two regions. It is found that the industrial heritages in both places are large in quantity and rich in types. The imbalance index of Walloon Region and Northeast China is 0.5001 and 0.6129. The closer the number is to 1, the more imbalance the distribution is. The distribution density of industrial heritages in Walloon Region is higher than that in Northeast China, and the types of industrial heritages in Northeast China are more than those in Walloon Region. The utilization rate of industrial heritage in the Walloon Region area is higher than that in Northeast China, and the main way of renewal is industrial tourism. The protection of industrial heritage in Northeast China is later than that in the Walloon Region, but it has the characteristics of sufficient stamina and rapid development. Strengthening the relationship between Europe and China in the protection of industrial heritage can provide reflection for the protection in Europe and experience for the protection in China.
Keywords: heritage category, industrial heritage conservation, industrial heritage, imbalance index, Northeast China, national industrial heritage, spatial density distribution, spatial distribution characteristics, Walloon Region.

1 INTRODUCTION

Along with the process of social development, an ocean of industrial facilities has been finished them historical duties. The research and conservation of them can contribute to the maintaining of historical memory and the sustainable development of urban construction [1]. Belgium, as one of the first countries, started the industrial revolution in Western Europe, has a great number of industrial heritage. As the leading region of the traditional industry, the industry has been developed in Walloon Region for more than 200 years. There is a huge number of industrial heritage and an affluent of experience in industrial heritage protection [2]. In East Asia, the traditional industrial bases in Northeast China are the cradle of China's industry. Like most of the traditional industrial region, with the deepening of reform and industrial update, the institutional and structural contradictions of the old industrial bases are increasingly apparent, and the further development is facing difficulties and problems [3]. The regeneration and aftermath of the industry in Northeast China is an important issue in the whole urban construction not only for China but also in East Asia.

On the one hand, the industry system in Walloon Region and Northeast China has similarities, for instance, both of them are the leading industrial region in their countries, both of them have a set of independent and complete industrial systems in their interior. Inside this system, most of the industrial facilities belong to the heavy industries that are facing elimination or transformation and upgrading. On the other hand, they have different cities, for instance, the industry in Walloon Region emerged earlier than in Northeast China. Therefore, put these two regions together to make a comparison, can not only burst out experience in industrial heritage protection but also offer an experience in urban regeneration in urban planning.

This paper is structured accordingly research the industrial category and spatial density distribution in Walloon Region and Northeast China: firstly, Section 2 introduce the data sources and research method, Sections 3 and 4 provide a presentation and analysis of the industrial heritage category and spatial density distribution in their individual region's; secondly, Sections 5 and 6 provide the main findings of the work the summary and discuss the perspectives for its continuity.

2 DATA SOURCES AND RESEARCH METHOD

2.1 Data sources

Thirty-six industrial heritages in the Walloon Region of Belgium based on the ERIH (European Route of Industrial Heritage) and 30 industrial heritages in Northeast China based on the List of the National Industrial Heritage of China (the first batch, the second batch, the third batch) were selected. Due to the analysis of spatial structure, the industrial heritages distributed in different regions are calculated according to different regions. The actual research objects are 38 industrial heritages in the Walloon Region of Belgium and 31 industrial heritages in Northeast China.

2.2 Research method

To provide strong support for scientific planning, classified protection, and effective utilization, this paper deeply analyzes the overall spatial distribution characteristics and an overall pattern of Walloon Region Belgium and Northeast China's industrial heritage applied through the spatial structure analysis that is one of the common methods in planning research. It can better analyze the spatial structure characteristics and spatial distribution law of the research object [4].

2.3 Imbalance index

The discrepancy of industrial heritage spatial structure presents discrepancy in different regional industrial heritage amount of distribution. The disequilibrium index is used to reflect the balanced distribution of industrial heritage in different regions, the equation is:

$$S = \frac{\sum_{i=1}^{n} Y_i - 50(n+1)}{100n - 50(n+1)} \tag{1}$$

In this equation: n presents the number of provinces and municipalities, in Walloon Region the $n = 5$, in Northeast China the $n = 3$; Y_i presents the cumulative percentage of the ith region in the total amount. Inside, $0 \leq S \leq 1$, if $S = 0$, it presents industrial heritage is evenly distributed in all provinces and municipalities in Walloon Region or Northeast China; if $S = 1$, it presents industrial heritage concentrating in one of the provinces or municipalities in Walloon Region or Northeast China; if the value of S is more approaching 1, it presents the more imbalanced the distribution of industrial heritage is.

3 THE INDUSTRIAL HERITAGE CATEGORY AND SPATIAL DENSITY DISTRIBUTION IN WALLOON REGION

Walloon Region is one of the three regions of Belgium which has 16,901 km^2 with 3.644 million of the population [5]. As the leading region in the industry system in Belgium, there are a huge number of industrial facilities and heritages. Based on the European route of industrial heritage, in the Walloon region, it has thirty-eight industrial heritage landmarks. The main categories of them are Mining, Transport, and Manufacturing (Table 1). The spatial distribution of industrial heritage landmarks mainly concentrated in Liege district (Liege province) and Mons – Charleroi district (Hainaut province) (Fig. 1).

Table 1: Industrial heritage category and amount in Walloon Region.

District distribution	Industrial heritage category					
	Transport	Mining	Manufacturing	Military	Others	Total
Liege	3	6	3	1	3	16
Mons-Charleroi	4	5	3	-	-	12
Namur	2	-	2	1	-	5
Arlon	1	2	1	-	-	4
Wavre	1	-	-	-	-	1
Walloon Region	11	13	9	2	3	38

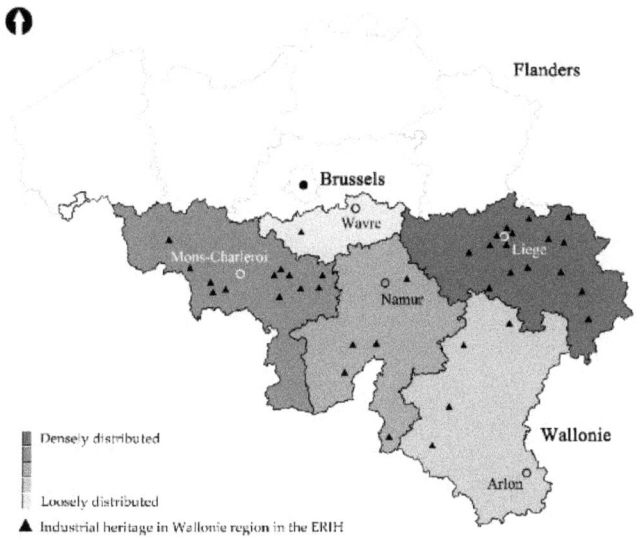

Figure 1: The industrial heritage landmarks spatial distribution in Walloon Region, Belgium.

According to the equation, after calculating and analyzing the data in the statistics on the distribution of industrial heritage in Walloon Region (Table 2), the imbalance index of industrial heritage in the Walloon Region is $S = 0.5001$. This shows that the distribution of industrial heritage in the Walloon Region is imbalanced. Besides, it can also be found that the number of industrial heritage in districts of Liege and Mons-Charleroi alone is 73.69% of the total number of industrial heritages in Walloon Region.

Table 2: Statistics on the distribution of industrial heritage in Walloon Region.

No.	Administrative division	Amount	Proportion (%)	Cumulative proportion (%)
1	Liege	16	42.11	42.11
2	Mons-Charleroi	12	31.58	73.69
3	Namur	5	13.16	86.85
4	Arlon	4	10.52	97.37
5	Wavre	1	02.63	100.00

From the angle of history, Liege district and Mons-Charleroi district are the most developed districts in Walloon Region in terms of economy and industrial construction since the beginning of the Industrial Revolution [6]. There are three main reasons. The first reason is geopolitics. Liege district borders Germany and Mons-Charleroi district borders France. The rapid development of the huge domestic markets of Germany and France has provided orders and markets for industrial products for Liege and Mons-Charleroi. The second reason is transportation, both Liege and Mons-Charleroi have well-developed railway and water transport facilities, which provides trade conditions for the expanding production capacity of the two places. To sum up, the industrial heritages created and left by these two districts are relatively large (Table 3) [7]–[9].

Table 3: Representative of industrial heritage in Walloon Region signed in UNESCO World Heritage site.

Name	Location	Category	Image
The strépy-thieu boat lift	Mons-Charleroi district	Transport	
Le bois du cazier world heritage site	Mons-Charleroi district	Mining	
Blegny mine world heritage site	Liege district	Mining	

4 THE INDUSTRIAL HERITAGE CATEGORY AND SPATIAL DENSITY DISTRIBUTION IN NORTHEAST CHINA

Northeast China is a region of the northeast part of China, that including Liaoning, Heilongjiang, Jilin, three provinces. It has 787,300 km^2 with 108 million of the population [10]. Like the Walloon Region in Belgium, Northeast China is the leading region in the industry system in history in China. The modern industry of China was born and raised here. Based on the List of the national industrial heritage of China (the first batch, the second batch, the third batch), in the Northeast China region, it has thirty-one industrial heritage landmarks. The main categories of them are Transport, Manufacturing, and Mining (Table 4). The spatial distribution of industrial heritage landmarks is mainly concentrated in Liaoning province (Fig. 2).

Table 4: Industrial heritage category and amount in Northeast China.

District distribution	Industrial heritage category						
	Transport	Mining	Manufacturing	Military	Energy	Others	Total
Liaoning	7	4	7	2	2	-	22
Heilongjiang	2	1	1	-	-	2	6
Jilin	1	-	-	-	1	1	3
Northeast China	10	5	8	2	3	1	31

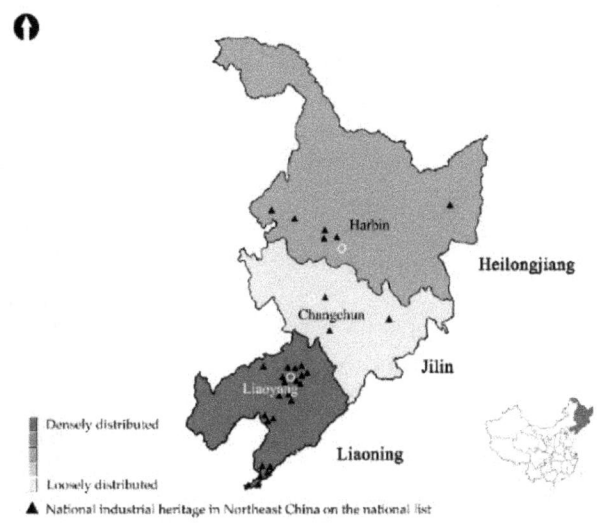

Figure 2: The industrial heritage landmarks spatial distribution in Northeast, China (PRC).

According to the equation, after calculating and analyzing the data in the statistics on the distribution of industrial heritage in Northeast China (Table 5), the imbalance index of industrial heritage in Northeast China is $S = 0.6129$. This shows that the distribution of industrial heritage in Northeast China is extremely imbalanced. It can also be found that only the number of industrial heritage in Liaoning province alone is 70.97% of the total number of industrial heritages in Northeast China.

Table 5: Statistics on the distribution of industrial heritage in Northeast China.

No.	Administrative division	Amount	Proportion (%)	Cumulative proportion (%)
1	Liaoning	22	70.97	70.97
2	Heilongjiang	6	19.35	90.32
3	Jilin	3	09.68	100.00

Due to its special historical background, at the end of the 19th century and the beginning of the 20th century, the textile, printing and dyeing, match and other light industry, machinery manufacturing, and even military industry rose in Liaoning province firstly.

Since the late Qing Dynasty in 1898, the germination of machine industry has been produced in Liaoning. After entering the period of China (ROC) in 1912, Liaoning's industrial development has entered a rapid period. Then Liaoning began in 1935 and was colonized by Japan for 14 years until the end of World War II [11]. After the founding of China (PRC), a national heavy industry base and a military-industrial base were first built here, giving birth to countless "firsts" in the industrial history of China (PRC).

At the end of the first 5-year plan, Liaoning's total industrial output value accounted for 16% of the country, ranking second in the country [12]. These special historical processes left rich industrial heritage not only for Liaoning province but also for the whole three northeast provinces of China (Table 6) [13].

Table 6: Representative of industrial heritage in Northeast China.

Name	Location	Category	Image
Yingkou paper mill	Yingkou city, Liaoning province	Manufacturing	
Daqing oilfield	Daqing city, Heilongjiang province	Mining	
Fengman hydropower station	Jinlin city, Jilin province	Energy	

5 RESULT

The imbalance index of Walloon Region's 0.5001 is much lower than Northeast China's 0.6129, it shows the distribution of industrial heritage in Northeast China is much higher imbalance than Walloon Region. The numbers of industrial heritage category in Walloon Region is less than in Northeast China, but the industrial heritage landmarks in Walloon Region is more than Northeast China. The main industrial heritage in Walloon Region is Mining, but it is Transport in Northeast China. The birth and development of the industry of these two regions mainly due to their geographical location and historical background. Therefore, the most density district distribution of their industrial heritage is located close to neighboring countries and the main categories are built by their historical stages of industrial development. In Walloon Region, Mons-Charleroi is close to France, Liege is close to Germany. In Northeast China, Heilongjiang is close to Russia, Liaoning is close to the Sea of Japan.

6 DISCUSSION

This research discusses the characteristics and causes of the spatial distribution of industrial heritage in Walloon Region, Belgium, and Northeast China from the regional scale, which is of great significance in scientific understanding and mastering the geographical spatial distribution of industrial heritage in Belgium and China and also has certain reference value for the protection and utilization of industrial heritage. The following aspects can be further studied in the future:

1. This research only discusses the current distribution characteristics of industrial heritage in Walloon Region and Northeast China from the perspective of space, and does not consider the internal differentiation of industrial heritage in China from the perspective of industrial categories. The influencing factors of distribution only consider the causes of the development of industrial heritage in the two regions from the perspective of history and geopolitics. The influence of industrial change, production division, and operation mode on the spatial distribution of industrial heritage is not considered.
2. This research only takes the industrial heritages of Walloon Region and Northeast China as the research objects. Although they are the leading industrial regions in their countries, however, this research did not put them into a broader regional scale to compared horizontally.
3. Industry is in a long-term changing process, social transformation and urbanization process will affect the industry. The post-industrialization degree of Walloon Region is higher than that of Northeast China. This study does not expand the concept of industrial heritage to a broader dimension to compare.

7 CONCLUSION

Both of the Walloon Region and Northeast China is the leading region in the industry in Belgium and China (PRC). However, there is no study to compare the two regions horizontally. This research fills this gap. In this research, the spatial distribution characteristics and influencing factors of 38 industrial heritages in Walloon Region of Belgium listed in the European Route of Industrial Heritage and 31 industrial heritages in Northeast China listed in the List of the National Industrial Heritage of China were analyzed in detail. From the scope of industrial heritage spatial density distribution, through geographic research method the imbalance index, after calculating and analyzing the data in the statistics on the distribution of industrial heritage. It found the imbalance index of industrial heritage in Walloon Region is much higher than that in Northeast China, the industrial heritages are mainly distributed in Liege and Mons-Charleroi districts, its distribution of industrial heritage category is less than Northeast China, only has 5 industrial heritage categories. Due to the special historical background, the industrial heritage of Northeast China has a strong colonial cultural color, the industrial heritages are mainly distributed in Liaoning province where is close to the sea of Japan. However, the same is that their main categories are concentrated in transportation and mining.

Through the longitudinal comparison of industrial heritage category and protection progress, the regeneration for industrial heritage in Walloon Region was started earlier than in Northeast China. Walloon region has entered the post-industrial era, with a large number of successful conservation cases and mature conservation experience. Northeast China is still exploring its path for the protection of industrial heritage. This can not only offer advanced experience in conservation from Walloon Region to Northeast China but also can present reflection for the protection of industrial heritage in Belgium.

The development of industry in Walloon Region was started earlier than in Northeast China, as well as the conservation of industrial heritage. However, the development speed has been slow down in the recent half-century gradually. On the contrary, the conservation of industrial heritage in Northeast China is getting attention incrementally. Landmark and social capital investment and social attention are increasing year by year. Keeping on eyes on industrial heritage conservation in Walloon Region and Northeast China will promote the two places to continuously stimulate endogenous power and radiate more vigor and vitality.

ACKNOWLEDGEMENTS
This research was funded and supported by the CoMod project (Compacité urbaine sous l' angle de la modélisation mathématique (théorie des graphes et des jeux)) and supported by the Faculty of Architecture and Urban Planning and the Faculty of Sciences, University of Mons.

REFERENCES
[1]   Glumac, B. & Decoville, A., Brownfield redevelopment challenges: A Luxembourg example. *Journal of Urban Planning and Development*, p. 146, 05020001-1-9, 2020.
[2]   Zhang, J., The overview of the conservation and renewal of the industrial Belgian heritage as a vector for cultural regeneration. *Information*, **12**, p. 27, 2021.
[3]   Gaofeng, W., The research status of industrial heritage in China. *Industrial Architecture*, **43**(1), 2013 (in Chinese).
[4]   Du Plessis, D.J., The evolving spatial structure of South African cities: A reflection on the influence of spatial planning policies. *International Planning Studies*, **20**(1–2), p. 87, 2015.
[5]   Statbel.fgov.be.   https://statbel.fgov.be/nl/themas/bevolking/structuur-van-de-bevolking. Accessed on: 01 Feb. 2021.
[6]   Zhang, J., The overview of the conservation and renewal of the industrial Belgian heritage as a vector for cultural regeneration. *Information*, **12**, 27, 2021.
[7]   The official website of the strépy-thieu boat lift. https://www.canalducentre.be/en/visits/the-boat-lift-of-strepy-thieu/. Accessed on: 26 Apr. 2021.
[8]   The official website of the le bois du cazier world heritage site. https://www.leboisducazier.be/. Accessed on: 26 Apr. 2021.
[9]   The official website of the Blegny mine world heritage site. https://www.blegnymine.be/. Accessed on: 26 Apr. 2021.
[10]   Yin, L., Study on the interprovincial population migration and its influencing factors in Northeast China. Jilin University, 2015 (in Chinese).

[11] Li, M., Advantages, problems and countermeasures of industrial heritage tourism development in Liaoning Province. *Journal of Capital University of Economics and Trade*, **16**(4), pp. 65–68, 2014 (in Chinese).

[12] Guo, M., Challenges and countermeasures of developing open economy in the northeast old industrial base – Taking Liaoning old industrial base as an example. *Economic Review*, **1**, pp. 65–69, 2016 (in Chinese).

[13] National industrial heritage list released, the official website of the ministry of industry and information technology website. https://www.miit.gov.cn/zwgk/wjgs/art/2020/art_dde0deaecfaf46b395e1de1ee401058c.html. Accessed on: 26 Apr. 2021 (in Chinese).

SECTION 7
LEARNING FROM THE PAST

TRADITION AND INNOVATION IN THE SCENERY CITY'S ARCHITECTURES: THE IMPACT OF FILIPPO JUVARRA IN CARLOS MARDEL'S 1733 PLAN FOR LISBON'S RIVERFRONT – A WATER-CITY PROPOSED DESIGN FOR THE ENVISIONED "ROME OF THE OCCIDENT"

ARMÉNIO DA CONCEIÇÃO LOPES & CARLOS JORGE HENRIQUES FERREIRA
CIAUD, Lisbon School of Architecture, Universidade de Lisboa, Portugal

ABSTRACT

The territory, as it was understood in the medieval period, undergoes a shift of paradigm in the early stages of the modern age. The growing importance of European capitals and city-states, as an expression of the court life of absolutist monarchies, and the advent of the Counter-Reform movement provide urban spaces with diverse moments and events, where architecture plays a decisive role in the assertion of different powers – notably within royal and religious elites. In the early 18th century, the highest power belonged to King João V, who wanted to transform Lisbon into the new "Rome of the Occident" in an effort to seek validation from the religious establishment of Rome. For that purpose, Filippo Juvarra was one of the talented architects brought to Portugal to apply the monarch's ideas. One of the least studied projects is the one signed by Carlos Mardel, which intended to significantly change the image of the city throughout several miles of its riverfront. By employing scenographic strategies from the Baroque period, the proposed plan reveals a very smart hydraulic technique which allows for the blending of the water and the "new city," forming a harmonious combination. The Portuguese model shows that the need to assert Lisbon as the capital of an overseas empire triggers changes in the architecture and the urban scenography, in order to feed new desires and ambitions. Taking into account that the most widespread images of Lisbon are its views from the river, the proposals for the regularization of the riverfront are now seen as an innovative and strategic motivation to recreate the city's image. As heirs of a strong tradition, based on a constructive praxis, engineers, architects, and construction masters develop innovative projects in response to new challenges, which will have considerable relevance in the international context.
Keywords: Baroque scenography, Carlos Mardel, Filippo Juvarra, Lisbon's riverfront theater curtain, Rome of the Occident, urbemarism, water-city.

1 INTRODUCTION

Time, shapes and places have very particular contexts, sometimes difficult to unravel, especially from a distant research perspective. Between the mid-17th century and throughout the 18th century, we identify a period of unique architectural production, when its relationship with the landscape reveals new developments with repercussions on project theory and practice.

The conflicts generated by the reformist ideas and the counter-reform response – primarily from Rome – the absolutism of the French monarchy centered on Louis XIV and the growing influence of the enlightenment thought, define the complexity of this period, which, in turn, underlines the need to search for alternatives to recover the idea of a unified society that supported the ancient world.

The mentality and innovations that emerge from this period – although pluralistic in nature, paradoxically long for the idea of unity. Amid religious convictions, political arrogance, and a growing awareness of free will (due to the enlightenment thought), compromises are established between faith, art, technique and economics that feed the creative spirit of the Baroque style, justifying the controversy and ambiguity that its expression originated.

A new trend, partly inherited and inspired by the rediscovery of classical architecture, is characterized by greater formal and conceptual complexity, in which several artists stand out, with emphasis on two major European locations, namely in the Italian atmosphere of papal influence and in the French monarchy.

In this context, a deeper commitment between architecture and other arts is considered, focusing on the design of gardens and the relationship between architecture and the landscape, within a new sense of place, with emphasis on its importance in urban development. The plans for Rome, guided by strategic axes that favor perspective effects referenced in notable buildings or sculptures and symbolic elements, accentuate a specific idea of monumentality, where architecture and place gain new meanings and new purposes within a certain scenography and theatrical exploration.

On the one hand, the Renaissance flourished linked to the interpretations of the classic works of imperial Rome, where Vitruvius' legacy remained a unique reference to the thinking and projecting process. On the other hand, the mannerist interpretations and innovations of its heritage contributed to new, bolder expressions, through Michelangelo or Serlio. These trends that marked architectural production throughout the 17th century – such as the Vitruvian and Serlian orientations, and the affirmation of a particular architecture designated as *Chã* (Plain Style) [1] – changed in the 18th century, due to a new dynamism associated with the Baroque spirit. In a new artistic and conceptual exploration, several authors stand out as main influencers in this period, such as Gian Lorenzo Bernini, Francesco Borromini, Guarino Guarini and Pietro da Cortona, to which we can add François Mansart, Christopher Wren, Andre Le Notre, among others.

Along these lines, architecture emerges within a singular synthesis between dynamism and systematization, in search for a significant completeness [2]. This synthesis of centralization and linear direction, reflected in the French royal squares or the axes suggested in the Rome of Sisto V, will find means of project exploration and unique achievements in several European cities.

Among the innovations of the Baroque period, specifically in the relationship between architecture, landscape and place, it should be highlighted the idea that space does not surround architecture but permeates its forms, it is imbedded in the design. Space is not an external element, that merely outlines architectural pieces. Space is a meaningful, inclusive aspect, generator of the spectacular perspective seen in the overall city and in the objects within (such as buildings, statues, staircases, streets, and so on). It is a glorious symbiosis of space and matter. This idea strengths the interpretation of a continuum fueled by perspective simulation and formal dynamism [3].

To these ideas, referenced by several authors who studied this period, it is important to add contributions that reveal the originality of this thought, free from the total control of papal Rome or French absolutism. In this regard, it is imperative to discuss the Iberian endeavors, distinguished by an overseas commercial and cultural context, and an unparallel relationship between architecture and waterfront landscapes. By this time it emerges a new data: the development of a city's project process has a great significance in the history of the city, as an object of architectural and urban reflection of the Baroque period.

2 A NEW IMAGE FOR LISBON IN THE REIGN OF KING JOÃO V

The decision to establish the view from the river – the *Ribeira* – as the physical support for a new image of the city, was based on the strong and ancestral relationship of the capital with a vast overseas empire. The main entrance to Lisbon, the waterfront, was the main attraction, the calling card of the city. Thus, those who arrived from the river needed to be amazed with the splendid, glamorous view, in aesthetic terms, of the capital of the empire.

The Class of the Sphere – in the College of *Santo Antão* – lectured by the Jesuits (1590–1759), contributed to the transmission of scientific and mathematical knowledge. The development of navigation in the age of exploration and discovery of new territories, added practical and theoretical knowledge to the traditional teachings of mathematics, geometry, cartography, algebra, and cosmography, as shown by Pedro Nunes (1502–1577), the high-cosmographer of the reign [4] (Araujo). This new knowledge advanced the materialization of a solid relationship between theory and constructive practice, particularly on the defense requirements of ancient and new overseas possessions.

The whole commercial, political and administrative dynamic surrounding the maritime Empire, was centralized in Lisbon – particularly, in the new downtown: the *Ribeira*. To support this new centralized dynamic, private palaces are built positioned along the riverfront, starting from *Terreiro do Paço* and *Ribeira das Naus*, to the west and east [5]. The gradual occupation of the land outside the walled system revealed a high sense of security in the defense. The sense of security and the new buildings outside the high walls, fostered even more the development of the riverfront.

The subject of security is relevant. Aside from people already living outside the walled system, there was a great development in pyro ballistics. Additionally, since Lisbon had an exceptional strategic position, the new palatine riverfront suppressed defense features. This was a major shift of paradigm regarding the capital's defense needs.

The idea of a "Rome of the Occident" filled the imagination of many influential people for many decades. For instance, Father António Vieira (1608–1697) encouraged the myth of a Fifth Empire [6]. The writings of Father António Vieira had a religious and mystical dimension surrounded by a holistic view of the capital of an overseas Empire: this was viewed as a way of spreading the Christian message to the world. King João V (1689–1750), a devout man with various cultural interests, probably knew his sermons as he was born during the life of the Jesuit priest. Most religious sermons in that era were written with such assumptions: it had become usual to compare Lisbon to Rome. Thus, it is not surprising that the monarch took a messianic strategy. The goal was to turn Lisbon into the successor of Rome, and the Portuguese Empire would be the replacement of the Roman Empire.

King João V's desire to elevate Lisbon as a new Rome was established at the beginning of his reign [7]. And it lasted until his death, to the extent that the eulogy in memory of King João V references the New Rome of the West, Lisbon, in comparison with the New Rome of the East, Byzantium [8]. King João V had an enormous need for approval by the European nations; he wanted to be seen as a powerful sovereign of an overseas empire. In order to achieve that, he decided to undertake a grandiose endeavor. The best place to show it, to provoke the most significant impact, is the riverfront, since the main images of Lisbon are views from the river. Thus, to create a new image for the city, it would be crucial the regularization of the riverside by conquering some land over the river. The sovereign had the intention to assert Lisbon as the capital of the Empire, with a new image as viewed from the river. Aided by religious devotion and, above all, by Brazil's recent gold and diamond discoveries, the absolute monarch stimulated the construction of an impressive riverfront. The king himself acquired six farms in the *Belém* area [9], in order to contribute to an increase in demand for this location by members of the royal court and other members of nobility. The option to expand the city to the west was an old desire that peaked in the period of João V. Rome and Italy were the great cultural and artistic references of the time: they were the major influential hubs for knowledge, culture, art and artists. Many artists and knowledge seekers from all over Europe were sent to study in Italy; and many established artists and experts in different fields left Italy to develop various parts of Europe. Since Sixtus V's plan for Rome (with the new requisites that emerged from the counter-reform), the principles applied to the embellishment of buildings and the city as a whole required an architectural component. The strong Baroque influence of the spectacle-city populated the imaginary of European courts.

In the 18th century, Portugal had great experience in waterfront urban settlements, in the many maritime cities of its vast overseas empire. In those places, in the middle of the Baroque period, autonomous models were developed in the occupation of waterfront territories. In the vast territory of the Portuguese empire, there was a need to consolidate squares, places, and cities – the technical drawing component, transmitted by treaties and writings, had become a critical tool, as seen in the developed projects. As a result of this systematic action, until the end of the 18th century, Portugal kept military architects and engineers developing public projects. Some of these experts, with military training, mutually shared their knowledge with architects, master builders, masons, and other craftsmen. Like in other European cities, Lisbon's riverfront established itself, starting to take shape with a renewed land project. The spectacle character of the Baroque established the physical reality of the riverside as a new way of creating waterfront architecture. The city sought to create moments for visual pleasure. It became a scenographic support, as a means to see the city beyond its scale (from within) and, simultaneously, to see it as an object of contemplation (from the outside).

During a Royal embassy, sent to Rome by King João V, an invitation was made to one of the most recognized Italian architects of that time – Filippo Juvarra (1678–1736) – to work in Lisbon. Juvarra's professional training as a goldsmith made him craft beautiful pieces of jewelry; these skills expanded into the design of creative pieces of architecture. Juvarra was born in Messina, which had been an important port for centuries and had a physiographic position akin to Lisbon, in terms of the dimension of the riverside. Both cities had similar characteristics, especially when viewed from the water (the ships arrived from the sea in analogous circumstances). Those assumptions are very clear in Juvarra's drawings, for Messina such as for Lisbon, with a real similarity in the approach to the territory [10]. At the Portuguese court, Juvarra developed designs for Lisbon's embellishment projects, such as a Royal Palace, a Patriarchal church and a lighthouse. All of the draws stablishing a scenic view to Lisbon, from the Tagus river. Maybe this was the main reason for Juvarra's criticism to Mafra complex (*Convent of Mafra*), from the architectural planes to the place – saying that it was an arid and desert land, with no water [11]. By using sketches and drawings to a great extent, the works of the Italian architect conveyed idyllic scenes. Juvarra was familiar with the *Teatro Marittimo di Messina* (also known as *Palazzata*), the magnificent water front where he also develops the project of restructuration and enlargement of the Royal Palace in 1714 [12]. Messina was known by is natural and architectural beauty, mainly due to the vast urban front of sumptuous palaces that faced the sea. Juvarra's personal view for Lisbon mirrored the spectacle-city where he was born, Messina (Fig. 1).

Furthermore, Juvarra designed a lighthouse for Lisbon (Fig. 2) extremely similar in structure and image to the one he also designed for Messina five years earlier (Fig. 3). He proposed the same elevation of the lighting point, through a circular tower, that simulates a pedestal in the form of a commemorative column (triumphal column), which resembles the Doric column ornamented with a human figure at the top [13]. Both proposals convey a strong relationship with the Trajan's column: the proposal for the Tagus lighthouse tower is the one that most resembles it, topped by a pedestal, with a human figure at the top – which is almost identical to the Roman column, surmounted by a pedestal with a statue of St. Peter, placed there in 1588 by order of Pope Sixtus V (Fig. 4).

Figure 1:   Juvarra's drawing – view of Messina and the plan for the new Royal Palace, 1714. *(Source: Turin National Library.)*

Figure 2: Juvarra's lighthouse design for Lisbon – 1719.

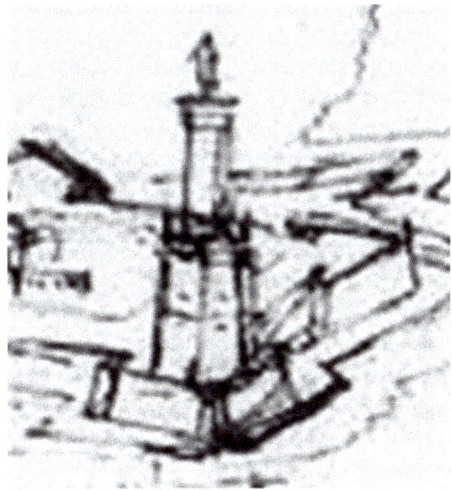

Figure 3: Juvarra's lighthouse at Messina (excerpt from Fig. 1).

Figure 4:   Trajan's column in Rome – illustration by Piranesi.

According to illustrations of that time, Messina already had a lighthouse. However, the Italian architect, certainly inspired by the emblematic columns topped by statues, redesigns the lighthouse, proposing to place at the top of Messina's lighthouse an identical image, in representative terms, to the one used in the proposal of the lighthouse for Lisbon's waterfront. The lighthouse would not only function as a warning for navigation, but also as an ornamental column – a scenographic image to embellish the riverfront. Thus, the grandeur aspect of the Baroque character that defines the thinking behind a whole proposal of city scenery is expressed in this new idea to beautify Lisbon's riverfront: the spectacle, the elegance, the exaltation of certain public figures, the grandiose style of territorial landmarks, along with other characteristics of similar nature.

After the departure of Juvarra from Portugal, no plans were developed to change the image of the city. After about a decade, the Hungarian architect Carlos Mardel (1695–1763) arrived in Lisbon from England. Between the departure of Juvarra and the arrival of Mardel in Portugal, no significant changes happened to the riverfront. During this interval, the king ordered a survey to get information in terms of geography, physiography and cartography, with the goal of providing groundwork for future developments. Even after Juvarra left Lisbon, the monarch's intentions remained unalterable: a letter from King João V to the Chief-Engineer of the kingdom, Manuel de Azevedo Fortes, references the geographical locations of *Pedrouços* and *Corte Real* (Palace) providing an indication for the size of the territory to be mapped, intended to be modified [14]. However, as mentioned, nothing relevant was done to the riverfront until Mardel's arrival.

Mardel arrived at Lisbon in 1733, roughly three years after the consecration of the Convent – Royal Palace of Mafra (October 22nd, 1730), whose works would continue until 1744. In all likelihood, Mardel visited Mafra's work, where he witnessed the royal aspirations at the time, in aesthetic terms. The building and its entire factory were established as champions of the national Baroque.

The military engineer and architect was valued in terms of regularizing waterfronts. Coincidentally, in the year of his arrival, a plan to regularize the riverfront appears with his signature. As for the contacts that take him to Lisbon and that make him develop a plan for the city, the context is not entirely clear. The records indicate that he immediately takes part in the building of the *Águas Livres* aqueduct and develops the project that ends the influx of water to the capital: the water reservoir of *Mãe de Água*, at *Amoreiras*. The professional connection to the waterfront is maintained, as he develops projects (examples: stone pier, *Lázaro Leitão* house, palace of the Eagles) located between *Belém* and *Alcântara* – along

Rua da Junqueira, which were on the riverside at the time. Moreover, Mardel develops several projects, in which the "water architecture" plays an important role, such as the urban embellishment through several fountains – among many other projects and plans, from his arrival in Portugal until his death, where he shows his skills as an architect and engineer [15].

3 A PLAN FOR A CITY: A SCENOGRAPHY BASED UPON WATER
Juvarra and Mardel may have never crossed paths, but the ideas of the former likely influenced the work of the latter. The king himself had, probably, some weight on this influence, since he had the opportunity to discuss ideas and drawings by Juvarra with Mardel, including the coastal image of the city of Messina and the *Palazzata* – named after the row of palaces placed on a continuous urban front, facing the sea (Fig. 5). The image of Messina was well known in Spain (Fig. 6) [16], even during the Iberian Union, wen Spain's empire had also the domain of Portugal and Sardegna Kingdom.

Figure 5: Painting depicting the city of Messina (18th century).

Figure 6: Gabriele Merelli, *Messina*, 1677 (24.6 × 37.5 cm), Madrid (Spain), Library Francisco de Zabálburu, Ms. 73-511, fol. 24r.

A similar idea was attempted for Lisbon: an idea for the entire city was based on a plan for Lisbon's riverfront, which would have a land extension bigger than ever before (Fig. 7). It establishes a whole new design for the western riverside front. It is through this riverfront scenography that a new image for the capital is sought.

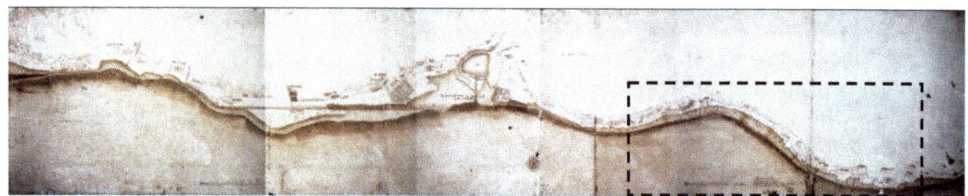

Figure 7:   Carlos Mardel's 1733 plan for Lisbon's riverfront (black rectangle is analyzed on Fig. 8). *(Source: Library and Historic Archive – Ministry of Public Works, Transportation and Communications. Carlos Mardel, Explicação do Caes Novo de Belém ao Caes de Santarém. Cota D 27C.)*

The proposal is loaded with a strong Baroque character, with an extreme appetite for the spectacular. There was a mixture of spectacle, fantasy and the sublime, in order to generate an innovative design. New ranges of Baroque urbanism are established, by defining a new scale and a new design, empowering the creation of a new city architecture. The end goal is to develop an "ideal city," a "scenic city."

Mardel showed that his combined capabilities as a military engineer and an architect, enabled him to excel as an expert, providing evidences of his boldness and of a grand vision for the capital of the kingdom, which was aligned with the King's expectations. Mardel reveals a great capacity for understanding and expressing the Baroque principles in his projects. As Hélder Carita defends, Carlos Mardel reveals another way of thinking about the city and its potential urbanism, opposed to the plain Cartesian rationalism and militaristic uniformity promoted by Portuguese engineers. Mardel's works displayed a scenic and festive sense of Baroque philosophy, and a wise and pragmatic adaptation to the realities of the natural landscape of a city cut between valleys and hills overlooking the Tagus river [17]. Later on, in the course of his work, the Hungarian architect would manifest talent and technical capabilities beyond the layout of the "new Lisbon," developing professional activity in various fields of architecture.

The plan was developed in the early 1930s of the 18th century (1733). It may have acted as a professional test to assert himself as an architect on his arrival at the kingdom. By combining aspects of constructive techniques by the water, Mardel gave evidences, as a military engineer, of a skill established by Vitruvius for the profession of architect, in the VIII book: to master the subject of hydraulics. This is, for G. Guarini, the 5th Factory: the Aquatic [18]. Analyzing the architectural project, we can see the intention of having water running through the city adapted to a new design. The area between tidal lines is intended for new buildings. A programmatic analysis concludes that, in fact, the improved city is drawn *in* water, advancing beyond the stream banks, onto the river (Fig. 7 and detail on Fig. 8).

One of the aims of the plan is to create a distinct urban moment, opening the city to the river: a "water square" is created, defined by two edified volumes, perpendicular to the river, without a building at the top [19]. It is called *Molho Grande*, located in the area called *Boa Vista*, between *Largo da Esperança* and *Casa da Moeda* (indicated by the white arrow in Fig. 8). Previously, in this area, facing the river, there was the intention of building a royal palace – this early idea probably justifies this exception in terms of river views. Delimiting the whole front, the continuous "river wall" serves as support for a wooded riverside promenade.

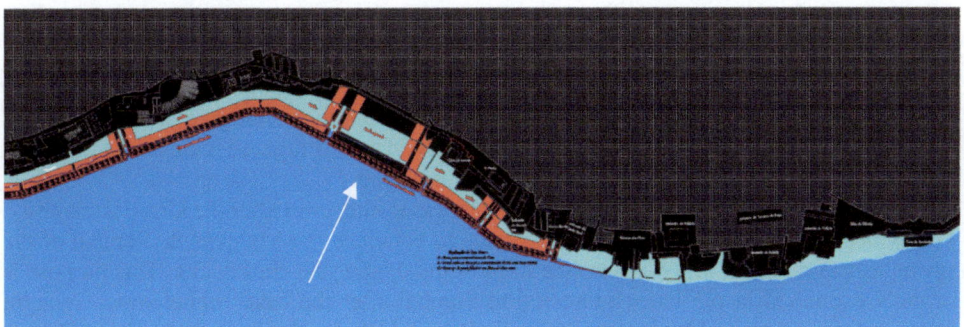

Figure 8:   Excerpt from Fig. 7 (area of the black rectangle). Mardel's plan for the area
             between *Tercenas* and *Terreiro do Paço*. Red: the buildings and the wooded
             riverside promenade; public walk with trees over the water. Blue: water. Cyan:
             the water between the new and the old buildings; tidal water entering through
             the canals. White arrow: interior water square, scenographic area open to the
             river. *(Source: Drawing made by the author of this article.)*

Juvarra's proposals essentially consist of embellishing the hillside facing the river, as well as entering Lisbon by sea. Mardel's proposal, however, is bold and progressive. Mardel plans the "new riverside city," with a linear layout and geometry, integrated with the environment – land and water. It is an example of sustainable development that is very up to date for the present – as if the architect himself could foresee centuries of distance, the very problem of rising sea waters.

One of the evidences of the programmatic assumptions of Mardel's plan, is surprisingly expressed in a letter addressed to Marquês de Pombal, sent by Lázaro Leitão Aranha [20]. The author of the letter is the owner of the palace at *Rua da Junqueira*, named *Casa Nobre – Lázaro Leitão*, projected by Carlos Mardel. This mansion, located on the riverfront at the time, was the first materialized project by Mardel in Portugal. Besides being a relevant ecclesiastical figure, among other titles, its owner was a teacher at the University of Coimbra. He was a person close to King João V – their relationship began in the first decade of the reign. He was aware of the monarch's desires; as a result, he was included in the royal group – called *embassy*, a diplomatic mission sent by the king – that travelled to Rome in 1712. Lázaro Leitão Aranha – Deputy of the Holy Office – was then 34 years old.

Lázaro Leitão is one of the main agents of the monarch's intentions and magnanimous aspirations. Among other aspects, Leitão visited Rome with the mission of obtaining a "Bula Aurea" from the Pope: the aim was to divide the city of Lisbon in two halves (Eastern and Western). Leitão also requested an authorization to build a new Patriarchal (the city's second cathedral). These authorizations, including the "Bula Aurea," were obtained on November 7th, 1716. Consequently, the city turned into a new administrative, religious and legal center [21].

The Royal expedition returns to Lisbon in 1718. In the following year, Filipe Juvarra – who, at the time, was considered the most famous Italian architect – arrives in Portugal. While in Rome, Lázaro Leitão presumably met Juvarra and learned about his importance as an architect. They were both ecclesiastical men, with Juvarra belonging to the Order of the Theatines. This religious organization had a great influence on the development of the Italian Baroque – they were the heirs of Mannerist principles and were committed agents of the aesthetic languages of the Counter-Reform movement. The most popular member of the order was Guarino Guarini, author of the project for the *Igreja da Divina Providência* (Church of Divine Providence) in Lisbon [22]. Guarini had developed projects in Messina, where he worked as a mathematics and philosophy teacher, from 1660 onwards. Juvarra was Guarini's disciple, although he became mainly known for working in Carlo Fontana's workshop and for the catafalque designed in memory of the Portuguese King Pedro II located at the Church of Santo António dos Portugueses, in Rome.

Lázaro Leitão remained loyal to his assignment in Rome, even after the king passed away. The evidence is in the efforts that Leitão made to convince the Marquês de Pombal of the advantages to redesign the Western side of the capital, after the earthquake. This purpose is strongly expressed in the letter addressed to Pombal after the catastrophe, which brought down, among other buildings, the Patriarchal for which Leitão had obtained the Pope's permission. The letter was written when the location for the reconstruction of the capital was being decided. According to Arthur Lamas, the text lists and corrects the main flaws of the old city. The message argued in favor of building to the West. The different reasons included beauty, functionality, security, commercial issues, good taste, and specifically, the view from the river: the image of Lisbon for those who enter or leave the capital by sea – by the main port of Lisbon. On those aspects he even appeals, as examples of beauty and public quality, to Italian cities like Napoles and Messina. Leitão even referred the great characteristics of the beaches in the western region (their length and terrain), although the beaches had two private owners – this disadvantage would be easily overcome, according to the letter, with the declaration that the site was "convenient to the public": thus, the individuals should give up the right to the land. The occidental area where the former King João V, had already acquire several parcels of land, previously to the earthquake.

The suggestion to redesign the Western zone was expected, since it was the most advocated hypothesis initially – right after the catastrophe, and fearing earthquake repetitions, the predominantly supported solution was the revamp of this area, since it was the section least affected by the disaster [23]. As França explained, in the first advanced document or "dissertation," as to the principle that the reconstruction should obey – among the five hypotheses (…); the fifth hypothesis proposes to abandon the city in ruins leaving the owners to do whatever they wanted, and to build a new Lisbon on the west side of the old one, on the land close to the Tagus river that the earthquake has saved – this was the most discussed idea, to rid the capital of similar disasters (…). Manuel da Maia declared his preference for such an idea that would allow him, without hindrance, to create a more solid and more beautiful city, but the choice belonged to the king – in fact, to Pombal, according to José-Augusto França.

When comparing Leitão's letter to Pombal and Mardel's project to regularize the riverfront, it is obvious that all the aspects listed as flaws in Leitão's letter were fixed in Mardel's project, and the several features considered as advantages by Leitão were highlighted in the Hungarian architect's plan. In regard to the flaws, these were qualities that the city lacked in contrast with other great European cities: the city lacked a public walk for the public to enjoy the view of the sea and lacked a wide street over the sea; Lisbon's view from the sea, up close, included only bald precipices and cliffs (not beautiful tall buildings); the beaches of Lisbon served for contraband; the people who traded did not have safe warehouses to store their possessions; and so on, as A. Lamas mentioned.

As mentioned before, the goal was to create an "ideal city," with Europeanized characteristics. The Baroque philosophy included components of exuberance and magnificence, that the capital of the empire lacked. The criticisms focused mainly on what was idealized, that did not exist in the city. Overall, the terrible earthquake was seen as an opportunity to provide dignity to the image of the capital.

Yet, some questions still remain: did Mardel share the ideas of the plan with his client Lázaro Leitão, who now sought to defend them by convincing Pombal? Or was King João V who introduced the ideas for Lisbon's great plan to the religious man, seeking to impress the Pope?

The design for the new urban front of Lisbon surpasses the vision of a basic façade. The Baroque tendency for the scenographic spectacle and opulence, pleased the magnanimous King. Due to the royal desire to turn the capital into a grandiose sight, the intended volume of construction was impressive – the image for the "new European capital" could not be limited to building a new patriarchal and a palace. The "new city," expressed in Mardel's plan, defined the new architecture of the city and simultaneously its own scenario. The entire western riverfront, from *Terreiro do Paço* to *Belém*, would include a vast housing complex, in several blocks, along the shore. It is a simultaneously regular and organic layout, which metamorphoses between the old riverside fabric and the tidal line – defining a new cityscape which "hides" the ancestral urban "mess" by creating a highly scenic built curtain. That's an example of the layout as a contribution of homogeneity and regularity, for the architectural unity of the whole, as a strategy to preserve the Baroque waterfront urban scenery.

Between the period of Mardel's scenography for the riverside front and the post-earthquake recovery, another Italian architect arrived in Lisbon, in 1752. Giovanni Carlo Galli-Bibiena arrived in Portugal by request of the King José I, and developed works with a heavy scenographic French for the capital: from the Tagus Opera to short-lived architectural pieces [24]. However, the magnificent plan outlined by Mardel would only be matched, in terms of spectacularity of city design and urban scenery, by the *Plano da Baixa Pombalina* – the plan for the downtown area named after Marquês de Pombal.

4 CONCLUSION

In the Baroque period, there isn't a specific established urbanism, but a set of Baroque principles applied in the cities. The design of the capital, by itself, would not turn Lisbon into a Baroque city. In other cities of the world, in the same period, there are some urban pieces, places, streets, squares, urban fronts, alignments, fountains, obelisks, and other elements with a great scenic effect, where there is an appeal to the spectacle and the view. The same would happen to Lisbon: the western section of the capital would be flooded with Baroque ingredients; however, we still could not call it a "Baroque city."

Mardel's project for the capital is a bold proposal, with a strong structural character for the western side of Lisbon. In the portions of the city that benefit from the Baroque style, most design proposals take advantage of the void (space) to create the moment of spectacle of the square, from a perspective point of view. A piece of an urban design project that, due to the nature of its connection to water and the definition of its layout, is completely unlike any other original design for Baroque cities of the time. In Mardel's proposal, this also happens, but it goes a step beyond: the empty space, the opening of the field where one seeks to establish and to create the expectation for the new view of the city, is located in the water. The river is the central feature of this landscape. And it is from the river, the best view of the city turned spectacle.

Through Mardel's proposal, Lisbon would also have its "edified theater curtain" – the *Teatro Ribeirinho Lisboeta*, similar to the *Teatro Marittimo di Messina* (*Palazzata*). The proposal takes advantage of the amphitheater position, the excellent physiography, favouring the view from the river, from South to North. The great view from the river, added to excellent views of the river, clean waters, healthy air and a substantial sun exposure, create a scenario of exceptional maritime and fluvial dynamics. This architecture never materialized

– it remained an idealization. However, the design drawn as a project, allows us to extract the ideas and ideals behind the plan – it can be seen as a preservation of the design process, as an urban architectural unit, which was established before the 1755 earthquake and materialized after it, with the development of the plan for the Baixa Pombalina area of the city. We would highlight the fact that Carlos Mardel, who drew up the 1733 Plan, was a member of the team of architects who won the competition for the *Baixa* plan.

Carlos Mardel's plan for Lisbon's waterfront was, perhaps, the last opportunity for a city design parallel to the river. This water-city proposed design, where the water is the pivotal element of the design, could be called "*urbemarism*": a term that results from the fusion of the words *urban* and *maritime* (meaning the water of the river and of the sea). The waterfront is, undeniably, the defining element of this projected new territory. The architect combined the traditional linkage between Lisbon and the water, with an extraordinary technical innovation – in the national context of the time – of great impact on urban morphology. The result would be a continuous front of buildings parallel to the water.

At the same time, there is a new image established for the city, along with an "event" – something unique in the urban context, a phenomenal scene, a *happening:* two built-up urban fronts, where the river component is defined as an intermediate element; keeping the riverside dynamics active between the two urban fronts – as if it was separating two realities of the same city, old and new. The old and the new Lisbon are connected by the dynamics of the tides.

Mardel's plan addressed the problems of the time, in light of the city's irregular waterfront situation, as well as the deficit of port infrastructures that hampered the huge flux of maritime trade into Lisbon; the sustainability of the old riverside was thus ensured and a new urban dynamic that was rather innovative and advanced for its time was created. This new approach to the problem of the water connection allowed the new urban environment – "the new Lisbon of the West" – to establish a form of urban sustainability consisting of different factors and functionalities which had thus far been lacking, both jointly and individually. Here, naval, military, commercial, residential, leisure and recreational aspects would be brought together and combined in a single plan.

The project reflects the understanding of the traditional city, its characteristics and its potentialities, as a scenographic support for a new image of the city. The innovation also lies in the form of appropriation of an erratic territory, transient in terms of its rhythms and its dynamics. As a solution, the plan assigns it a uniform, ruled, clear reading character, supported by a geometrized, linear and organized form; which, in turn, enables contemplation and an urban spectacle through a new front of palatine water – worthy of the capital of a Maritime Empire.

REFERENCES

[1] Kubler, G., *Portuguese Plain Architecture – Between Spices and Diamonds, 1521–1706*, Wesleyan University Press: Middletown, Connecticut, 1972.

[2] Norberg-Schulz, C., *Baroque Architecture*, Fantonigrafica – Industrie Grafiche Editoriali S.p.A.: Venice, Italy, 1971.

[3] Argan, G.C., *La Europa de las capitals*, Skira-Carrogio Ediciones: Barcelona, Spain, 1965.

[4] Araujo, R., *As cidades da Amazónia no Séc. XVIII – Belém, Macapá e Mazagão*, FAUP Publicações: Porto, Portugal, pp. 31, 39, 1992.

[5] Carita, H., *Lisboa Manuelina e a formação da Época Moderna (1495–1521)*, Livros Horizonte: Lisbon, Portugal, p. 193, 1999.

[6]   Pereira, P.J.G., *Lisboa – Porta do Mundo*, p. 56. https://www.academia.edu/
        7423872/LISBOA_PORTA_DO_MUNDO.
[7]   Rossa, W., *Fomos condenados à cidade – uma década de estudos sobre património
        urbanístico*, Imprensa da Universidade de Coimbra: Coimbra, Portugal, p. 365, 2015.
[8]   Sylva, F.X., *Elogio Funebre de D. João V.*, Regia Officina SYLVIANA e da Academia
        Real: Lisbon, Portugal, p. 85, 1750.
[9]   Rossa, W., *Além da Baixa, indícios de planeamento urbano na Lisboa setecentista*,
        Ministério da Cultura – IPPA: Lisbon, Portugal, p. 30, 1998.
[10]   Manfredi, T., *Prospettive dal Tejo. La nuova Lisbona di Giovanni V in tre vedute di
        Filippo Juvarra*, ArcHistoR, No. 7, Università degli Studi Mediterranea di Reggio
        Calabria, pp. 10–15, 2017. http://pkp.unirc.it/ojs/index.php/archistor/article/view/
        231/207. Accessed on: 15 Jan. 2021.
[11]   Raggi, G., *Filippo Juvarra in Portogallo: Documenti inediti per i progetti di Lisbona
        e Mafra*, ArcHistoR, No. 7, Università degli Studi Mediterranea di Reggio Calabria,
        p. 45, 2017. http://pkp.unirc.it/ojs/index.php/archistor/article/view/222.
[12]   Sutera, D., LEXICON: Storie e architettura in Sicilia (Messina tra Seicento e
        Settecento), No. 1. *L'Iconografia del Palazzo Reale di Messina*, Edizioni Caracol:
        Palermo, Italy, pp. 49–51, 2005.
[13]   Sansone, S., La reggia di João V di Portogallo. Il progetto per Buenos Aires a Lisboa.
        Filippo Juvarra 1678–1736, architetto in Europa, vol. 2, Campisano Editore: Rome,
        Italia, p. 205, 2014.
[14]   Ribeiro, D.M., *A Formação dos Engenheiros Militares: Azevedo Fortes, Matemática
        e ensino da Engenharia Militar no século XVIII em Portugal e no Brasil*, Tese de
        Doutoramento, USP-FE: São Paulo, Brazil, p. 72, 2009.
[15]   Vasconcelos, F.T. de M., *Carlos Mardel – elementos para a história da arquitectura
        portuguesa do Século XVIII*, Tese de Licenciatura: Lisbon, Portugal, 1955.
[16]   Manfrè, V., La Sicilia de los cartógrafos: Vistas, mapas y corografías en la Edad
        Moderna. *Anales de Historia del Arte*, **23**, pp. 79–94 and 81–87, 2013.
        DOI: 10.5209/rev_ANHA.2013.v23.41903.
[17]   Carita, H., *Dois Alçados Inéditos Do Palácio Real De Campo De Ourique*, Estudos de
        Lisboa, Revista de História da Arte n.º 11, Instituto de História da Arte, Faculdade de
        Ciências Sociais e Humanas – FCSH-UNL: Lisbon, Portugal, pp. 192–194, 2014.
[18]   Toussaint, M., *Da Arquitectura à Teoria – Teoria da Arquitectura na Primeira Metade
        do Século XX*, Caleidoscópio: Lisbon, Portugal, pp. 28–29.
[19]   Lopes, A. & Ferreira, C., *The Planned and Untold Story of the City's Architecture –
        the Pre-industrial Plan for the Riverside Boundary of Lisbon: Materialized and
        Remaining Aspects of Carlos Mardel's Plan from 1733*, vol. 191, WIT – Structural
        Studies, Repairs and Maintenance of Heritage Architecture XVI: Brussels, Belgium,
        pp. 91–92, 2019.
[20]   Lamas, A., *A Casa-Nobre de Lázaro Leitão, no Sítio da Junqueira (extra-muros da
        antiga Lisboa)*, Imprensa Lucas: Lisbon, Portugal, pp. 56–59, 1925.
[21]   Murteira, H., *Lisboa, da Restauração às Luzes*, Editorial Presença: Lisbon, Portugal,
        p. 103, 1999.
[22]   Tavares, D., *Guarino Guarini – Geometrias Arquitectónicas*, Dafne Editora: Porto,
        Portugal, p. 107, 2010.
[23]   França, J.-A., *História da arte em Portugal – o Pombalismo e o Romantismo*, Ed.
        Presença: Lisbon, Portugal, pp. 13–14, 2004.
[24]   Garcia, A., *Espaço Cénico, Arquitectura e Cidade – Guimarães, Um Modelo
        Conceptual*, Caleidoscópio: Lisbon, Portugal, 2016.

SOCIAL ROLE OF THE WALL: THE DOMESTIC VERNACULAR ARCHITECTURE OF SOUTH INDIA

ANJALI SADANAND[1], RAMASAMY VEERANASAMY NAGARAJAN[2] & MONSINGH DEVADOSS[1]
[1]Measi Academy of Architecture, India
[2]Hindustan University, India

ABSTRACT

Meaning and symbolism consecrate the vernacular house. The vernacular house is a response to the environment visible in the investment of knowledge systems imbedded in its architectural fabric. The house becomes an agency of socio-cultural norms through the architecture language and materiality of its spaces and elements and in doing so the domestic house acts as a repository of intangible and tangible cultural heritage. The objective of this paper is to explore the role walls play in conserving this heritage. The paper will study the manner in which walls construct social realities by looking at the different roles they play, through an exploration of their character, materiality spatial, structural and social function. Theoretical frameworks espoused by Simon Unwin, Yatin Pandya and Julienne Hanson will be used to support the discussion. In order to illustrate the context of the wall in domestic vernacular architecture. The discussion will focus on a comparison of walls across four typologies of vernacular houses of varying size, spatial organization and materiality from Tamil Nadu, South India. It will be argued that in cases where change is present, modern materials have shifted the emphasis of a value system based on the significance of socio-cultural norms to that of socio-economic considerations which have resulted in subtle transformations. In other situations vernacular traditions as architectural strategies and devices are repeated in modern houses with the intention of continuing tradition. The wall is objectivized and finds its way in contemporary architecture as an artefact which through memory of association assures identity and continuity.

Keywords: wall, vernacular, gender, status, symbolism, layering.

1 INTRODUCTION

Architecture it is widely accepted creates and shapes relations between people. Vernacular architecture is embedded with socio cultural knowledge systems as intangible and tangible heritage present in tradition and beliefs and manifest in their physical realities. Glassie elaborates on this with respect to walls and technology. "With the act of physical alteration that calls into space, implying a past and a future, and with walls that divide space, at once including and excluding, architecture has happened" [1]. "Technology is a corollary of human existence. As life unfolds, every technological act brings changes in two great relations: the one that always connects the humans and nonhuman spheres and, the other that is built to connect people with one another" [1]. "Vernacular technology depends on direct connections: direct success to materials and direct connections amongst suppliers, suppliers and consumers who simultaneously shape landscapes, social orders and economic arrangements, while wealth circulates in the vicinity" [1]. A study of walls in domestic vernacular architecture will inform us about local traditions and lifestyle patterns and hint at the direction taken in terms of transformations due to the transmission of other technologies through modernization. On the one hand, there is a desire to inculcate new systems and on the other hand a desire to keep certain traditions alive. The paper will demonstrate this aspect through case studies, in a vernacular setting where there has been an adaption of "new" technologies, relative to time, and as a response to needs. It will show that in an urban setting vernacular heritage is objectified and features as motifs and artefacts.

WIT Transactions on The Built Environment, Vol 203, © 2021 WIT Press
www.witpress.com, ISSN 1743-3509 (on-line)
doi:10.2495/STR210261

2 DEFINITIONS

The word "Vernacular" derives from the Latin "vernaculus" meaning domestic, "native" so the definition "native science of building" is really quite appropriate [2]. Oliver further elaborates his definition to say "in using the generally accepted phrase 'vernacular architecture,' I am embracing all the types of building made by people in tribal, folk, peasant and popular societies where an architect or specialist designer, is not employed" [2]. Included in the scope of vernacular architecture he continues to say "Although traditional village building has declined and the barriadas of Peru, the bustees of Calcutta or the favelas of Brazil are made from salvaged and scrap material, some architects have seen in the peri-urban squatter's passionate desire to build their own" [2]. "Vernacular architecture studies may in this way defined as the study of those human actions and behaviors that are manifest in commonplace architecture. "Vernacular architecture is a set of objects, the common buildings of a given place and time: as ensembles of buildings or vernacular landscapes, the products of a particular architectural community: as vernacular architecture studies, an approach to studying buildings as cultural manifestations" [3].

2.1 Building type definitions based on materials

"Those composed of short-lived materials-mud, sticks, grass-are defined as kaccha. (hindi for unripe raw incomplete) They contrast with pukka (proper, ripe, cooked. structures made to last, using more tenacious material-worked stone or timber, burnt bricks, lime plaster. Many combine several ingredients: Indian architects refer to them as semi-pukka" [4].

3 THEORETICAL FRAMEWORK

Paul Oliver and Rapoport put forth the connection between culture and material and construction as significant to vernacular Architecture. In Built to meet needs, Oliver states "The main purpose of the present collection is to consider the cultural factors that bear upon the subject" [2]. He makes references to material, construction techniques, plans etc. Talking about heritage he considers traditions which have been retained and are still in use. He comments, "Cultural traits and environmental contexts constituted the focus of vernacular tradition in buildings, which have often existed for centuries" [2]. "Undoubtedly, physical cultural and perceptual factors affect the degree of significance of certain features in form, structure, space use, or detail in buildings although changes over time are to be seen in most building traditions, the persistence of distinct building types and forms, of material resources and methods of construction and of space use and of associated value is undeniable" [2]. Oliver points out to aspects of tradition and transmission. "Traditions are sustained if they have meaning: they may be practical or symbolic" [2]. In reference to materials and technology, he says" innovation and change result from diffusion and experiment rather than from inducement and intervention" [2].

Rapoport in House Form and Culture in agreement with Oliver's, in House, Form and Culture suggests that socio-cultural factors override climatic factors and factors related to material and construction. Rapoport lists five socio cultural factors as "basic needs, family, position of women, privacy and social intercourse" [5]. He stresses on the significance of time, meaning and communication. Rapoport says that "vernacular design is achieved through the application of a system of shared rules. In effect vernacular design is best defined as being based on the use of a model with variations and differing from primitive design in the extent of variations. Since the model is shared and widely accepted, the resulting environment communicates clearly to their inhabitants" [6]. "Since in humans symbolic behavior generally is central and since artifacts, including buildings and settlements are one type of symbol which make concrete the immaterial, space less, timeless nature of values, meanings and life" [6]. Rapoport extends this understanding to the environment and provides us with a set of cues which he names as fixed and non-fixed elements to establish a system of non -verbal communication in which walls are considered as fixed elements.

Yatin Pandya in 'Elements of Spacemaking' suggests a framework for looking at architectural elements. He comments that walls are protective barriers and used in vernacular architecture also as storage. With respect to openings, he comments, "The door connotes an act of passage between two realms" [7]. Yatin Pandya in 'Concepts of Space in traditional Indian Architecture' describes strategies used in Indian vernacular architecture. He elaborates on sequential space, as articulated in architecture, to engage with the body in the form of movement through a system layering which create spatial narratives and employ columns and walls as elemental to cater to their purpose. Layering, Yatin Pandya comments is a spatial strategy for preserving social hierarchy and ensuring security and privacy.

McMurtrie states "Semioticised spaces are spaces that we can reach and use as resources for a specific social purpose and as such they are social constructs. Only structured semioticised spaces are relevant to the semiotics of movement in space" [8]. Space is text and the readings and experience of space, as meaning, with respect to volume, solid, as in wall, materiality, light and texture create the narrative which concertize spatial experience.

Kulbushan Jain in 'Thematic Spaces' in the Indian context, reinstates the significance of the threshold as an important aspect of Indian traditional architecture and element of the façade wall.

Simon Unwin in 'Analysing Architecture' and the Wall talks about the different roles the walls play by illustrating different types of walls. Examples are seen in the wall as a marker, as a form of enclosure, loadbearing wall and the inhabited wall which contains space and, wall as frame.

Julienne Hanson in Decoding Houses comments that "The important thing about a house is not that it is a list of activities or rooms, but that it is a pattern of space, governed by intricate conventions about what spaces there are, how they are connected together and sequenced, which activities go together and which are separated out, how the interior is decorated, and even what kinds of household objects should be displayed in the different parts of the home" [9]. Hanson uses many parameters such as, relationship of house to street, control of entrance, placing of objects within the interior, relationships of specific activities to rooms and spatial variables such as visibility/permeability, insulation/sequencing categoric differentiation/relative position. These are used for analysis within the domestic interior and between the interior and exterior of the house. Social norms that dictate behavior patterns and privacy are significant to understanding traditional architecture and the three spatial variables enlisted by Hanson in decoding houses suggest a way of analysis. The three significant spatial variables she postulates are as follows. "Visibility/permeability: Visibility refers to whether or not the interior of the dwelling can be seen from the street, or to whether it is possible to see clearly from one part of the domestic interior into another" [9]. "Visibility is about whether space is used to manifest objects and behaviors or to conceal them. It tells us about the relative transparency or opacity of the domestic setting, a permeability, which refers to the amount of control exercised over the way in which it is possible to move from one space to another" [9].

Insulation/sequencing are the second pair of spatial variables. "By insulation is meant the degree of discontinuity, that is, the strength of the boundary, between rooms. Where insulation is plus, rooms may be separated by a partition, or perhaps face each other across an intervening space. A railing or line of columns, a change of floor level or ceiling height, or even by differences in surface appearance. Cupboards or stores may be used to add mass and to emphasize the boundary wall" [9]. Hanson argues that a low insulation value is attributed to spaces demarcated by "a railing or line of columns, a change of floor level or ceiling height, or even by differences in surface appearance" [9]. "Sequencing is defined as the way in which spaces are connected together into chains" [9]. Hanson continues to say "Where sequencing is plus, it is always necessary to go through one space to reach another, and minus sequencing means that spaces are one cell deep from a central circulation space" [9]. Categoric differentiation/relative position looks at "aspects of spatial organization which are not so much morphological – to do with the internal logic of the physical arrangement – as microcosm effects – to do with the way in which spaces acquire particular social identities which particular functions are assigned unambiguously to specific spaces within the home" [9].

A review of parameters for analysis can be listed as:

- Perception of the wall

  — Size, form and material, elements
  — Aesthetics
  — Spatial arrangement – layering
  — Structural aspects
  — Openings
  — Social and Symbolic function as material culture
  — Aesthetics

- Types of walls

  — Boundary, marker
  — Inhabited wall
  — Structural
  — Barrier
  — enclosure

- Parameters based on spatial variables
- Visibility/permeability
- Insulation/sequencing

4 SCOPE AND LIMITATIONS

The four house samples are of varying sizes and typology and restricted to Tamil Nadu. They include kaccha, and pukka houses of different time periods and settings.

5 METHODOLOGY

The houses to be studied will be studied in a specified order based on the understanding of aspects of vernacular architecture explained at the beginning. The first house is a traditional four hundred year old house, located in an Agraharam, which will introduce the social system related to houses. The second example will show a larger house of a different typology built with the fusion of colonial and the vernacular tradition, in a town, indicating the transmission of technology into the form and aesthetic detail of building elements, the third house will depict a 'progressive' vernacular house, but built in modern materials and with the vernacular tradition relating to measurements and perceptions and the last house will be an example of a tribal Toda hut illustrating the vernacular tradition in kacha construction in a natural setting. References will be made to other examples for points of comparison. The examples will be analyzed with respect to the above listed parameters. A table showing spatial variables will indicate values for spaces in all four types. 0 indicates the space is not associated with any activity while 5 will indicate the space is used only for a particular activity. A mid-value of 3 will indicate the space has many activities which change during different times of the day.

6 DISCUSSION

6.1 House 1 – Type 1

House 1 is located in a village close to Tanjore. It is an example of a typical Agraharam model house. Andre Beteille describes an Agraharam as "the Agraharam is where all the Brahmin houses are located" [10]. It is the center of their social life. Agraharams are usually built around a temple either by encircling it on three sides like a garland or form a linear street pattern with houses on either side facing east west and leading to a temple. Houses share a common wall with their neighbor. Every part of the house responds to a functional need which caters to their relevant social norms. The typical house is deep, spatially organized around a single or multiple courtyards and has a back door leading to an alley or land. The house is linear and space is sandwiched between two walls. The street houses only the same caste. Houses directly interface with the street. The street is an extension of the house. The street view is of a string of similar house facades which differ not in architectural language but in size as in their respective widths construct a wall enclosing the street which in its original context was an extension of the house. The house directly interfaces with the street. The street is an extension of the house. The facades of the houses construct a wall enclosing the street which in its original context was an extension of the house Caste separation was strictly maintained in traditional India and people of other castes communicated through the rear door. In a social milieu of strong hierarchies and differentiation the Agraharam house displays a spatial program which preserves social differentiation based on gender and hierarchy through a strategy of layering of space using walls to articulate reveal and conceal public, semi public and private spaces.

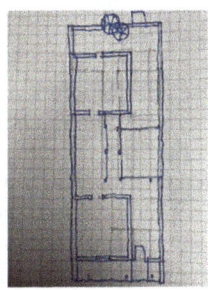

Figure 1: Plan of Type 1 house. *(Source: Authors.)*

Figure 2: Elevation of Agraharam house. *(Source: Authors.)*

Figure 3: Interior wall showing deity. *(Source: Authors.)*

6.1.1 Layering with respect to exterior
The house façade consists of a Thinnai area. A break in the Thinnai facilitates a point of entry. Layering is seen in two locations: (1) with respect to the street and (2) with respect to

the courtyard (Fig. 1). The house interfaces with the street via a Thinnai (Fig. 1). The first Thinnai, layer is demarcated by a raise in plinth and two low walls which protrude from the house façade and act as territorial markers and function as seats. They are called Thinnai. Social custom permits any passerby to sit on these low walls and the Thinnai demarcate public space. They are made of brick with plasterwork. The façade is divided into a wall which has one window, closed and shuttered and a second Thinnai which is semi-public in character and is covered and shaded, visible and permeable with a raised seating area. The raised area is designed for someone to sit on the floor from the street. Timber columns with carving on brackets support the sloped and tiled roof. As a boundary wall of the house, it corresponds to Simon Unwin's description of an inhabited wall. It is also a barrier as it insulates the house from the street with respect to socially screening the interior and in addition climatically creates a buffer zone to reduce the absorption of heat from the street. There is a haptic function to building elements. This space is multifunctional. Gender differentiation is followed as in the afternoon women visit each other using this space for social interaction while during the day men use the space and a bed can be placed here to catch the breeze. Beyond this space, internal layering is experienced in sequential placement of transition.

6.1.2 Elements of facade

Craftsmanship aside from being invested on the columns in Thinnai is focused on the front door. The architectural character in terms of carving and detail of columns, wall surface color and detail of the door give identity and reflect status of the house. The Thinnai rear wall is a backrest as is the columns. In some instances, a dado of paint on the wall corresponds to the portion of wall used as a backrest. Color and paintwork are used to create identity. The front door is aligned to the entrance of the second space and to the exit door beyond which leads to the alley. In the façade only one window is visible. The window is rarely opened ensuring privacy to the room. The front door is usually only five feet in height and one bends one's head when one's enters. The door frame is detailed with a carved overmantel over the door shutter usually decorated with the symbol of a deity. The carving and woodwork of the door follow a vernacular language of plank construction in locally available timber and with traditional craftsmanship and detail. Thresholds are symbolic and manifest fetishes, symbols which endow prosperity and keep off the evil eye. On either side of the door are two niches made into the wall which house lamps which will be lit when darkness falls. The front room, called camera is provided with a small window, closed at most times with timber shutters, and is dark. It the only room in the house with a small window overlooking the street and kept closed. In the interior, the wall of the main hall which has a picture ledge along which ancestral photographs are hung and in one area religious pictures of reigning deities serve as an altar for worship (Fig. 3). The wall itself is painted with iconography of symbols to which offerings and prayers are made. This imparts a sacred quality to the hall which results in taboos being maintained towards entry to the space for women during menstruation. Space is text. The wall itself in part becomes an object of ritual and symbolic in its use. The Hall is a primary social gathering space of the house.

6.1.3 Layering in the interior

An ante space leads to a colonnaded passage open to a courtyard. There are two rows of columns and as metaphoric walls the columns demarcate different zones ranging from public to private, profane to less profane and more important. Spaces in the houses are designated according to Hindu building norms called 'vastu.' There are two rows of columns which between them bound a transition space of low categoric value. The inner row of columns border a hall the family space of the house which is used for social functions and has therefore more significant social value attached to it. The columns are short with decorated capitals richness of detail in wood craftsmanship and reflects status (Fig. 1). The columns are more articulated than the outer row close to the courtyard reflecting the status of the thalavaram which is for general use. The spatial narrative created in the form of a sequential journey from public to semi public and private, through spaces of differing volume materiality light and articulation established through a system of layering characterize the architectural character of the Agraharam house. Walls are loadbearing and made of bricks set in lime mortar which keeps the house cool. The front door is aligned to a rear opening which leads to an alley and fields at the back of the house.

6.1.4 Spatial variables
Visibility is ranked low as the interior is not displayed to the street through alignment of front door (see Table 1). Categoric differentiation of spaces value is a neutral 3 (Table 1), as only kitchen and the front room have designated functions. The interior wall of the main hall which has a picture ledge along which ancestral photographs are hung in one area with religious pictures of reigning deities serve as an altar for worship (see Fig. 3). The wall itself is painted with iconography of symbols to which offerings and prayers are made. This imparts a sacred quality to the hall which results in taboos being maintained towards entry to the space for women during menstruation. Wall and Space are text. The wall itself in part becomes an object of ritual and symbolic in its use. The Hall is a primary social gathering space of the house. The house presents cues which are universal to the members and can be read and followed rigorously. In Rapoport's terminology walls act as fixed elements delivering a message which determines social behavior. In Gibson's terminology of affordances the wall in an Agraharam house is negotiated, adapted and transformed into every usable aspect to satisfy social needs. It is an architectural structural element, a component of enclosure, marker of territory, furniture, provider of social interaction and neighborliness and an object of ritual. It enacts its role within a set of codes acknowledged, accepted and read and understood by the people of its environment. Areas of taboo are maintained and respected even if they are not physically marked by territorial walls. The effect of the wall as a social barrier is felt even in its absence. The interior colonnades act as metaphoric walls dividing more profane areas from spaces of higher status. Common walls between courtyard and neighbor are barriers and high and do not encourage social interaction between neighbors but insulate the houses socially.

6.2 House 2 – Type 2

Type 2 house is the house of a gentleman of Marathi origin, who was adviser to the Thanjavur royalty and represents the bungalow type residence built in a town, more than 100 years ago, during the colonial period. It is set in a large ground of land, set back from the road and represents a fusion of colonial and the traditional model. It has a large hall typical to such houses of the period as suggested by Ananthalwar, in his treatise on Indian architecture.

The house can be divided into two zones, a public formal zone in front and a rear private zone organized around a courtyard.

6.2.1 Layering
A compound wall (Fig. 4), demarcates the property and an ornate entry set in the wall marks the entry point. There is a clear definition of public and private. The house unlike the

Figure 4: Entry to House 2. *(Source: Authors.)*

Figure 5: Thinnai in two layers. *(Source: Authors.)*

Figure 6: Showing arched entry, niches and colonial light fitting. *(Source: Authors.)*

Figure 7: Interior hall showing colonial columns and clerestory. *(Source: Authors.)*

Figure 8: Niche with painting of deity and fire altar on ground. *(Source: Authors.)*

Figure 9: Columns in rear and private area. *(Source: Authors.)*

Agraharam house is not permeable to the street. The compound wall delineates boundary as in territory and acts as a barrier protecting the house through which there is a single point of access. Unlike as in the Agraharam house, security is demanded by its setting as the cultural landscape is diverse. Type 2 demonstrates the use of layering as a tool for creating social differentiation primarily based on gender and status. There are two levels of Thinnai (Fig. 5), incorporated into the bungalow and separated by a change of level. Through a sequence of layers one enters the main house interior. The first layer is two storey high and is composed of structural brick piers. The second layer is made of arched openings set in a wall. Technology is used here to differentiate the two environments and reflect status, the higher Thinnai being reserved for people of higher rank. This space is a transition to the interior and a space restricted to men and outsiders of a rank who cannot be entertained in the reception hall in the interior. The front door (Fig. 6), aligns to the interior kitchen window on one side and the front entry in the compound wall on the other side. The window is a point of control. Past the front Thinnai one enters another transition space, before entering the hall. Windows flanking the entry open into the second Thinnai area. Windows are large and disclose the interior. The rooms and transition spaces 'insulate' the main central hall, which the most significant space and accessible from the outside through three intermediate layers The reception hall has a large central area covered by a clerestory roof which is delineated by a colonnade of circular columns with capitals which reference colonial buildings and further indicates transmission of colonial technology (Fig. 7). It can be mentioned here about the similarity of the columnar framed space used as a reception area to that in Vishrambaugh Wada in Pune of Maratha origin where the columns are made however, of timber. The front part of the house which represents the public image is detailed in a colonial architectural language.

6.2.2 Elements

The front door is a double door with a deity in a panel above the door frame. Traditional elements such as niches and motifs of deities contribute to the resilience of belief systems (Fig. 6). The window is a point of control. Private rooms are spatially organized around the inner courtyard which holds a tulsi plant. Unlike the Agraharam house visibility is increased here of the interior and its objects and permeability is controlled. Two layers of Thinnai offer insulation from the outside in addition to the house being set at a distance from the road. In the Interior the main hall is enclosed by walls from which portraits of ancestors hang. One wall has a niche and for images of local deities form an altar.

6.2.3 Layering in the interior

The inside is for women and family and treated simpler and in a different scale with different finishes seen also in the rudimentary columns (Fig. 9), around the courtyard which follow the similar system of layering and spatial hierarchy as in the Agraharam house.

6.2.4 Spatial variables

Type 2 maintains status and gender differentiation is used by tools of layering resulting in high insulation of the hall and private spaces followed by categoric differentiation and high values of visibility and permeability from inside the compound showing a tendency to discreetly reveal the interior to the chosen few who are allowed to enter.

6.3 House 3 – Type 3

Type 3 house is represented by the potter's Kasi Rajan's house in Chettinad. Type 3 offers an example of the new "vernacular" with industrial materials to form a new vernacular language. Type 3 is representative of his choices within the context of the vernacular. Type 3 shows the aspect of vernacular architecture which constitutes it as a process. The potter is an established craftsperson known for making Ayyanar horses. Ayyanar is a deity specific to Chettinad. The house evolved from a semi-kaccha structure of local materials of thatch roof and brick walls in mud mortar to a concrete and brick house (Fig. 11). The first house of the potter comprised of three spaces. A front ante space with a cooking alcove and two rooms behind. As his livelihood relies on pottery keeping his pottery safe and dry is his main concern. As the thatch proved to be not watertight the new structure was built with a concrete flat roof, with a terrace used for gatherings and corresponds with the advancement in his stature to becoming a prominent potter. Houses in Chettinad are built for security as often women are left behind while husbands and menfolk travel overseas on business. The Chettinad community has a definitive house form visible in the plan in Fig. 13. The new house of the potter is an adaption of this model on a smaller scale.

Figure 10: Interior wall with window. *(Source: Authors.)*

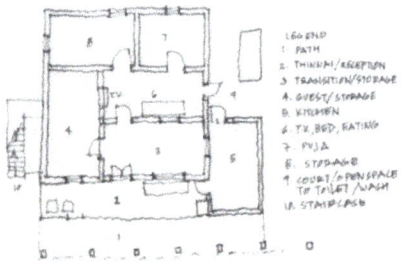

Figure 11: Kasi Rajan house. *(Source: Authors.)*

Figure 12: Thinnai area showing window. *(Source: Authors.)*

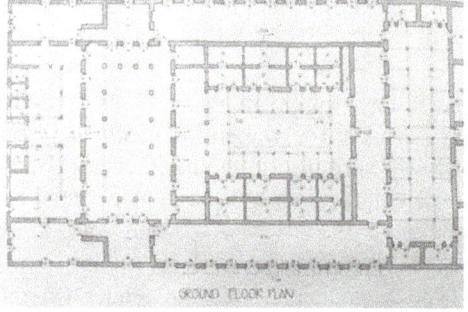

Figure 13: Plan of Chettinad house. *(Source: Measi academy of architecture student documentation.)*

6.3.1 Layering with respect to exterior

The house interfaces with the street through an open Thinnai unlike the standard Chettinad model of Fig. 13, where security and privacy are paramount. The potter is assisted by his wife and the Thinnai functions as a space for social interaction with outsiders and a space for work.

6.3.2 Layering with respect to the interior

The house retains the essential spatial elements of the Chettinad house seen in layering and the use of long rectilinear halls which are transition spaces and makes provisions for small transformations determined by the potter's lifestyle. Layering continues beyond the front door, and the next space in sequence is a transition space of no specific function, followed by the living space and then a series of rooms. Layering of walls insulate through the creation of transition spaces and enforce privacy to the interior and family and private zone. In the potter's house all spaces are used for storage of pots.

6.3.3 Elements

The façade wall of his new house has a window and a grille which is painted in many colors (Fig. 12). The Thinnai roof is supported on a concrete column. The doorway is embellished. The façade wall is decorated with colourful paintwork. A painted grille reveals the interior transition space. Windows are present but kept closed (Fig. 11). They have glazed shutters and are designed in a contemporary idiom, unlike Agraharam windows with wooden shutters. They offer no visibility to the interior but suffice as elements of status and are replicated to emulate superficially the style of more affluent Chettinad homes and in this sense are symbolic. They have glazed shutters and are designed in a contemporary idiom, unlike Agraharam windows with wooden.

6.3.4 Spatial variables

Categoric differentiation is low. Visibility and permeability are low and transition spaces insulate the interior living spaces. All efforts to afford privacy to the main central living area is manifest in the plan. The Thinnai roof is supported on a concrete column. Technology is used to suggest stature seen in concrete column and roof and the window shutters.

6.4 House 4 – Type 4

Type 4 is a Toda hut, the abode of Tribal, buffalo herdsman in the Nilgiris in Tamil Nadu. The Toda hut is a kuccha structure (Figs 14 and 15). The hut is parabolic in shape, it contains a single volume which is entered through a small three feet by three feet opening. The roof and wall merge and the wall can be redefined as a building skin and not comprising the planarity of the other three typologies. Cultural heritage invested in building technology is present in the constructional technique of the roof cum wall and in the front wall which is made of granite slabs.

6.4.1 Layering

Toda huts are single volume structures within a compound bordered by a low stone compound wall which defines territory and keeps animals out. A Thinnai area opens to the environment and is the transition between outside and inside.

Figure 14: Toda hut. *(Source: https://www.ijitee.org/wp.)*

Figure 15: Toda hut showing raised platform for visitors. *(Source: http://psrcentre.org/images/extra images/15.%201312074.pdf.)*

6.4.2 Layering in the interior

Toda huts are single volume structures and space internally is divided into three main spaces. A raised platform on one side and equipment for milk churning, an activity performed by men which screens a cooking area, reserved for women, in the rear of the hut.

6.4.3 Elements

The roof cum wall in bamboo reflects a skill in crafting which has been compared to embroidery for which Toda women are well re-known. The hut facades are often painted with symbols for protection such as the sun, moon floral motifs and buffalo horns. The hut is built by the community and in this respect a social artifact.

6.4.4 Spatial variables

Categoric differentiation is specific in the interior in the rear reserved for women. There is no visibility and little permeability in the façade as the huts are located in natural landscape of the Nilgiris Hills and need protection from wild animals.

7 FINDINGS

Table 1 show the spatial character of the four houses and further illustrates the significance of the wall in establishing certain preset systems of social relations. The relationship of house to street or environment is seen to be more open in the Agraharam and potters house and more hierarchical in the Marathi house, where it uses depth and extensive layering to maintain exclusivity. Categoric differentiation which is dependent more on gender separation than can be attributed to activities is displayed in all cases. It can be seen that there is a significance of privacy and culture over technology, with the maintenance of certain fundamental value systems. Categoric differentiation is not invested in different types of activity but invested in social hierarchies. The wall acts as a frame for establishing ancestry and thereby constituting status (Fig. 8). Four kinds of walls can be seen-the inhabited wall seen in the niche for deities, the permeable columnar wall and the impermeable compound wall. Colonial construction technologies introduced the masonry pier and high flat ceilings on timber beams called madras terrace roofing. Technological innovation is used to emphasize stature, and security measures and durability as in the case of the potter's house. The potters house shows the vernacular interpretation of modern construction technology within a modified traditional spatial matrix. Modifications are visible, seen in the provision of a terrace and its social function and the use of concrete. The hut is built by the community and in this respect a social artifact. The Toda hut exemplifies Rapoport's dictum that vernacular architecture is both process and product. The hut presents an authenticity and cultural rootedness which has been passed on by generations and preserved. As local indigenous materials are used there is constant need for rebuilding and this continues the tradition and reinforces the essence of vernacular architecture. The tectonics of its structure are based on a knowledge system handed down by generations and realized by the community. The door remains a significant element withstanding change and retaining its status as a threshold between realms. Windows are secondary. Changing conditions in the settlements, have resulted in transformations for example in Fig. 19, a compound wall is added to an Agraharam house for security and in Fig. 2 the front wall is used as a surface for political graffiti (Fig. 17), is converted to a shop. There has been a loss of social meaning to the Thinnai. While the vernacular man takes what is available to modify his conditions to survive, the urbanite uses the vernacular architectural language as a source of patterns, which can be used to adorn his house evoking a past in a Post-Modern language of semiotics (Fig. 18), where vernacular motifs are used symbolically to highlight entry and create stylistic statements.

8 CONCLUSION

Change and transformation is happening at the village level. Demolitions and change have brought new environment and changed the cultural landscape and setting seen in the additional need for security and a degree of inclusiveness (Fig. 19). In the city two directions reference the vernacular. The first as Fig. 18, show contemporary houses where the vernacular element is used to suggest an association of roots and tradition as an attempt to recreate the past with a sense of romanticism and to retain its histories through association.

Table 1:   Relation of house to street via its façade through visibility and permeability and spatial social function.

Spatial variables	Type 1	Type 2	Type 3	Type 4
Visibility-facade	5	2	3	5
Interior	1	3	3	1
Permeability	5	2	2	1
Sequencing	5	5	5	1
Insulation	4	4	4	1
Categoric differentiation	4	4	3	4

Figure 16: Grilles in thinnai. *(Source: Authors.)*

Figure 17: Shop. *(Source: Authors.)*

Figure 18: Modern house with traditional entry. *(Source: Authors.)*

Figure 19: Agraharam house with compound wall. *(Source: Authors.)*

Figure 20: Thinnai in modern bungalow. *(Source: Authors.)*

Figure 21: Traditional door. *(Source: https://property.sulekha.com/in dividual-houses-villas-for-lease/madhavaram-chennai.)*

The second as in Fig. 20, shows the vernacular element as motif. Words disjointed from a vocabulary used not in own grammar but symbolically used as a reminder of the past as fragments of the wall. The Thinnai no longer affords a view of the street, its social meaning is modified as a message of welcome, perhaps alluding to its past. The context has changed and its purpose is modified, to serve as a form of imagery of the vernacular. There is a transfer and not transmission of knowledge. Tradition is still held in respect though its manifestations are different and in the current form, as an artifact, social value is removed and what was part of a system of cues in a fabric of tradition woven according to need and symbolic in function now becomes a commodity of economic value representing material value. In this respect heavily carved doors and capitals and columns find their way into newly constructed villas (Fig. 21). What was part of a cultural landscape as material culture lives now in its new form, as a fragment as an *object d'art* and as a metonym.

REFERENCES

[1]   Glassie, D., *Vernacular Architecture*, Indiana Press: USA, p. 37, 2000.
[2]   Oliver, P., *Built to Meet Needs*, Architectural Press: UK, pp. 4–18, 2006.
[3]   Carter, T. & Elizabeth, C.C., *Invitation to Vernacular Architecture, A Guide to the Study of Ordinary Buildings and Landscapes*, The University of Tennessee Press: USA, p. 18, 2005.

[4] Cooper, I. & Dawson, B., *Traditional Buildings of India*, Thames and Hudson: London, p. 11, 1998.
[5] Rapoport, A., *House Form and Culture*, Prentice Hill: USA, p. 12, 1969.
[6] Rapoport, A., *Vernacular Architecture and the Cultural Determinants of Form, Buildings and Society*, ed. K. Anthony, Routledge Kegan and Paul: UK, p. 286, 1980.
[7] Pandya, Y., *Elements of Space Making*, Mapin Publishing Private Ltd.: India, p. 14, 2007.
[8] McMurtrie, R.J., *The Semiotics of Movement in Space; A User's Perspective*, Routledge, 2017.
[9] Hanson, J., *Decoding House*, Oxford University Press, 1998.
[10] Beteille, A., The Economic weekly annual number February 1962, Sripuram: A village in Tanjore district. https://www.epw.in/system/files/pdf/1962_14/4-5-6/sripuram_a_village_in_tanjore_district.pdf. Accessed on: 15 Mar. 2021.

TOWARDS DEVELOPING AN ECOLOGICAL TOURISM SETTLEMENT IN SIWA OASIS, EGYPT: CASE STUDY OF BABENSHAL ECO-LODGE

OLA ALI BAYOUMI[1], MOHAMED ABDELALL[2] & MOHAMED ANWAR FIKRY[2]
[1]Department of Architecture, Alexandria University, Egypt
[2]Alexandria University, Egypt

ABSTRACT

This paper is aiming to revive the real meaning of an "Eco-lodge" in the Egyptian western desert. Siwa was a totally isolated oasis centered at the North of Sahara desert and connected to the Egyptian Western desert, and thus it has a unique environmental features developed for thousands of years ago. Unfortunately, the new roads established by the Egyptian government to connect between Siwa oasis, Matrouh and Giza cities lead to the transformation of lots of new ideologies and innovative building materials. Beside the previous obstacles, foreign architects ignore most of the building strategies and restrictions created by local residence during establishing new "Siwan-tourism settlements." According to Dr. Emad's words, the designer of Babenshal eco-lodge, Siwan builders exchange their experience to future generations in order to respect Nature; the reason is they are totally aware that Nature's punishment is totally destructive and this is the truth that most architects forget about. For that the researcher will try to continue their research on giving a sample weighting for any future eco-lodges that are or will be built in Siwa oasis. According to their previous paper under the name "Developing an Ecological Assessment Tool for Siwan eco-lodges in the Egyptian Western Desert (EWD)" the researcher had created a criterion specified for local Siwan eco-lodges giving *weighting coefficient values* for Environmental and Social items, indicators and their parameters. By field work study and the usage of air quality multi-meter device they can find the mean value of the highest three indicators of IEQ which are CO_2 in ppm, temperature in degree Celsius and air flow in m/s, multiplying them with their weighting coefficient = 0.14. This method will be applied on one of the most famous eco-lodges "Babenshal Eco-lodge" as a case study to scientifically rate its environmental indoor quality.
Keywords: air quality measurements, ethnographic approach, Babenshal eco-lodge, Siwa oasis.

1 INTRODUCTION

Siwa oasis is the heart of the Egyptian western desert. It has a completely isolated culture and unique topographical features, the first is due to the different cultures that passed through it without sustaining their existence starting from the old Egyptians, Romans, and Greeks and ended with the Islamic conquest and the "*Amazigh*," while the second is due to its location where it locates in between 15 and 20 m below sea level and surrounded by high lands with the height 30 to 40 m above sea level [1]. On other hand its geological lands are concentrated by highly salty soil and shallow saline lacks that are used for medical purposes. Although the high salinity of the soil, it is so reach with minerals and mud that help palm trees to grow up (see Figs 1 and 2).

All of these features created unique architectural characteristics for Siwan lodges. Siwans in return respects Nature and fortune their private life. On other hand their Social life has a direct impact on creating those unique Siwan buildings to serve their "Indoor life quality"; this is because the harsh environmental characteristics as the oasis locate in the heart of the Egyptian western dry desert. After the researchers multiple visits to understand all those above features, they noticed the extreme and dangerous changes that happened to Siwan eco-lodges over the last 10 years. And thus, they decided to continue developing their research to calculate the values of the highest environmental aspects indicators of their previous research

WIT Transactions on The Built Environment, Vol 203, © 2021 WIT Press
www.witpress.com, ISSN 1743-3509 (on-line)
doi:10.2495/STR210271

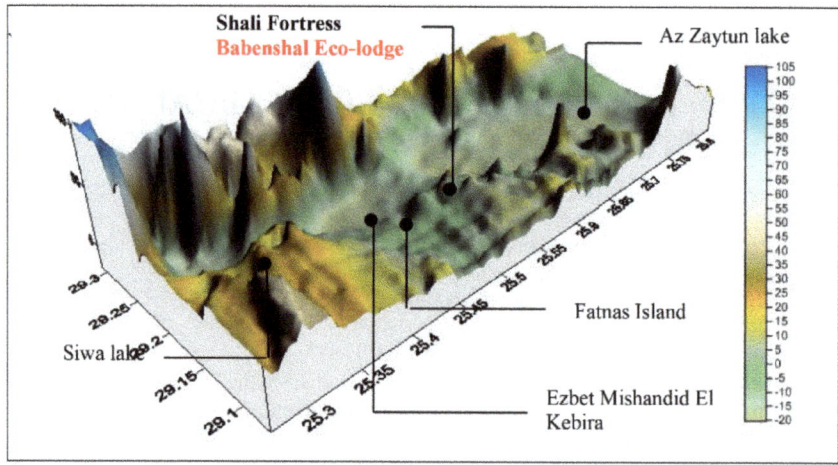

Figure 1: Topographical perspective of Siwa oasis. *(Source: Authors.)*

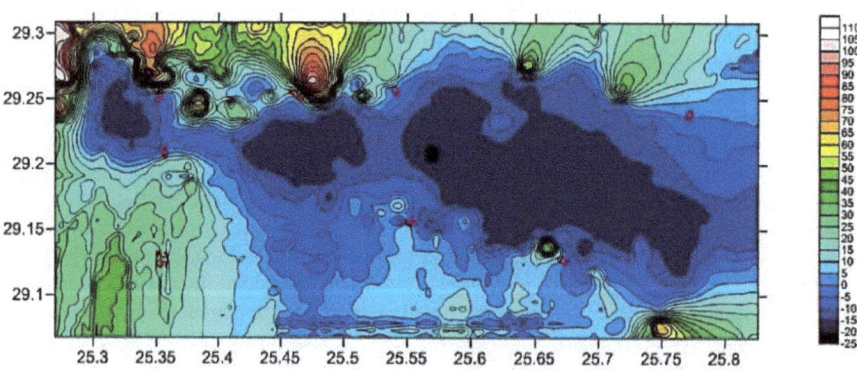

Figure 2: Topographical map of Siwa oasis. *(Source: Authors.)*

which are (life quality/CO_2, thermal comfort/Temperature and Ventilation quality/Air flow). Those values will be multiplied with their previous-established weighting coefficient = 0.14 that is built for only Siwan eco-lodges [2] and the result will be the environmental rating of the Babenshal eco-lodge. This method is a key method that can be used to measure any Siwan eco-lodge's IEQ.

Moreover, those environmental aspects indicators have been chosen by Babenshal's designer and its construction engineer during their interview with the researchers. This interview has been recorded and transformed into a written index to show their opinion on the way in which Siwan eco-lodges should be developed [3].

2 SIWAN ECO-LODGE DEVELOPMENT

Siwan eco-lodges are the key-projects and the clear image that reflects Siwan culture and their architectural unique features. Due to economic, political and limited environmental variations, the main properties of Siwan accommodation and EWD Eco-lodges in specific have significantly been changed [4].

According to Table 1, Siwan eco-lodges have been changed in their size, way of construction and the distribution of internal spaces and those variables have been connected to the different locations and the construction method they are related to. However each case carries its pros and cons, but unfortunately the recently developed eco-lodges where related to the short term benefits. According to Dr. Emad words; "Due to the invasion of new chemical materials in the oasis, lots of local labors lost their mechanism and their experience in how to build with local provisions and lots of them already had stopped using their traditional techniques in their buildings and thus it was essential to encourage people financially and scientifically to build in their old ways" [3].

Table 1:   A comparison shows the significant variations happened to Siwan eco-lodges since the beginning of its establishment till now. *(Source: By the researcher.)*

	"Abu Muslim" village accommodation	"Shali Village"	Residence descended to the plain	Siwa now
Siwan building development				
Main properties	Primitive shelter	• Accommodation that suits Shali Site topography • Vertical extension • The roof is alternative to open courtyards	• Accommodation that suits the plain below Shali fortress • Horizontal extension • Open courtyard appeared	• Governmental new buildings that develop vertically with no courtyards
Openings % of the shelter elevation	Tent structure	Kershef material (mud–salt–straw)	Kershef material (mud–salt–straw)	Limestone or red bricks are covered with Kershef as clay or paste
Mass exposure % Elevation exposure % Roof exposure % Building mass/land	10% 30% 75% 30% 100%	5% 12% 47% 12% 100%	10–15% 11–14% 27–33% 12–14% 70 (Gov.)–74% (local people)	15–20% 10–15% 33–36% 5–15% 100 (Gov.)–787.4% (local people)
Images of developed eco-lodges				

3 RESEARCH METHOD FOR MEASURING ENVIRONMENTAL QUALITIES OF FUTURE BUILT ECO-LODGES

3.1 Indicating the highest items

This research will try to use the "*how method*" to measure the highest two items and their maximum indicators that had been valued in the "*What method*" see Table 1 and Fig. 4, and had been indicated by professionals (Babenshal designer and its construction engineer) during their structural interview with the researcher. Those items are the indoor environmental quality and its relation to the site. The "*how method*" is done by using work field study and its aim is to insure the quality and the accuracy of the results.

Table 2: The second highest environmental items. *(Source: The researchers in the paper "Developing an Ecological Assessment Tool for Siwan eco-lodges in the Egyptian Western Desert (EWD)".)*

Items	Assessment categories	Their weights
S	Site	0.14
E	Energy efficiency	0.288
W	Water efficiency	0.066
M	Material	0.275
IEQ	Indoor environmental quality	0.14
W&P	Waste and pollution	0.067
E&C	Economic and cost	0.025

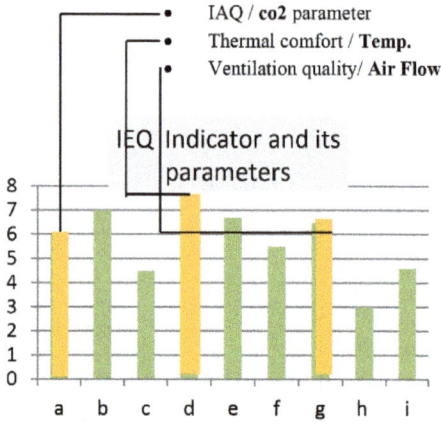

Figure 3: The researcher visiting Babenshal eco-lodge for field work study. *(Source: Authors.)*

Figure 4: IEQ highest indicators (IAQ/CO$_2$, thermal comfort/ temp, ventilation quality/air flow). *(Source: Authors.)*

3.2 Questionnaire contribution

Engineer Emad Farid (the Architect), Eng. Ramiz Azmi (the construction engineering) and Raies Abdalla Hamoda (Siwan Builder of the lodge).

3.3 Questionnaire structure

The Questionnaire structure was summarized as the following: At first, the researchers tried to understand the criteria took by the architect and the construction engineering during developing their project (Babenshal eco-lodge), so they started asking about the site obstacles and what are the main criteria took in order to choose the target project location? According to Eng. Emad; in the past local people used to build on high lands in order to overcome the underground water humidity, but recently it is not an obstacle. However, architects must comprehend the nature of the project lands with the help of local labors and scientists must respect local builders experience and provide them with scientific knowledge respectively. Then the second question was based on the nature of the material that should be used in all design process and whether there is any contribution of advanced building materials to the traditional Kershef substance?

Dr. Emad answer was that Kershef material made of mud straw and salt is an excellent choice and perfect substance for isolation of sound, heat during summer and cold during winter, but the idea is that eco-lodges built with Kershef need large land space compared to contemporary eco-lodges, moreover, although environmental material's life cycle is not long as chemical materials, finishing materials made of white mud as shown in Babenshal lodge is more durable and can stay much longer than chemical pastes that should be renewed annually, on other hand, according to various scientific researches, chemical constructing materials have multiple negative health impact on humans. Then Eng. Ramis added "*There are environmental solutions that can solve the pros related to this traditional material as using olive leaves to reduce cracks exist several years after the eco-lodge is constructed*" then he added "*A good example for this is Shali lodge that is built of raw Siwan rocks since 22 years ago and no cracks have appeared till now*" [3]. Finally the questionnaire was ended by asking whether there is a criterion developed by officials or by the Government to control the system of new or previous established Siwan-lodges.

Eng. Emad said: "*of course we need to build our own criteria for building any eco-lodge in Siwa oasis, we have already talked to coordinating system here to build one to make the concept of Environmental Eco-lodge in Siwa real. This system will aim to preserve the inheritance and construction coordination of the oasis on both building and urban scales*" [3].

3.4 Field work study: Babenshal eco-lodge

3.4.1 Coding tracing points for Babenshal internal spaces

As an initial step, the researcher had to code the most usable internal spaces. The Ground floor has the codes starting with (IT = Restaurant) and (IR = open courtyard), the First floor has the codes starting with (FC = cafeteria and sitting area) and the Second floor has the codes starting with (RO = Roof Restaurant) (see Fig. 5).

3.4.2 CO_2 tracing points in ppm

By using CO_2 measuring tool in the field/Babenshal eco-lodge depending on the codes given in Fig. 6, the researcher found that the first floor restaurant carries the highest % of CO_2 ppm

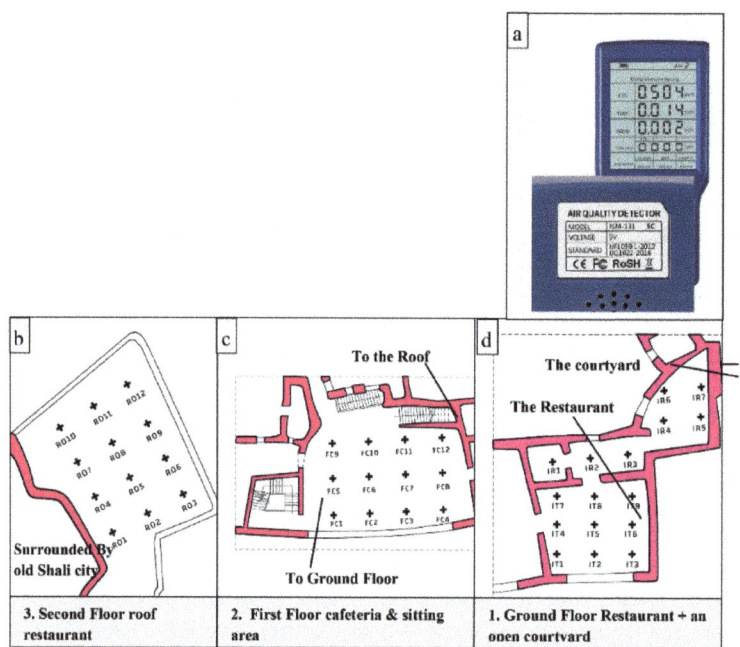

Figure 5: (a) Air quality detector used by the researcher in Babenshal eco-lodge; and (b)–
(d) Coding tracing points for Babenshal internal spaces: Ground, first and second
floors. *(Source: Authors.)*

although the good cross ventilation occurred in this place (due to the wide window a crossed
to the narrow stairs), on other hand the roof restaurant on the roof carries the lowest CO_2
ppm %.

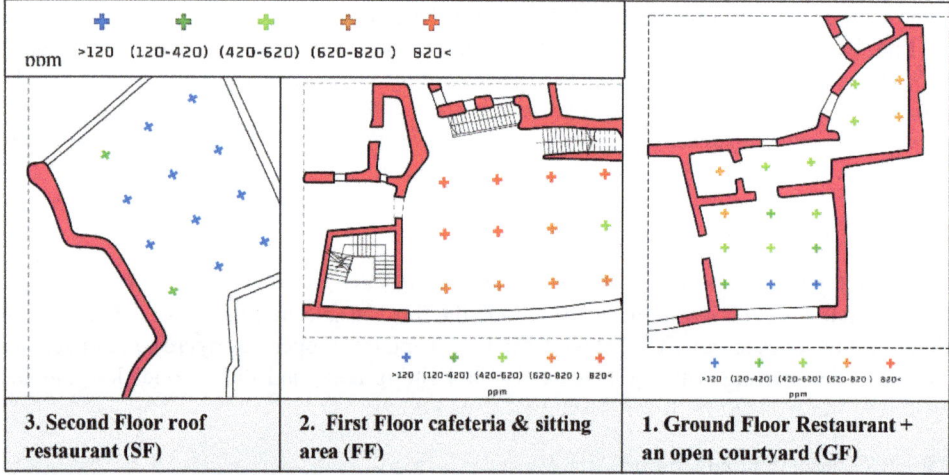

Figure 6: CO_2 measurements of ground, first and second floors of Babenshal eco-lodges
in ppm. *(Source: Authors.)*

- Mean CO_2 tracing measurements in Babenshal eco-lodge = [220 (Mean of GF) + 637 (Mean of FF) + 170 (Mean of SF)]/3 = 343 ppm.

3.4.3 Air speed tracing points (m/s)

By calculating the average of the measured air flow the researcher found that the minimum value for air flow presented in the first cafeteria floor, while the roof floor carries the highest value. This result can be concluded and easily predicted when tourists visit the place. This is because although the roof is not surrounded by built blocks, it was disappointing that the first floor roof is poorly ventilated and this is contradicting to the aim of the eco-lodge designer which is creating comfortable cross ventilation in this area (see Fig. 7).

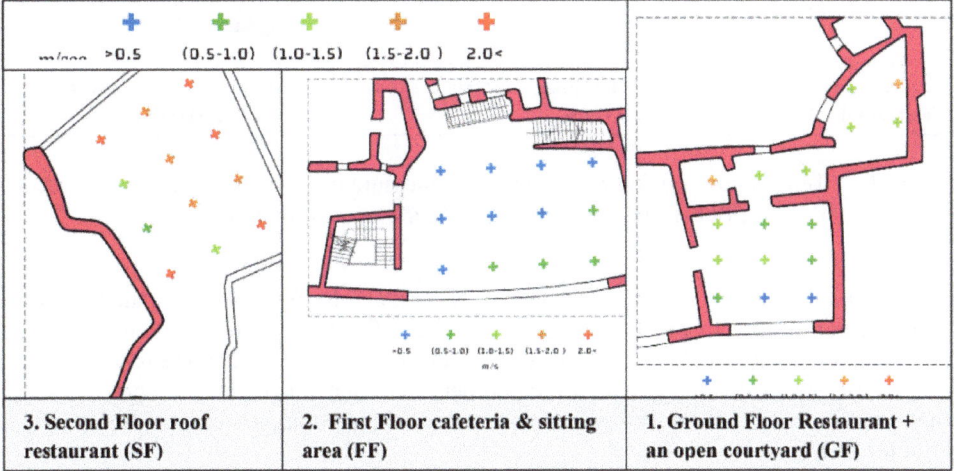

3. Second Floor roof restaurant (SF)	2. First Floor cafeteria & sitting area (FF)	1. Ground Floor Restaurant + an open courtyard (GF)

Figure 7: Air Flow measurements of Ground, first and second floors of Babenshal eco-lodges in m/s. *(Source: Authors.)*

- Mean air flow tracing measurements in Babenshal eco-lodge = [0.75 (Mean of GF) + 0.2 (Mean of FF) + 1.2 (Mean of SF)] / 3 = 0.72 m/s.

3.4.4 Temperature tracing points (degree Celsius)

The below pictures show that the roof has the highest average temperature relative to the rest of the eco-lodge's floors. This result is variable through the day hours but the aim of the researcher is to identify the mean value of the air quality in each point of the eco-lodge's internal spaces including the roof (see Fig. 8).

- Mean of temperature tracing measurements in Babenshal eco-lodge = 16.5 [(Mean of GF) + 15 (Mean of FF) + 20.5 (Mean of SF)]/3 = 17.34°C.

3.5 Calculating the weighting of each value to obtain the credible weighting of the lodge environmentally

In this paper the researcher had only measured and calculated part of the equation that was developed in her paper under the name *"Developing an Ecological Assessment Tool for*

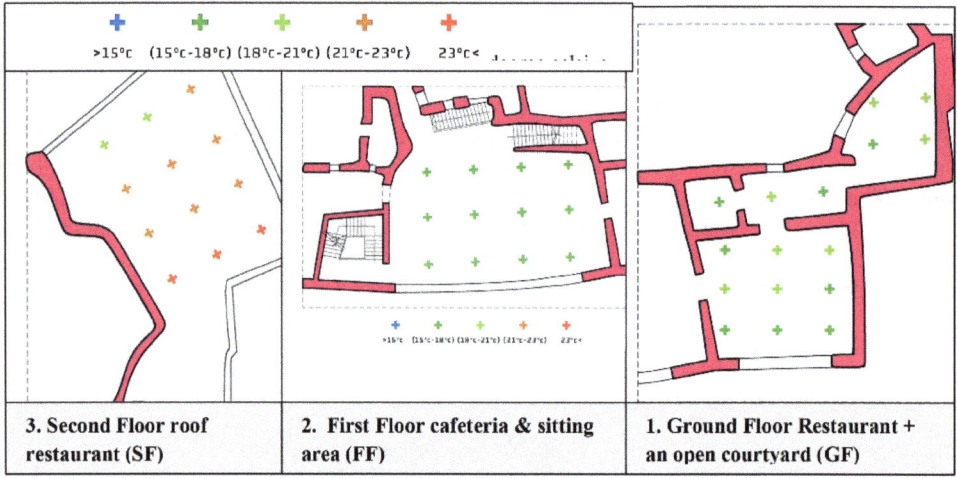

3. Second Floor roof restaurant (SF)	2. First Floor cafeteria & sitting area (FF)	1. Ground Floor Restaurant + an open courtyard (GF)

Figure 8:   Temperature tracing points and their measurements of ground, first and second floors of Babenshal eco-lodges in m/s. *(Source: Authors.)*

Siwan eco-lodges in the Egyptian western desert (EWD)" by finding the parameters of the IEQ indicators multiplying them by IEQ weighting coefficient (0.14) to find their real values. Those values are then added to the rest of the Indicators related to gain the real IEQ mean value of Babenshal eco-lodge. Moreover in order to calculate the total Environmental mean quality of the lodge, all Environmental Categories should be added repeating the previous steps for each category.

Each category has its own parameter measuring system in order to gain there values. In this paper the researcher only choose to measure the highest indicators of the IEQ Category.

4 RESULTS

This paper is defining the system of finding the results for the criterion equation which is built to serve Siwan eco-lodge in the EWD. This process is complicated and must be accurately defined by the mean required tool that suits the items and categories that were defined in previous researcher papers. In this paper the researcher shows an accurate way that can be used for weighting existed eco-lodges. Moreover the tools used are not suitable to measure unbuilt eco-lodge otherwise computational methods as simulation are suitable in such model. And finally and not last, each Category as Economic Category or Social one has its specific accurate process and thus the researcher is aiming to continue to search for the most suitable way for each Category of the Environmental Aspect to be use in any Siwan eco-lodge in the future.

5 CONCLUSION AND RECOMMENDATIONS

- An ecological assessment model is a cyclic process that must be developed and adapt to periodical variations and challenges.
- There must be an intellectual understanding and respect between architects and local labors during their contribution for developing Siwan eco-lodge.
- Nature has the first priority among all phenomena that directly affects Siwan eco-lodges during the total design processes.

➢ **Score of measured Parameter (1 or 0.5 or 0) * Parameter weighting Coefficient = Parameter Result (of next indicator)**

➢ **Total result of previous indicator's parameters * Indicator weighting Coefficient = Indicator Result (of next Category)**

1. Mean Babenshal seasonal co2 production (IAQ Indicator) * Indicator weighting= 343*0.14 = **48.02**.

2. Mean Babenshal seasonal Air Flow (Ventilation quality Indicator)* Indicator weighting Coefficient= 0.72 *0.14 = **0.1008**

3. Mean of Babenshal seasonal Temperature (Thermal Comfort Indicator)* Indicator weighting Coefficient= 17.34 * 0.14 = **2.427**

➢ **Total Result of previous Category's Indicator * Category weighting coefficient = IEQ Category result (CR)**

(48.02 +0.1008+2.427+ life quality value +Day light + Visual quality + Occupant Quality+ Safety + Acoustic Control) * 0.14 = ***The real value of Babenshal IEQ Category***.

Final Result of Environmental Assessment Tool = CR (site) + CR (Energy) + CR (Water) +CR (Material) + CR (IEQ) + CR (Waste) +CR (Economic).

- The weighting process of an eco-lodge in specific Siwan eco-lodges must be accurate and specialized to the required eco-lodge meant to be measured and getting its Environmental or Social real weighting.
- The weighting system is a system built to serve certain eco-lodges of specific location and shouldn't be applied globally to serve its real purpose.
- Scientists and architects are required to visit the eco-lodge and live among local people for a period not less than 3 month in order to conclude and summarize all required data for measuring the real environmental weighting of this eco-lodge.

REFERENCES

[1] Fakhry, A., *Siwa: The Oasis*, The American University in Cairo Press: Siwa Oasis, Cairo, 1973. ISBN 9774241231.

[2] Bayoumi, O.A. & Bayoumi, A.A., *Developing an Ecological Assessment Tool for Siwan Eco-lodges in the Egyptian Western Desert (EWD)*. Wessex Institute: Valencia, Spain, 2020.

[3] Azmi R., Raies, A.H. & Emad, F., An interview with the architect, construction engineer and the builder of Babenshal Eco-lodge [مقابلة]. *Babenshal Ecolodge*, Ola Ali Bayoumi & Amr Ali Bayoumi Recorder: Siwa Oasis, 25 (4.32 pm), Jan. 2021.

[4] Vivian, C., *The Western Desert of Egypt*, vol. 1, The American University: Cairo, Egypt, 2002.

CRISIS OR OPPORTUNITY: LOOKING AT THE PAST FOR THE RESILIENCE OF SETTLEMENTS

GULIZ OZORHON & ILKER FATIF OZORHON
Faculty of Architecture and Design, Özyeğin University, Turkey

ABSTRACT
This study focuses on learning from the examples of traditional architecture for planning/designing resilient living environments. It is important for the living environments of the future to take advantage of the ancient knowledge of traditional settlements, which are handed down from generation to generation, centring on "human" and "environment" and knitted with the dynamics of life. Within the scope of the study, 7 traditional settlement examples selected from Anatolia were examined under the titles of settlement, building and technology. The characteristics of these settlements are revealed with their ecological, economic, social and spatial dimensions in relation to their resilience. As a result of the research, all settlement examples were evaluated in an integrated manner, and some determinations were made for future resilient settlements.
Keywords: traditional settlement, resilience, learning from past.

1 INTRODUCTION
Today, the whole world is trying to adapt to pandemic conditions while trying to develop theories and produce projects on how the life and its dynamics will be shaped after the pandemic. What should the planning principles of post-pandemic living environments be? How should public spaces or housing units transform? Pandemic also made us question whether our living environments are resistant to the crises we may face or not.

The pandemic has shown us how weak we are or can be in the face of unforeseen situations. We came across the reality of how fragile our lives and the spaces we live in can actually be. Moreover, the pandemic was only one of the possible crisis scenarios. With this confrontation, we once again remembered how vulnerable our living spaces are against environmental problems, natural disasters, floods, earthquakes, and climate crisis.

At this point, the eyes of the researchers turned to traditional settlement examples once again after many times, and the main question of the research in this study emerged in this way: *Can traditional architectural examples be used for planning/designing resilient living environments?*

What is the resilience? According to Ward [1]; "*A resilient system is adaptable and diverse. It has some redundancy built in. A resilient perspective acknowledges that change is constant and prediction difficult in a world that is complex and dynamic ... Resilience thinking is a new lens for looking at the natural world we are embedded in and the man-made world we have imposed upon it.*" Resilience is the capacity of a system to absorb disturbance and reorganize while undergoing change, so as to still remain essentially the same function, structure, identity, and feedbacks [2] Resilience is mainly studied in the context of urban planning in the literature. According to Yaman Galantini [3], defining urban resilience with its ecological, social, socio-ecological, spatial, economic and institutional/governance dimensions is important in terms of constructing the relationship between resilience and urban planning.

On the other hand, traditional principles have evolved over a long period of time all countries of the world. People have developed building techniques excellently adapted to the building materials available and local conditions such as the climate [4]. Recent research has emphasised the potential contribution of traditional ecological knowledge to cope with challenges from global environmental change [5]. Since traditional settlements are created based on environmental data, they are easier to adapt to environmental changes.

Anatolian architecture offers an important and rich resource in this respect. Anatolia has been home to many civilizations over the course of history and all of them developed by adding their expertise and knowledge to the total accumulation pool of the landscape they inherited. Every civilization has developed its own building principles by drawing on accumulated knowledge and taking them one step further. Learning from the past and from past expertise and the transfer of knowledge have been key factors in the building traditions of Anatolian civilizations [6]. This study asserts that the argument for learning from the past is still valid. As noted by McIntosh et al. [7] *"by ignoring the great laboratory of millennia of responses to environmental change, we condemn ourselves to reinventing a very complex wheel in the face of one of humanity's greatest challenges"*.

1.1 Methodology

This study focuses on learning from traditional settlements for the resilience of existing and new living environments. In this context, the method of the research (Fig. 1) is based on the examination of selected traditional settlement examples with a holistic perspective and with different layers under the headings of settlement (pattern), building (unit) and technology (component). With this examination, the characteristic features of these living environments, which could protect itself against many years and various difficulties, shaped by environmental data, socio-cultural dynamics, habits and the identity of the place, have been tried to be understood. In the study, instead of revealing the settlements one by one, it was preferred to examine them as a whole, so that it was aimed to reveal the strategies repeated in all of them.

In the study, the relationship between the ancient knowledge of traditional settlements and resilience (adapted from Yaman Galanti's [3] approach) is systematically revealed in its ecological, economic, social and spatial contexts.

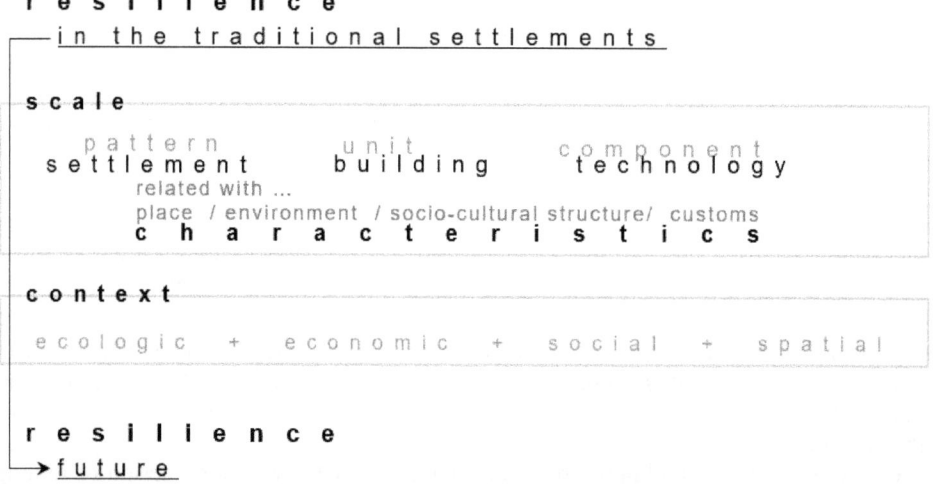

Figure 1: The framework of the study.

2 CASE STUDY: INVESTIGATION OF TRADITIONAL SETTLEMENTS IN THE CONTEXT OF RESILIENCE

In this study, which focuses on the issue of resilience, the knowledge of traditional settlement examples selected from Anatolia is referred. Within the scope of the paper, 7 traditional settlement examples (*Antakya, Mardin, Taraklı, Göynük, Cumalıkızık, Uçmakdere, Sille*) from Anatolia with varying climatic characteristics were examined (Fig. 2) and these settlements were revealed with their characteristic features. While doing this, an integrated approach was preferred instead of examining the settlements separately, and the settlements were examined under the headings of settlement, building, technology.

Figure 2: The map that shows the locations of the traditional settlements.

2.1 Settlements

Traditional Anatolian settlements have centred nature. For example, an attitude compatible with the topography was displayed in these settlements. Generally, the settlements were established on a sloping area, and the flat land was left to agriculture/production. In the example of Sille, in order to adapt the settlement to the land conditions, vineyards and gardens were built on the flat lands in the valley, and residential settlements on the sloping lands. Settlements in Göynük (Fig. 3), one of the examples that have survived to the present day with little change, and in Taraklı (Fig. 4), a typical Ottoman settlement have been shaped parallel to the sloping structure of the land [8]. The narrow streets, which show an organic formation, are sometimes quite inclined due to the topography. In Cumalıkızık, the squares, streets and houses of the village nestling at the foot of the mountain are in harmony with the natural topography [9].

On the other hand, one of the most important features of these settlements is that they are formed according to the climate. For example, the strong climatic conditions of Mardin (Fig. 5), played key roles in the way the architecture was developed. The hot and dry climate ranks at the top of these climatic conditions. Therefore, the pattern of the city developed to harmonize with this hot climate [9]. The narrow streets, where only one car can pass, are shaped according to the climate and prevailing winds. For example, in Antakya (Fig. 6), the streets are formed in such a way that the wind blowing from the Asi River refreshes the city in the summer months while they receive sun light in the winter [10]. In hot climatic regions (*in the examples of Mardin and Antakya*), narrow streets create sheltered and shady transition areas for the hot and overwhelming effect of summer.

Figure 3: Göynük. (*Source: Authors.*)

Figure 4: Taraklı. (*Source: Authors.*)

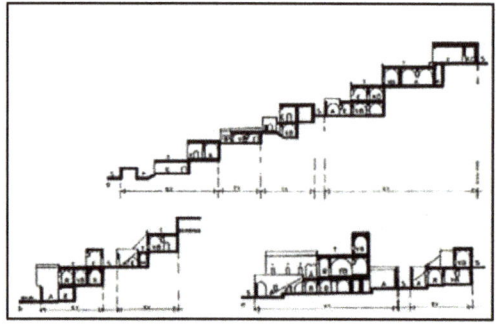

Figure 5: Mardin. (*Source: F. Alioğlu.*)

Figure 6: Antakya. (*Source: C. Bektaş.*)

Streets are also narrow in Uçmakdere (Fig. 7) which was established on a sloping land towards the mountain foot. Some houses were built in adjacent order, while some houses were built individually in a garden. The middle of the cobbled streets is inclined like a canal, and both sides are left high. Thus, in the village, which was established on a sloping land, when it rained, the water flowing from the midway was transferred directly to the creek. There is no accumulation of water in any part of the village [11]. It is possible to see a similar approach in many traditional settlement examples. For example, streets in Sille are extremely narrow. The middle of the stone-paved streets are arranged in the form of gutters to channel the water coming from above [12]. A similar attitude can be seen when looking at the cross sections of the streets in Antakya. The gutters in the middle of the streets allow the rain water flowing down the mountain skirts to reach the Asi River in a controlled manner [10].

2.2 Buildings

In traditional settlements, houses show a similar approach in terms of scale, system and material. In the settlements, the houses are mostly located adjacent to the street boundary or in a separate order in the gardens. Due to the topographical structure in Sille; The privacy of

Figure 7: Uçmakdere. *(Source: Visuals and drawings – T. Çınar.)*

the houses is also provided by the gradual settlement of traditional houses, streets and neighbourhoods. An architecture in harmony with nature was created while ensuring that the buildings benefit from the sun and the landscape equally [13]. The fact that Göynük houses do not block each other's outlook, sun and even air due to respect for the environment and neighbours can be associated with social life discipline. In Göynük, the organic street (Fig. 8) texture shaped by the topography and the parcel layouts of the houses have been created, and the fact that the streets are perpendicular or parallel to the slope has been determinant in the settlement of the houses within the parcel [8]. Traditional houses, which make up the urban texture of Göynük, can be accessed from different streets and levels thanks to their appropriate use of topography.

Figure 8: Göynük. *(Source: Authors.)*   Figure 9: Sille. *(Source: Authors.)*   Figure 10: Göynük. *(Source: Authors.)*

These houses have developed in harmony with the life culture, habits, social structure and human scale. Natural events that have been going on for centuries in traditional life are an important part of the life culture. The process of production that progresses with the soil, sun and rain has shaped the daily life and the places where this life takes place (Fig. 9). The production that started in the field continues at home as well. For this reason, the formation of open/semi-open spaces that will allow this, gains importance in shaping the houses.

These places are shaped according to the climate and take local names (hayat, taşlık, ...). For example, in Taraklı, courtyards are behind the house and these courtyards are surrounded by mud brick, stone or wooden walls. In the courtyards there are areas not only for daily life activities, but also for warehouses, haystacks, ovens and poultry [14]. In Göynük houses, stone-paved courtyards, which are surrounded by walls and also known as "taşlık", are open spaces that are used extensively. The bakery, where meals are also cooked on feast days take place in the courtyards, in parallel with the economic situation of the owner of the house. In some cases it is seen that the gusulhouses (the place where Muslims have a bath) on the living floor are also located in the courtyards. In some courtyard examples there is a pool for cooling, and in a small number of examples there is a toilet placed in a corner of the courtyard [8]. In Mardin, courtyards have fountains or small pools that support the microclimate. One of the features of the traditional houses is that they have iwans situated between the rooms. An "Iwan" (eywan) is a roofed space that is usually three-walled, affording open access to the courtyard. These spaces are very functional for the residents, especially in summer [9]. When the plans of Antakya houses are examined, the courtyard is always in the center. The rooms are located on one, two or three sides of the courtyard. It is important that the courtyards are functional, restful and simple. Landscape elements that create shade and coolness are also important [10]. Another important element of traditional Anatolian architecture is the "sofa". Just as the houses open to a street, every room in the house opens to sofa. The Sofa is an access to the rooms, as well as a space where all households gather and ceremonies are held [15]. In the spatial organization of Göynük houses, residential types with internal and external halls (sofa) which are shaped according to whether the hall is open or closed to the outside, stand out more. In the living floors of the houses that have survived to the present day in Göynük, plan schemes with internal hall ("cleavage" plan) and external hall has been used [8].

One of the most important principles of traditional houses is flexibility. *"The houses are capable of growing with the family; one unit at a time. Or that can be divided up later on ..."* [16]. Therefore, one of the most important elements of these houses is the room. Each room in the house is equipped to accommodate a family (husband and wife). Here you can sit, sleep, cook, eat and even have a bath. One of the rooms in the houses is different from the others. In this place, which is called as the head room, guests are welcome. This most equipped room is the closest one to the stairs and it has the best view.

2.3 Technology

The basic parameters that shape the façade in traditional houses are entrance, projections, doors and windows, the number of floors and the roof. The diversity seen in the plans of the houses is also reflected in the exterior façades. The arrangement of the windows on the street façades, which enable them to orient themselves to the view from a wide angle, is shaped by the size and spatial function of the space they are in [8].

Natural and local materials were used in traditional settlements. As Bektaş [16] pointed out *"the materials used can be obtained easily from nature, the building method are not pushed to their limits…one always seeks solutions that achieves more with less, are smarter are simpler…"*. For example, in Göynük houses, the ground floor is made of stone masonry on the stone foundation walls, and the upper floor is made of timber carcass, brick or mud brick filling material (Fig. 10) The use of wood in buildings is common due to the availability of wood in the region, its abundance and low cost. Similarly in Cumalıkızık, The structures of most Cumalıkızık are formed with timber as its basic timber construction material. In Cumalıkızık houses, the construction system is a hybrid system that includes a ground floor setting and a carcass system for upper floors. Stone, adobe bricks and regular bricks are among the construction materials used [9].

When the traditional structures in Taraklı (Fig. 11) are examined in terms of construction system and material; it is seen that the buildings were built in adobe or brick filled wooden carcass system. Since the region is covered with forests, the easiest building material to obtain is wood. The timber pillars resting on the stone foundation walls create a rigid system with the floor covering and the upper floor frame. The biggest reason why Taraklı, which is in the first degree earthquake zone, survived the earthquakes with insignificant damages is shown as that the majority of the buildings in the city have a traditional timber carcass system [15].

In Mardin, limestone found in deposits near the city is the basic construction material. While limestone is used for the primary supporting element of the structure, it is also used to ornament the buildings. The types of limestone used for these diverse purposes are different from each other. The stones used for the building material are clear yellow and hard, while the stones used as ornamentation are still soft when they are removed from the stone quarry [17]. in the micro-climate of the Mardin region, with hot and dry summers and significant temperature changes between day and night, the customarily thick walls perform both as heat reservoirs and insulators. During the hot day, the heat flow from the exterior due to the solar radiation to the inside is retarded and stored. Part of the heat is gradually released during the cooler hours, especially at night. In the cold winter months these thick walls reduce heating requirements because the heat stored in the walls is radiated during the night [9].

Figure 11: Taraklı. *(Source: Authors.)*

Figure 12: Uçmakdere. *(Source: Authors.)*

Figure 13: Sille. *(Source: Authors.)*

The dominant building material in the buildings in Sille (Fig. 13) is stone. Sille stone, which is easy to obtain and use, is a type of volcanite and andesite characteristics extracted from the quarries 2–3 km south of Sille [18]. It is used as both wall and flooring material. The structures in Sille were built by the craftsmen of Sille. As a result of being as strong and durable as possible, they have survived until today with various repairs [19]. Antakya houses are generally two-storey, made of stone and mud brick. The thick main walls on the ground floor are generally covered with hammer-dressed stones, and the upper floors are a light construction, which is mostly made with a bagdadi technique [10].

3 HOLISTIC EVALUATION

What Velinga says about vernacular architecture is also a pertinent remark about learning from the past: "*If the aim is truly to learn from vernacular architecture, what is needed is a*

Table 1: The features of the traditional settlements in the context of the resilience.

	Ecologic	Economic	Social	Spatial
Settlements	Settlement compatible with topography			Settlement compatible with topography
	Settlement compatible with climate			Settlement compatible with climate
	Settlement compatible with sun			Settlement compatible with sun
	Protection of green areas			
	Integrated-compact settlement			
	Settlements on the slopes, leaving the flat areas for agricultural land			
			Taking into account each other's possibilities in the formation of the building	
			Individuals with different beliefs and habits are together	
	Streets compatible with topography and flow of water			Streets compatible with topography and flow of water
Buildings	Low rise buildings compatible with human scale			
	Façade openings (windows) compatible with climate conditions			
	Building size to fit family needs (whatever is needed, neither more nor less)			
	The shaping of the house to be able to be divided and articulated			
			Possibility of using both private (room) and collective (sofa) space	
			Housing unit suitable for family structure and needs	
			Open and semi-open spaces are shaped according to the requirements of social life and place (hayat, iwan, courtyard)	
				The rooms and the equipment in the rooms are designed to allow this space to respond to many functions (sitting, eating, washing, sleeping, storage).
				Planning of furniture integrated with the walls
Technology	Structural system and material selection compatible with geological features			Structural system and material selection compatible with geological features
	Usage of natural materials			
		Usage of materials chosen from nearby areas		
	Façade openings (windows) compatible with climate conditions			
			Façades compatible with living habits and traditions	

holistic, integrated and critical approach that complements the study of the environmental qualities and performance of vernacular architecture with an examination of its social, political and economic aspects" [20].

Similarly, here, the resilience of the settlements is examined with their ecological, economic, social and spatial contexts based on the approaches of [21]. Ecological resilience refers to develop a model that describes changes in structure and function; social resilience refers to the ability of society to cope with threats as a result of social, political and environmental change; spatial resilience refers to the provision of infrastructure requirements for socio-economic resilience, while economic resilience refers to the capacity of an economy to recover from or adapt to the negative effects of external shocks and to take advantage of positive shocks [3], [22]–[26].

In this part of the study, the common characteristics of the settlements are represented in ecologic, economic, social and spatial contexts under the headings of settlement, buildings and technology (Table 1).

Bektaş [16] listed the characteristics of the traditional Turkish house as being compatibility with nature, environmental conditions and lifestyle, reality and rationalism, the inside-out solution, harmony between the interior and exterior, frugality, dimensions to be based on human body, ease of use, choosing the materials and construction techniques from nearby areas and flexibility. Traditional settlements compatible with topography and climate are instructive in terms of energy efficiency and conscious resource use. The establishment of the settlements in parallel with the topography enabled the construction of buildings with minimum intervention to the soil/land and thus to the nature [15]. The fact that the buildings are oriented to the south (to the sun) provided the opportunity to make maximum use of daylight and solar energy, thus it was possible to meet some of the energy use for lighting and heating naturally. Especially in the houses, it is seen that the living spaces are located to the south and the service volumes to the north. The houses were built in the required size and compactly, which prevented material and energy consumption. The buildings were created with natural/local materials (such as wood, stone) that can be easily obtained from the immediate surroundings (structure, shell and spaces), thus both healthy spaces were created and material procurement was easily provided [15]. The use of natural/local materials in the immediate surroundings of the buildings has had many advantages. In this way, healthy and qualified spaces have been designed and logistical convenience has been provided while, on the other hand, the characteristic identity of this place has been formed spontaneously.

4 CONCLUSION

According to the results of the study conducted by Gomez Traditional ecological knowledge and shared systems of beliefs can facilitate collective responses to crises and contribute to the maintenance of long-term resilience of social–ecological systems [5]. Indeed, traditional settlements have been established by reading/understanding nature, with the strong knowledge that has been articulated over the centuries. In addition to this knowledge, today there is a huge accumulation of science and technology that can serve human beings. Despite all of this, settlements are damaged, buildings collapse and people die due to disasters such as earthquakes or floods in many parts of the world today. Moreover, most of these disasters are not caused by earthquakes or floods – they are cities and buildings created by ignoring the knowledge of the past and the present. We should take the traditional settlements that have survived for hundreds of years and that are aware of the structural features of the soil/place they are connected to, as an example. The streets shaped according to the flow of water in these environments also allow the flow and continuity of nature. While planning the development routes and transportation axes of cities, we should include nature, and be sensitive and responsible for the ecological impact of the spread of our living spaces.

We should rethink the building envelope construction of traditional architecture, which is oriented according to the climate/sun, transformed according to the seasonal characteristics of the location, and shaped with natural/recyclable material. Undoubtedly, there is a lot to learn from this attitude in terms of energy use. We should rethink the culture of space that develops in parallel with life. While the pandemic pushed the whole world to live within their own homes, we mostly felt the lack of open living spaces. Our huge buildings, shopping centers, restaurants, and office towers became unusable when the air conditioners did not work. We remembered that the places we live in also need fresh air and sunlight. However, in traditional houses, there were courtyards, hayat, trees in these places, surrounding these spaces were healthy breathing shells made of stone, adobe or wood. We should expand the scope of sustainable architecture research and applications. We should raise awareness for all actors of building production that the features of the place should be included in every phase of the architectural design process.

As Bektaş [16] stated, now "we have to change our living culture." While planning our living environments at different scales (from home to street to city) from the smallest to the largest, we should turn to the concept of space that puts nature and people at the centre. Instead of struggling with nature, we should develop strategies that are in harmony with nature and according to the current dynamics of life. Problems such as climate crisis, global warming and famine are always on the agenda. However, we must be aware that we cannot produce real solutions to these problems without questioning our living habits. Undoubtedly, life continues to change at full speed. Of course, we cannot reproduce traditional settlement examples as they are today, but we must not forget that there is an important accumulation of knowledge in the strong relationship they establish with their own contexts. While we are constructing our living environments, we have to be consistent in our contextual approach, including the necessities of time and the dynamics of life, and we should think in a multifaceted way. With the guidance of science and this source of information provided by traditional settlements, we have to say new things and develop new strategies for our living spaces.

REFERENCES

[1] Ward, C., *Diesel-Driven Bee Slums and Impotent Turkeys: The Case for Resilience.* TomDispatch.com. Accessed on: 20 Mar. 2021.

[2] Walker, B., Holling, C.S., Carpenter, S.R. & Kinzig, A., Resilience, adaptability and transformability in social–ecological systems. *Ecology and Society*, 9(2), p. 5, 2004.

[3] Yaman Galantini, Z.D., Belirsizliklere karşı kurumsal dayanıklılık ve beş bileşenli kent planlama süreci (Institutional resilience against uncertainties and five-elements urban planning process). *Planlama, Kavramlar ve Arayışlar, 43rd Annual World Urbanism Day Colloquium*, Ankara, 7–9 Nov. 2021.

[4] Oliver, P., *Encyclopedia of Vernacular Architecture of the World*, Cambridge University Press: Cambridge, 1997.

[5] Gómez-Baggethun, E., Reyes-García, V., Olsson, P. & Montes, C., Traditional ecological knowledge and community resilience to environmental extremes: A case study in Doñana, SW Spain. *Global Environmental Change*, 22(3), pp. 640–650, 2012.

[6] Ozorhon, G. & Ozorhon, I.F., Learning from vernacular architecture in architectural education. *Megaron*, 15(4), pp. 553–564, 2020.

[7] McIntosh, R.J., Tainter, J.A., McIntosh, S.K., Climate, history and human action. *The Way the Wind Blows: Climate, History, and Human Action*, eds R.J. McIntosh, J.A. Tainter & S.K. McIntosh, Columbia University Press: New York, pp. 1–42, 2000.

[8] Dikmen, Ç.B. & Toruk, F., Spatial structure and proposals of protection of traditional houses Göynük. *Afyon Kocatepe University Journal of Social Sciences*, 17(1), pp. 99–128, 2015.

[9] Ozorhon, G. & Ozorhon, I.F., Learning from Mardin and Cumalıkızık: Turkish vernacular architecture in the context of sustainability. *Arts*, 3(1), pp. 175–189, 2014.

[10] Arıman, B., Typological analysis in the ancient texture of the City of Antakya. Unpublished Master's thesis, Istanbul Technical University, Institute of Science and Technology, 2002.

[11] Kurtuluş, I.H., *Tekirdağ* coastal line survey (Mürefte Yukarı Kalamış, Tepeköy, Çınarlı, Hoşköy, Güzelköy, Gaziköy, Uçmakdere, Yeniköy, Kumbağ, Barbaros). Unpublished master thesis, Istanbul University, Institute of Social Sciences, İstanbul, 1992.

[12] Tapur, T., Konya'da Tarihi Bir Yerleşim Merkezi: Sille (A historical settlement in Konya: Sille). *Türk Coğrafya Dergisi*, **53**, pp. 15–30, 2009.

[13] Keleş, R., An assessment of the sustainability of the relationship between traditional craft and interior space: The case of Sille village. Unpublished master thesis, Selçuk University, Institute of Social Sciences, Konya, 2019.

[14] Davulcu, M., Sakarya yöresi kırsal yerleşmelerinde konut mimarisi ve ustalık geleneği üzerine bir inceleme (A study on house architecture and master tradition in the rural settlements of Sakarya province). *Kastamonu Eğitim Dergisi*, **17**(2), pp. 687–706, 2009.

[15] Ozorhon, G. & Ozorhon, I.F., Rural architecture and sustainability: Learning from the past. *Journal of Asian Rural Studies*, **5**(1), pp. 30–47, 2021.

[16] Bektaş, C., *Türk Evi* (Turkish House), Yapı Endüstri Merkezi Yayınları, 2012.

[17] Alioğlu, F., *Mardin Şehir Dokusu ve Evler*, Tarih Vakfı Yayını: İstanbul, 2000.

[18] Aklanoğlu, F., Sustainability of traditional settlements and ecological design: A case study on Sille. Unpublished PhD thesis, Ankara University, Ankara, 2009.

[19] Tazefidan, C., Evaluation of Sille Houses by space-syntax method. Unpublished master thesis, Selçuk University, Fen Bilimleri Enstitüsü, Institute of Science and Technology, Konya, pp. 50–53, 2018.

[20] Vellinga, M., The noble vernacular. *The Journal of Architecture*, **18**(4), pp. 570–590, 2013.

[21] Yaman-Galantini, Z.D., Urban resilience as a policy paradigm for sustainable urban planning and urban development: The case of Istanbul. Unpublished PhD thesis, Istanbul Technical University, 2018.

[22] Holling, C.S., Understanding the complexity of economic, ecological and social systems. *Ecosystems*, **4**, pp. 390–405, 2001.

[23] Abesamis, N.P, Corrigan, C., Drew, M., Campbell, S. & Samonte, G., Social resilience: A literature review on building resilience into human marine communities in and around MPA networks. MPA Networks Learning Partnership, Global Conservation Program, USAID, 2006.

[24] Adger, W.N., Social and ecological resilience: Are they related. *Progress in Human Geography*, **24**(3), pp. 347–364, 2000.

[25] Keck, M. & Sakdapolrak, P., What is social resilience? Lessons learned and ways forward. *Erdkunde*, **67**(1), pp. 5–19, 2013.

[26] Gotham, K.F. & Campanella, R., Toward a research agenda on transformative resilience: Challenges and opportunities for post-trauma urban ecosystems. *Critical Planning Summer*, pp. 9–23, 2010.

Author index

Abdelall M. PI-327
Abdelmonem M. G. PI-185
Ahimbisibwe A. PI-29
Aldreghetti I. PI-133
Andrić D. PI-161

Ball S. PI-201
Bayoumi O. A. PI-327
Becue V. PI-285
Bienvenido-Huertas D. PI-89
Bohackova T. PI-123
Boscato G. PI-133

Cascone S. M. PI-39
Cascone S. PI-39
Cenci J. PI-275, PII-285
Chávez J. L. G. PI-101
Chiken Soriano A. PI-101
Chu W. PI-215
Citron R. S. PI-263
Coll-Pla S. PI-115
Costa Jover A. PI-115
Costantino C. PI-173
Cyril S. PI-249

da Conceição Lopes A. PI-297
de Smet I. PI-275
Demirbag R. PI-249
Desai J. PI-185
Devadoss M. PI-311
Devilat B. PI-185

Eltarabishi F. PI-237

Ferreira C. J. H. PI-297
Fialova P. PI-123
Fikry M. A. PI-327
Fiorin M. PI-133

Galić J. PI-161
Gallo D. PI-5
Garzia F. PI-149
Guerrero Baca L. F. PI-101
Gulli R. PI-173

Hejny L. PI-77
Higgins D. PI-201

Ibrahim I. PI-237

Jigyasu R. PI-185

Kelly T. PI-63
Koutra S. PI-285
Kreislova K. PI-123
Kroftova K. PI-77
Kühtz S. PI-229

Lanuza F. PI-185
León-Muñoz M. PI-89
Liu Y. PI-249
Longhitano G. A. PI-39
Longhitano L. PI-39

Mane M. PI-185
Massafra A. PI-173
Massaria L. PI-133
Mateu G. PI-115

Nagarajan R. V. PI-311
Nash H. A. PI-201

Osasona C. O. PI-15
Ozorhon G. PI-337
Ozorhon I. F. PI-337

Pagliuca A. PI-5
Prati D. PI-173
Predari G. PI-173

Ramos A. P. PI-101
Roset-Calzada J. PI-115
Rubio-Bellido C. PI-89

Sadanand A. PI-311
Sánchez Espinosa M. Á. PI-101
Saverimuttu V. PI-249
Scafuri V. PI-133
Seddiki C. PI-275
Shaheen W. PI-51
Stepinac L. PI-161

Trausi P. P. PI-5

Vukić H. PI-161

Wako A. K. PI-29

Zhang J. PI-285

Part II

Earthquake Resistant Engineering Structures XIII

WIT *eLibrary*

Home of the Transactions of the Wessex Institute.
Papers published in this volume are archived in the WIT eLibrary in volume 202 of
WIT Transactions on the Built Environment (ISSN 1743-3509).
The WIT electronic-library provides the international scientific community with immediate and
permanent access to individual papers presented at WIT conferences.
Visit the WIT eLibrary at www.witpress.com.

Preface

This volume contains some of contributions to the 13th International Conference on Earthquake Resistant Engineering Structures (ERES), organised by the Wessex Institute. The meeting was sponsored by WIT Transactions on the Built Environment. The initial venue of the conference was the city of Rome but the situation created by the COVID-19 pandemic made it necessary to transform the in-person event into an online forum.

This series of conferences began in Thessaloniki, Greece in 1997, followed by Catania, Italy (1999); Malaga, Spain (2001); Ancona, Italy (2003); Skiathos, Greece (2005); Bologna, Italy (2007), Cyprus (2009), Tuscany, Italy (2011), A Coruña, Spain (2013), Opatija, Croatia (2015), Alicante, Spain (2017) and Seville, Spain (2019).

The meeting provides a unique forum for the discussion of basic and applied research in the various fields of earthquake engineering relevant to the design of structures.

Some papers of the book deal with the topic of aseismic designs able to undergo the expected earthquakes events in their location. Others relate to the evaluation of seismic risk due to the separation between buildings or due to the typology of the construction.

The evaluation of the seismic demand on the structures can be expressed by deterministic or probabilistic theories, both formulations are present in the book and the concept of structural reinforcement to resist earthquakes and the several strategies to achieve such objectives appear in several papers, describing their application to multistory buildings with reinforced concrete structural framework.

An important issue is the definition of the value of the seismic hazard in different zones of a vast territory. Such an approach was the objective of a number of papers in the book, with emphasis on South American countries such as Peru.

The problem of protecting the built environment in earthquake-prone regions involves not only the optimal design and construction of new facilities but also the upgrading and rehabilitation of existing structures including heritage buildings. The type of highly specialized retrofitting employed to protect the built heritage is an important area of research.

The conference addressed these problems expanding on the development of previous meetings in the series. Papers presented at ERES are a relevant addition to the state of the art in this field. All of them are now freely available on the Wessex Institute eLibrary (www.witpress.com/elibrary) where they are a permanent record demonstrating the quality of the research presented at the ERES conference series.

The Editors are grateful to the members of the International Scientific Advisory Committee and the reviewers that made an outstanding job selecting the papers in this book, as well as to all authors for their collaboration.

The Editors
Ashurst Lodge, 2021

FRAGILITY ASSESSMENT OF THE INTER-STORY POUNDING RISK BETWEEN ADJACENT REINFORCED CONCRETE STRUCTURES BASED ON PROBABILISTIC SEISMIC DEMAND MODELS

MARIA G. FLENGA & MARIA J. FAVVATA
Department of Civil Engineering, University of Patras, Greece

ABSTRACT

The aim of this study is the probabilistic evaluation of the seismic performance of a multistory reinforced concrete (RC) frame structure due to the inter-story pounding effect. The assessment is performed through fragility curves at different performance levels. For this purpose, different probabilistic seismic demand models (PSDMs) are developed based on the real seismic response of the RC structure as a function of the spectral acceleration (Sa). In this direction, the inter-story (floor-to-column) pounding between an 8-story RC frame structure and a 3-story rigid barrier (very stiff structure) is examined. Three different initial gap distances (d_g) between the adjacent structures are considered. The seismic fragility assessment of the 8-story RC structure without the inter-story pounding effect is also incorporated. Results indicate that the local performances of the columns of the 8-story RC structure are crucial demand parameters for the probabilistic assessment of the inter-story pounding risk. The fragility curves are shifted to lower values of Sa due to the pounding effect in comparison to the corresponding cases without pounding, while the probability of pounding between the examined structures is increased as the separation gap distance d_g decreases. Nevertheless, the more exigent the performance level is the fragility curves move towards greater values of earthquake intensity.

Keywords: reinforced concrete frame, inter-story pounding, probabilistic assessment, fragility curves, performance levels, Eurocode 8, nonlinear dynamic analyses, PSDMs.

1 INTRODUCTION

In modern seismic engineering, over the last two decades, several researches have been undertaken to address the seismic performance of buildings through probabilistic procedures providing solutions and a better insight into seismic risk aspects of structures.

An important parameter for the evaluation of the seismic performance of structures is the pounding phenomenon between adjacent buildings that have been constructed in contact or with insufficient gap distance. Although interaction effects are independent from the initial structural design provisions it has been proved to be crucial for the integrity of the structural stability. Thus, the problem of the structural pounding between adjacent buildings has received substantial attention over the last two to three decades [1]–[10] and numerous results based on the deterministic assessment of the pounding effect on the seismic performance of real multistory structures have been reported in the literature [7]–[10]. However, review of the literature indicates that the probabilistic evaluation of the pounding problem is still in an early state of knowledge. So far, single-degree of freedom (SDOF) and/or multi-degree of freedom (MDOF) linear elastic systems, as well SDOF nonlinear systems have been used. In most of the cases, the peak relative displacement between the adjacent structures has been examined as key parameter of the pounding risk, while only the case of floor-to-floor pounding type has been considered [11]–[21]. Lack of relation attempts between the seismic performances of a real structure with the probabilistic assessment of the pounding effect still exists and more limitations are identified. In existing probabilistic methodologies, the local

WIT Transactions on The Built Environment, Vol 202, © 2021 WIT Press
www.witpress.com, ISSN 1743-3509 (on-line)
doi:10.2495/ERES210011

inelastic demands of the critical structural members have not been incorporated. The necessity for the verification of the local demand parameter is pointed out since it has been proved that this parameter is the most crucial issue on the seismic performance of structures due to pounding effect [9].

Nevertheless, in 2020, Flegga and Favvata [22], evaluated the floor-to-floor structural pounding effect on the seismic performance of an 8-story RC frame structure at different performance levels based on probabilistic assessment methods. Initial results that are based on the local demands of the structural members have also been presented. The probabilistic evaluation of the pounding effect is performed through fragility curves in terms of global and local engineering demand parameters (EDPs) as a function of peak ground acceleration (PGA) and Sa. Finally, in 2021, Kazemi et al. [23], studied the floor-to-floor structural pounding between equally heighted 3, 5 and 9 story RC and steel moment resisting frames (MRFs) at different seismic performance levels. The seismic limit states' capacities of the structures are defined through the median incremental dynamic analysis (IDA) curve using the global engineering demand parameter of maximum interstory drift. For the evaluation of the pounding effect, fragility curves based on IDA have also been developed in terms of maximum interstory drift as a function of the Sa.

Based on the above discussion, in this study a probabilistic assessment of the seismic inter-story (floor-to-column) pounding risk is attempted taking into account both global and local seismic responses of the examined RC structure.

2 EXAMINED INTER-STORY POUNDING CASE

The examined inter-story pounding case is an interaction case between an 8-story RC frame and a 3-story rigid barrier (very stiff structure). The story levels of the adjacent structures are not equal, so the slab of the shorter and stiffer structure hits the external column of the taller building. The top level of the potential contact point is located at the 1/3 of the height of the 4th floor level of the 8-story RC frame. The examined pounding case is shown in Fig. 1. Based on Eurocode's 8 provisions, a minimum gap distance equal to 9.0 cm between the 8-story RC frame structure and the 3-story rigid barrier is required to prevent pounding at the 4th floor level of the RC frame. Nevertheless, for the needs of this study, three different initial gap distances (d_g) are considered, namely d_g = 0.0 cm (structures in contact from the beginning), d_g = 4.5 cm and d_g = 9.0 cm (Eurocode's requirement). The seismic performance of the 8-story RC structure without the pounding effect is also included for comparison reasons.

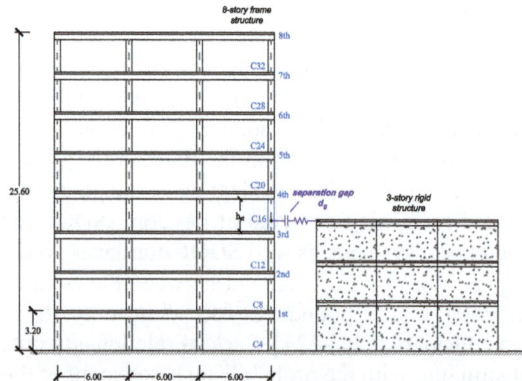

Figure 1: Examined 8-story RC frame structure.

3 DESIGN AND MODELLING ASSUMPTIONS

The 8-story RC frame structure was designed according to Eurocodes 2 and 8, meeting the Ductility Capacity Medium (DCM) criteria of the codes. The seismic behavior factor for the frame was q = 3.75. The mass was taken equal to M = (G + 0.3Q)/g (where G gravity loads and Q live loads) and the design base shear force was equal to V = (0.3 g/q)M. The computer program package Drain-2dx is used. Beams of the structure have been simulated through a common lumped plasticity model and columns by a special purpose element of "distributed plasticity" type accounting for the spread of inelastic behavior both over the cross-sections and along the deformable region of the member length. Collisions are simulated using contact elements that become active when the corresponding nodes come into contact. The response of the contact element is described by: a) the negative direction of the X-axis that represents the condition that the buildings move away from each other and b) the positive direction of the X-axis that simulates the actual behavior of the structures in case there is a small gap distance (d_g) between them. More details regarding the structural design characteristics and the modelling assumptions of the 8-story RC frame can be found in Favvata [9].

4 KEY ISSUES FOR DEVELOPING THE PSDMS

4.1 Examined EDPs-IMs

A probabilistic seismic demand model (PSDM) is a mathematical relation between the structural response and the earthquake intensity measure (IM). The structural response is described through engineer demand parameters (EDPs) representing the global or the local responses of a structural system while the intensity measures (IMs) refer to ground motion characteristics that may be structural independent (e.g. PGA) or structural dependent (e.g. Sa). In this study, the necessary EDPs samples are accomplished as a function of the Sa and the probabilistic seismic demand models are developed considering the following EDPs: (a) maximum displacement (δ_{max}) at the top level of contact point with the 3-story rigid structure, (b) maximum interstory drift (IDR_{max}), (c) maximum top drift (TDR_{max}), (d) maximum curvature ductility demands $\mu_{\varphi,max}$ and (e) maximum shear demands V_{max} of the external columns at the pounding side of the frame.

4.2 Probabilistic seismic demand models

In order to develop a PSDM, EDP|IM pairs should be generated through nonlinear dynamic analysis. In this direction, IDA method [24] has been used to define the seismic demands of the 8-story RC structure due to the pounding effect. Both components of seven different seismic excitations (totally 14 records) extracted from the PEER's database are used with Sa to be scaled in the range of 0.005 g to 1.4 g. Thus, a total number of 616 inelastic dynamic analyses have been performed. More details regarding the characteristic of the seismic excitations can be found in Flegga and Favvata [22]. Statistical process of EDP|IM pairs synthesizes the PSDM. The mathematical representation of median structural demand response $\widehat{EDP}$ and the IM can be approximated by a two parameters power law model [25], [26]:

$$\widehat{EDP}|IM = a\ IM^b. \tag{1}$$

The coefficients α and b are calculated through linear regression analysis of logarithm of IM and EDP, so the eqn (1) is transformed to the following expression:

$$ln\ \overline{EDP}|IM = b\ ln\ IM + ln\ a + \varepsilon|IM, \tag{2}$$

where $\varepsilon|IM$ is the random error with mean zero and variance σ^2.

The structural response demand is assumed to follow lognormal distribution [25] with logarithm standard deviation $\beta_{EDP|IM}$ which is calculated by the following equation:

$$\beta_{EDP|IM} = \sqrt{\frac{\sum_{i=1}^{n}(ln\ EDP_i|IM - ln\ \overline{EDP}|IM)^2}{n-2}}. \tag{3}$$

The mathematical representation of the PSDM leads to a closed form solution that permits the definition of the fragility curves. Each fragility curve describes, the probability an EDP to exceed the capacity $\hat{C}$ for a given IM and can be calculated as:

$$P[EDP|IM \geq C|IM] = \Phi\left(\frac{ln\ \overline{EDP}|IM - ln\ \hat{C}}{\beta_{EDP|IM}}\right). \tag{4}$$

$\Phi(.)$ denotes the standard normal cumulative function, $\hat{C}$ the median value of the capacity and $\beta_{EDP|IM}$ the logarithm standard deviation (eqn. (3)).

Nevertheless, linear representation of PSDM is not always accurate to describe the structural response for the entire range of IM [27]. The adoption of a bilinear regression model especially for local EDPs seems to be more valid to capture the nonlinear behaviour of the structural members. The bilinear regression model can be described as:

$$ln\ \overline{EDP}|IM = (a_1 + b_1\ ln\ IM)(1 - H_1) + [ln\ EDP|IM^* + b_2(ln\ IM - ln\ IM^*)]H_1 + \varepsilon|IM. \tag{5}$$

Coefficients α_1, b_1 and b_2 are defined through linear regression analysis. H_1 is a dummy variable is equal to $H_1 = 0$ for IM $\leq$ IM* and $H_1 = 1$ for IM $>$ IM*. The parameter IM* repressents the intersection of the two linear branches.

In this study, as it can be observed in Fig. 2, the global EDPs of IDR$_{max}$ and TDR$_{max}$ seem to fit better the linear regression model. On the other hand the local demand of the external column C16 that suffers the impact seems to be expressed more accurate through a bilinear model. In Figs 3 and 4 the PSDMs in terms of $\mu_{\varphi,max}$|Sa and V_{max}|Sa using linear and bilinear regression model are presented for all the examined pounding cases.

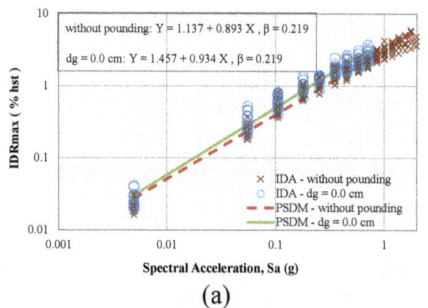
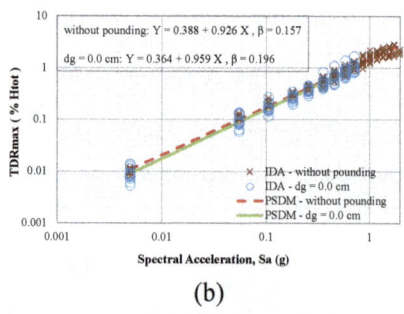

(a)   (b)

Figure 2:   Linear PSDMs in log-log space for the global EDPs of (a) maximum interstory drift IDR$_{max}$; and (b) maximum top drift TDR$_{max}$, as a function of Sa.

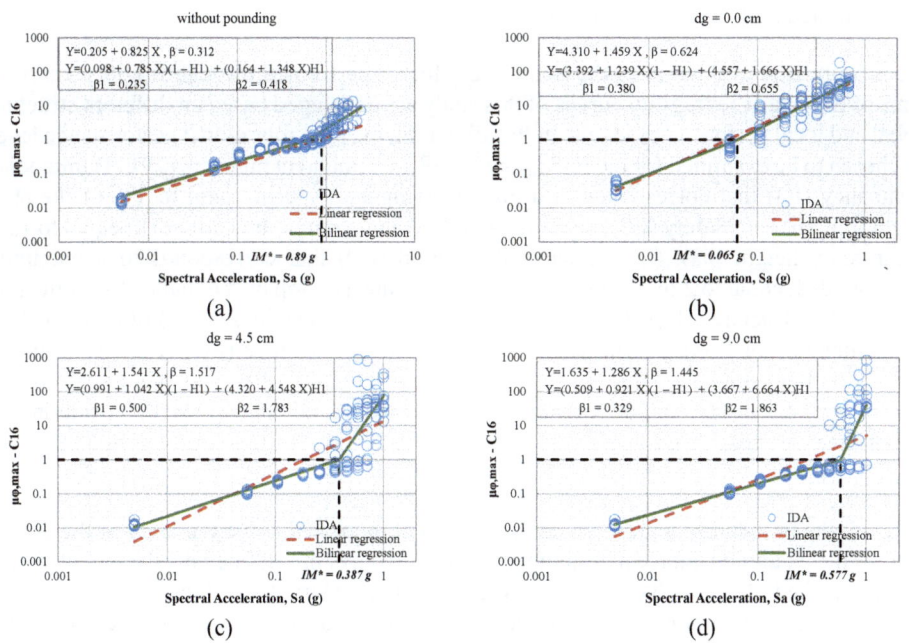

Figure 3: Linear and bilinear PSDMs in terms of $\mu_{\varphi,max}|$Sa (in log-log space). Examined cases. (a) Without pounding; (b) $d_g = 0.0$ cm; (c) $d_g = 4.5$ cm; and (d) $d_g = 9.0$ cm.

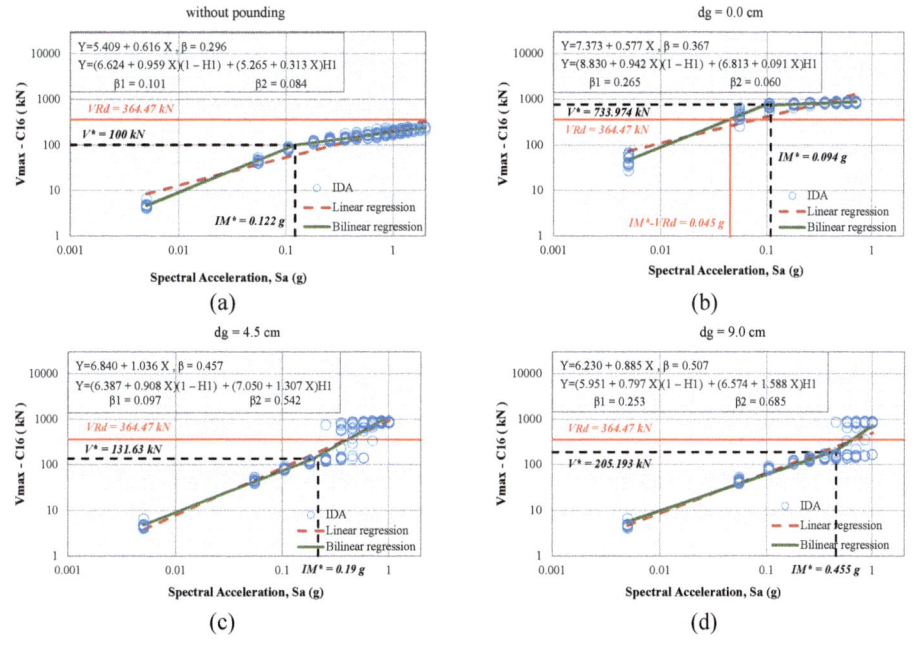

Figure 4: Linear and bilinear PSDMs in terms of $V_{max}|$Sa (in log-log space). Examined cases. (a) Without pounding; (b) $d_g = 0.0$ cm; (c) $d_g = 4.5$ cm; and (d) $d_g = 9.0$ cm.

4.3 Examined performance levels – PLs

The probabilistic evaluation of the pounding effect is performed through fragility curves in terms of EDP|Sa. The fragility curves of this study are developed based on different structural global and local performance levels. In this direction, the next four global performance levels are chosen to be examined for the IDR_{max} and TDR_{max} demand parameters [29]: (i) Immediate Occupancy (IO) that corresponds to a maximum interstory drift equal to 1% of the story height (h_{st}), (ii) Life Safety (LS) that corresponds to a maximum interstory drift equal to 1.5% of the story height (h_{st}), (iii) Collapse Prevention (CP) that corresponds to a maximum interstory drift equal to 2.5% of the story height (h_{st}) and (iv) top drift equal to 1% of the total height of the structure (H_{tot}). The local curvature ductility capacities ($\mu_{\varphi\text{-PL}}$) of columns have been estimated for Damage Limitation (DL), Significant Damage (SD) and Near Collapse (NC) performance levels based on Eurocode 8-part3. Finally, the local shear performance (*maximum shear force V_{max}*) of columns against pounding has been verified considering the design shear strength (V_{Rd}).

5 RESULTS

In Fig. 5, the probability a pre-defined displacement to exceed at the top level of the contact point of the examined adjacent structures, is evaluated. As pre-defined displacement limit is considered the initial gap distance d_g between the adjacent structures. It can be observed that the 8-story RC frame is more vulnerable to pounding with the adjacent shorter and stiffer structure as smaller the initial separation gap distance between them is. For a probability of pounding equal to 60%, an earthquake intensity of 0.35 g is critical when $d_g = 4.5$ cm, while in the case of $d_g = 9.0$ cm the critical value of IM is 0.75 g. A different approach provides that for an earthquake intensity of Sa = 0.3 g the probability of pounding between the examined structures is about 43% for $d_g = 4.5$ cm while in case of $d_g = 9.0$ cm this probability is less than 10%.

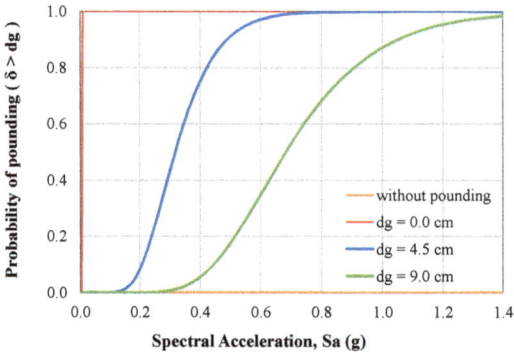

Figure 5:   Probability of inter-story pounding between the 8-story RC frame and the 3-story rigid barrier structure, when $d_g = 0.0$ cm, $d_g = 4.5$ cm, and $d_g = 9.0$ cm.

In Figs 6 and 7, the influence of the inter-story pounding effect on the fragility assessment of the 8-story RC frame in terms of $IDR_{max}|Sa$, and $TDR_{max}|Sa$ is presented and discussed. Fig. 6 shows that in the case of earthquake-induced pounding, the 8-story RC frame has exhibited the same probability of exceeding a particular performance level (PL) of IDRmax

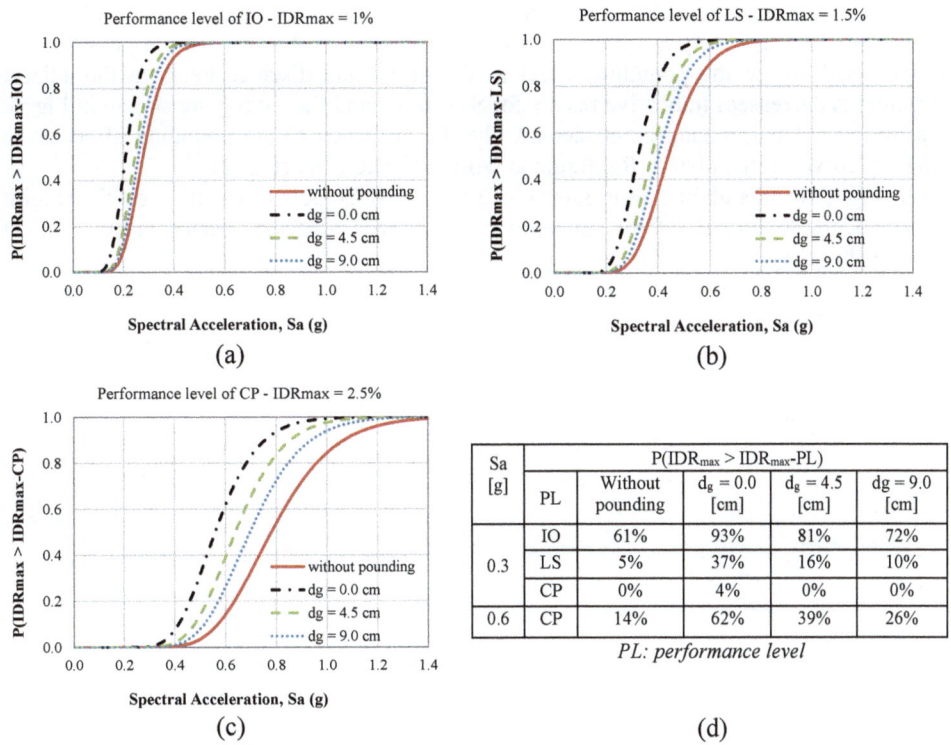

Figure 6: Fragility assessment of the 8-story RC frame against the inter-story pounding risk, in terms of IDRmax|Sa at different performance levels, when dg = 0.0 cm, dg = 4.5 cm, and dg = 9.0 cm.

The table in (d):

Sa [g]	PL	P(IDR$_{max}$ > IDR$_{max}$-PL)			
		Without pounding	$d_g = 0.0$ [cm]	$d_g = 4.5$ [cm]	$d_g = 9.0$ [cm]
0.3	IO	61%	93%	81%	72%
	LS	5%	37%	16%	10%
	CP	0%	4%	0%	0%
0.6	CP	14%	62%	39%	26%

PL: performance level

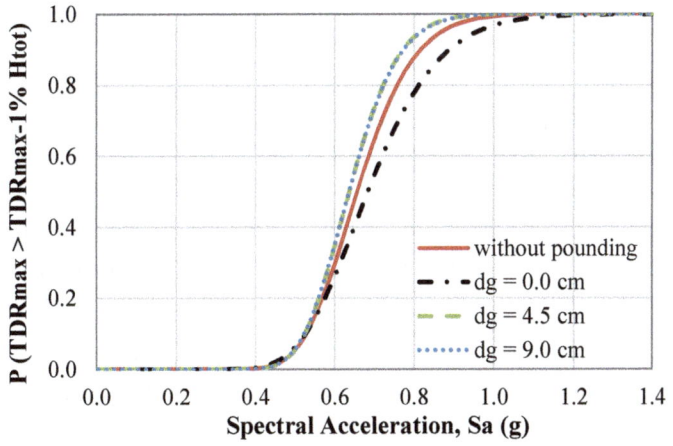

Figure 7: Fragility assessment of the 8-story RC frame against the inter-story pounding risk, in terms of TDR$_{max}$|Sa, when $d_g = 0.0$ cm, $d_g = 4.5$ cm, and $d_g = 9.0$ cm.

(%hst) at a lower value of Sa in comparison to the case without the pounding effect. The probability for the 8-story RC frame to exceed a specific PL of IDR_{max} at the same Sa value is increased due to the pounding effect as the initial gap distance between the adjacent structure is decreased. Indicative results for Sa = 0.3 g and Sa = 0.6 g are shown in Fig. 6d. On the other hand, meaningless seems to be the influence of the pounding effect on the fragility curves of the 8-story RC frame in terms of $TDR_{max}|$Sa (Fig. 7).

In Fig. 8, results about the influence of the inter-story pounding effect on the fragility curves of the 8-story RC frame in terms of $\mu_{\varphi,max}|$Sa at different performance levels are presented.

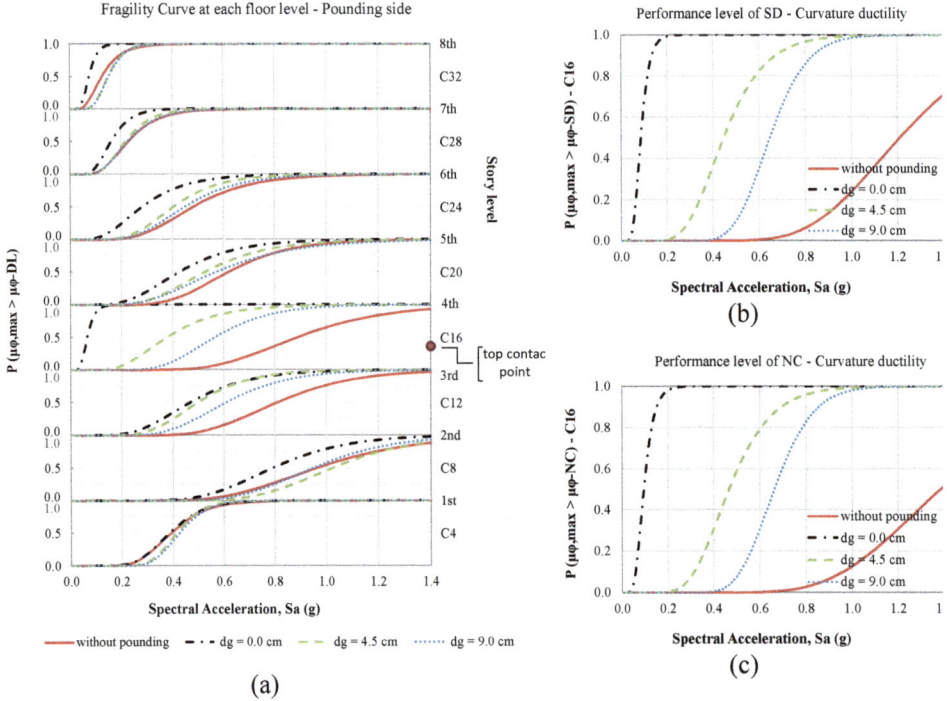

Figure 8:   Fragility assessment of the external columns (at the pounding side) of the 8-story RC frame against the inter-story pounding risk, in terms of $\mu_{\varphi,max}|$Sa, at different performance levels, when d_g = 0.0 cm, d_g = 4.5 cm, and d_g = 9.0 cm.

Based on these results it can be observed that critical due to the pounding effect is the local response of the external column C16. So, for example, when an earthquake intensity equal to 0.4 g is assumed, the probability the external column C16 to exceed the curvature capacity at the performance level of DL is 13% when d_g = 9.0 cm. This probability is increased to a value of 55% when the initial gap distance between the adjacent structures is equal to 4.5 cm and to a value of 100% for d_g = 0.0 cm. In all the examined cases of "without pounding effect" the column's response is elastic one (probability equal to zero) for Sa = 0.4 g. Also, it can be observed that in all the examined cases when the separation gap distance between the structures is decreased the fragility curves shift towards a lower value of Sa. In the case where the adjacent structures are in contact from the beginning (d_g =

0.0 cm), the critical flexural performance of the column C16 is evaluated even for values of Sa lower than 0.2 g.

In Fig. 9, the fragility assessment of the inter-story pounding risk based on the local shear performances of the column C16 is presented. It can be observed that this column is in critical condition due to shear capacity demand when the initial gap distance between the adjacent structures is equal to 0.0 cm even for values of Sa less than 0.1 g. In the case of considering a value of Sa = 0.4 g the probability of exceeding the shear capacity V_{Rd} is 46% and 9% for $d_g = 4.5$ cm and $d_g = 9.0$ cm, respectively. On the other hand, the probability is zero when the 8-story RC frame is vibrating freely.

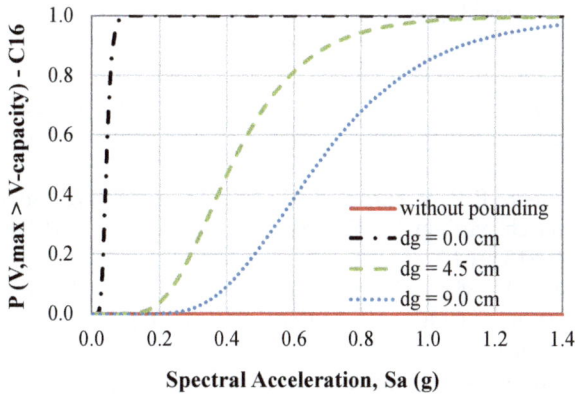

Figure 9: Fragility assessment of the critical column C16 of the 8-story RC frame against the inter-story pounding risk, in terms of V_{max}|Sa, when $d_g = 0.0$ cm, $d_g = 4.5$ cm, and $d_g = 9.0$ cm.

6 CONCLUSIONS

Based on the results of this study the following concluding remarks are noted:

- The 8-story RC frame is more vulnerable on suffering the inter-story pounding from the adjacent 3-story rigid structure, as smaller the initial separation distance (d_g) between them is.
- The fragility curves are shifted to lower values of Sa due to the inter-story pounding effect in comparison to the corresponding cases without pounding. Nevertheless, the more exigent the performance level is the fragility curves move towards greater values of earthquake intensity.
- The probability of the 8-story RC frame to exceed a specific global or local performance level at the same value of intensity measure is increased due to the inter-story pounding effect.
- The local performances of the columns of the 8-story RC structure proved to be crucial demand parameters for the probabilistic assessment of the inter-story pounding risk. Thus, it can be stated that the local fragility curves in terms of flexural and shear capacity of the most crucial structural member should be incorporated on the evaluation of the seismic performance of structures against inter-story pounding risk.

- Finally, PSDMs that are used for the fragility assessment of the inter-story pounding effect should be based on bilinear regression models in the case of evaluating the local EDPs, while global EDPs can be described based on linear PSDMs.

REFERENCES

[1] Anagnostopoulos, S.A. & Spiliopoulos, K.V., An investigation of earthquake induced pounding between adjacent buildings. *Earthquake Engineering and Structural Dynamics*, **21**, pp. 289–302, 1992.

[2] Karayannis, C.G. & Favvata, M.J., Earthquake-induced interaction between adjacent reinforced concrete structures with non-equal heights. *Earthquake Engineering and Structural Dynamics*, **34**, pp. 1–20, 2005.

[3] Karayannis, C.G. & Favvata, M.J., Inter-story pounding between multistory reinforced concrete structures. *Structural Engineering and Mechanics*, **20**(5), pp. 505–526, 2005.

[4] Jankowski, R., Assessment of damage due to earthquake-induced pounding between the main building and the stairway tower. *Key Engineering Materials*, **347**, pp. 339–344, 2007.

[5] Cole, G.L., Dhakal, R.P. & Turner, F.M., Building pounding damage observed in the 2011 Christchurch earthquake. *Earthquake Engineering and Structural Dynamics*, **41**(5), pp. 893–913, 2012.

[6] Abdel Raheem, S.E., Mitigation measures for earthquake induced pounding effects on seismic performance of adjacent buildings. *Bulletin of Earthquake Engineering*, **12**(4), pp. 1705–1724, 2014.

[7] Favvata, M.J., Interaction of adjacent multistory RC frames at significant damage and near collapse limit states. *WIT Transactions on the Built Environment*, vol. 152, WIT Press: Southampton and Boston, pp. 47–59, 2015.

[8] Jankowski, R., Pounding between inelastic three-storey buildings under seismic excitations. *Key Engineering Materials*, **665**, pp. 121–124, 2016.

[9] Favvata, M.J., Minimum required separation gap for adjacent RC frames with potential inter-story seismic pounding. *Engineering Structures*, **152**, pp. 643–659, 2017.

[10] Abdel Raheem, S.E., Alazrak, T., Abdel Shafy, A., Ahmed, M. & Gamal, Y., Seismic pounding between adjacent buildings considering soil-structure interaction. *Engineering Structures*, **20**(1), pp. 55–70, 2021.

[11] Hong, H.P., Wang, S.S. & Hong, P., Critical building separation distance in reducing pounding risk under earthquake excitation. *Structural Safety*, **25**(3), pp. 287–303, 2003.

[12] Lin, J.H. & Weng, C.C., Probability analysis of seismic pounding of adjacent buildings. *Earthquake Engineering and Structural Dynamics*, **30**(10), pp. 1539–1557, 2001.

[13] Lopez-Garcia, D. & Soong, T.T., Assessment of the separation necessary to prevent seismic pounding between linear structural systems. *Probabilistic Engineering Mechanics*, **24**(2), pp. 210–223, 2009.

[14] Lopez-Garcia, D. & Soong, T.T., Evaluation of current criteria in predicting the separation necessary to prevent seismic pounding between nonlinear hysteretic structural systems. *Engineering Structures*, **31**(5), pp. 1217–1229, 2009.

[15] Tubaldi, E. & Barbato, M., Stochastic analysis of the risk of seismic pounding between adjacent buildings. *Computational Methods in Stochastic Dynamics*, **26**, pp. 309–326, 2013.

[16] Tubaldi, E., Freddi, F. & Barbato, M., Probabilistic seismic demand and fragility assessment for evaluating the separation distance between adjacent buildings. *Proceedings of the 11th International Conference on Structural Safety and Reliability*, New York, USA, pp. 1641–1648, 2013.

[17] Barbato, M. & Tubaldi, E., A probabilistic performance-based approach for mitigating the seismic pounding risk between adjacent buildings. *Earthquake Engineering and Structural Dynamics*, **42**(8), pp. 1203–1219, 2013.

[18] Wu Q., Yan H., Zhu H. & Ding L., Probabilistic performance-based assessment for critical separation distance of adjacent buildings: Theoretical analysis. *ASCE Journal of Performance of Constructed Facilities*, **34**(4), pp. 1–12, 2020.

[19] Chase, J.G., Boyer, F., Rodgers, G.W., Labrosse, G. & MacRae, G.A., Probabilistic risk analysis of structural impact in seismic events for linear and nonlinear systems. *Earthquake Engineering and Structural Dynamics*, **43**(10), pp. 1565–1580, 2014.

[20] Tubaldi, E., Freddi, F. & Barbato, M., Probabilistic seismic demand model for pounding risk assessment. *Earthquake Engineering and Structural Dynamics*, **45**(11), pp. 1743–1758, 2016.

[21] Naeej, M., Amiri, J.V. & Jalali, S.G., Probabilistic evaluation of separation distance between two adjacent structures. *Structural Engineering and Mechanics*, **67**(5), pp. 427–437, 2018

[22] Flegga, M.G. & Favvata, M.J., Global and local performance levels on the probabilistic evaluation of the structural pounding effect between adjacent RC structures. *Proceedings of the 11th International Conference on Structural Dynamics, EURODYN*, vol. 2, Athens, Greece, pp. 3762–3779, 2020.

[23] Kazemi, F., Miari, M. & Jankowski, R., Investigating the effects of structural pounding on the seismic performance of adjacent RC and steel MRFs. *Bulletin of Earthquake Engineering*, **19**, pp. 317–343, 2021.

[24] Vamvatsikos, D. & Cornell, C.A., Incremental dynamic analysis. *Earthquake Engineering and Structural Dynamics*, **31**(3), pp. 491–514, 2002.

[25] Shome, N., Probabilistic seismic demand analysis of nonlinear structures. Report No. RMS-35, Department of Civil Engineering, Stanford University, 1999.

[26] Cornell, C.A., Jalayer, F., Hamburger, R.O. & Foutch, D.A., Probabilistic basis for 2000 SAC federal emergency management agency steel moment frame guidelines. *ASCE Journal of Structural Engineering*, **128**(4), pp. 526–533, 2002.

[27] Ramamoorthy, S.K., Gardoni, P. & Bracci, J.M., Probabilistic demand models and fragility curves for reinforced concrete frames. *ASCE Journal of Structural Engineering*, **132**(10), pp. 1563–1572, 2006.

[28] PEER ground motion database. 2011. http://peer.berkeley.edu/peer_ground_motion_database.

[29] Applied Technology Council (ATC), Seismic evaluation and retrofit of concrete buildings, vols 1 and 2. Report No. ATC-40, Redwood City, CA, 1996.

[30] Eurocode 2, Design of concrete structures. Part 1: General rules and rules for buildings. EN 1992-1-1, European Committee for Standardization, Brussels, 2004.

[31] Eurocode 8, Design of structures for earthquake resistance. Part 1: General rules, seismic actions and rules for buildings. EN 1998-1, European Committee for Standardization, Brussels, 2004.

[32] Eurocode 8, Design of structures for earthquake resistance. Part 3: Assessment and retrofitting of buildings. EN 1998-3, European Committee for Standardization, Brussels, 2005.

PROPOSAL OF IN SITU PARAMETERS FOR THE ASSESSMENT OF PHYSICAL VULNERABILITY TO SEISMIC EVENTS: A PERUVIAN CASE STUDY

LUIS IZQUIERDO-HORNA & ANDREA GALVÁN
Universidad Tecnológica del Perú, Perú

ABSTRACT

After the last seismic event that occurred in Pisco, Peru in 2007 (7.9 Mw), the fragility of the structural conditions of the existing houses in the Peruvian territory became evident. Therefore, it is important to be able to anticipate or recognize those dwellings that are more physically vulnerable in order to preserve the livelihoods and living conditions of the residents. In this sense, the instruments currently used for this purpose do not guarantee an adequate assessment of reality. Thus, the objective of this research is to propose a set of parameters to determine the level of physical vulnerability present for a given sector based on its structural and constructive characteristics. For this purpose, it was necessary to determine the structural typology of the sector based on the predominant material, number of floors, structural damage and the construction process stage. This information will allow the selection of a set of parameters potentially capable of adequately identifying the level of physical vulnerability in the dwelling. The district of Los Olivos was chosen as a case study. This research resulted in a set of parameters grouped into the following categories: construction system, irregularity and soil condition, roofing system, structural interaction and state of conservation. These categories and their descriptors are applicable and reproducible at different territorial scales.

Keywords: *physical vulnerability, in situ parameters, earthquakes, risk and disaster management.*

1 INTRODUCTION

Over the years, the multiple earthquakes that have occurred in Latin America have had a huge impact on the gross domestic product (GDP) [1]. For example, in 2007 there was an earthquake in Pisco with a magnitude equal to 7.9 Mw, which destroyed, disabled and affected 48,208, 45,500 and 45,813 homes, respectively, and seriously affected 14 health facilities [2], resulting in an economic impact of more than 139 million dollars [3]. These consequences can be explained through the poor awareness of the inhabitants and the high vulnerability of the built environment [4]. Therefore, in order to reduce the effects caused by this type of phenomena of natural origin, it is necessary to have a tool that allows an adequate identification of the conditions of the present houses in order to mitigate the seismic risk [5].

On the other hand, considering seismic risk as the possibility of altering a system, product of the interaction of hazard and vulnerability of the same system [6], this research will focus on analyzing this last component because it is determined by man [7]. Although, the vulnerability component can be studied in multiple dimensions (i.e., social, economic, environmental, etc.) [8], this research will focus on analyzing the physical component of vulnerability. This dimension will focus on analyzing the built environment of the end user, through in situ parameters and event characteristics [9]. The lack of attention to this dimension could be observed after the earthquake in Chile (6.5 Mw), where the different damages suffered by buildings that shared similar characteristics were evident [10]. Given this diversity of effects, it is necessary to design an assessment instrument that allows a comparative recognition between buildings within the same urbanization, allowing the proposal of useful measures that can reduce the seismic risk of the same [11], [12].

WIT Transactions on The Built Environment, Vol 202, © 2021 WIT Press
www.witpress.com, ISSN 1743-3509 (on-line)
doi:10.2495/ERES210021

With this in mind, Ordaz et al. [1] proposed assessing the physical vulnerability of the buildings in Toluca based on three parameters (i.e., type of material and construction design, age of the buildings and construction period) resulting in that 1,430 and 866 dwellings presented, respectively, a high and very high vulnerability showing an adequate classification, even when general and applicable parameters were considered as long as the area of interest is made up of correctly constructed structures. Under this same approach, in Venezuela, López et al. [11] considered 6 parameters that emphasized the differences between buildings within the study area (i.e., regulations, building age, irregularity, structural type, degree of deterioration, drainage, topography and depth of the deposit) to carry out an earthquake response prioritization study. On the other hand, Zora and Acevedo [13] propose to consider some additional parameters so that their results can demonstrate the conditions of the area, such as the existence of short columns, number of longitudinal axes in plan, year of construction, etc.

With the aim of achieving a simple and effective assessment, methods such as the Benedetti–Petrini or commonly called the vulnerability index emerged, which incorporates parameters that facilitate the inspection of each home [14]. Gent et al. [10] improved this methodology by applying it to Chilean confined masonry constructions, adding as a parameter the occurrence of an earthquake of magnitude equal to or greater than that of 1985. Similarly, Maldonado and Chio [12] adapted the method based on the characteristics of the sector (i.e., slope and topography). Similarly, Ortega et al. [15] combined the Benedetti–Petrini method with expert judgment and static modeling. Similarly, researchers like Formisano et al. [16], Kassem et al. [17], Chieffo et al. [18], Mosoarca et al. [19], [20], and Kappes et al. [21] updated the Benedetti–Petrini method considering the construction not only individually, but also as part of a whole. Therefore, parameters such as: structural heterogeneity, elevation interaction, number of staggered floors, building position and percentage difference in opening areas between adjacent facades were added.

For all the aforementioned, it should be emphasized that, even when the uncertainty or randomness of the parameters and phenomenon of natural origin is not considered [22], assessing structural vulnerability based on parameters is very useful, since it is considered within them, the most determining characteristics that influence the physical vulnerability of the construction to a seismic event [23]. Thus, this research will focus on developing a proposal for in situ parameters that allow determining the level of physical vulnerability in the study area with the greatest possible precision. Section 2 will describe the methods used. In addition, relevant information about the study field is also presented. The results will be described in Section 3 and discussed in Section 4. Finally, Section 5 shows our conclusions.

2 METHODOLOGY

Within the different methodological proposals to determine the level of physical vulnerability, the parameters used were chosen based on the opinion of experts regarding the factors that influence a construction [24]. However, in order to have an adequate assessment, it is important to know not only the available literature, but also the case study. In this sense, the A.H. (Human Settlement) Laura Caller – Los Olivos was chosen as an area of interest. Thus, from the analysis of the most prominent and characteristic parameters of physical vulnerability in homes, those that directly influence the area of interest will be identified. For this, through the virtual tour on Google Maps, the classification and description of the structural typology in the study area was obtained. Then, with the data previously obtained in conjunction with the literary review, the selection of the set of parameters that influence and classify the level of physical vulnerability among housing constructions is made. Fig. 1 summarizes the methodological proposal.

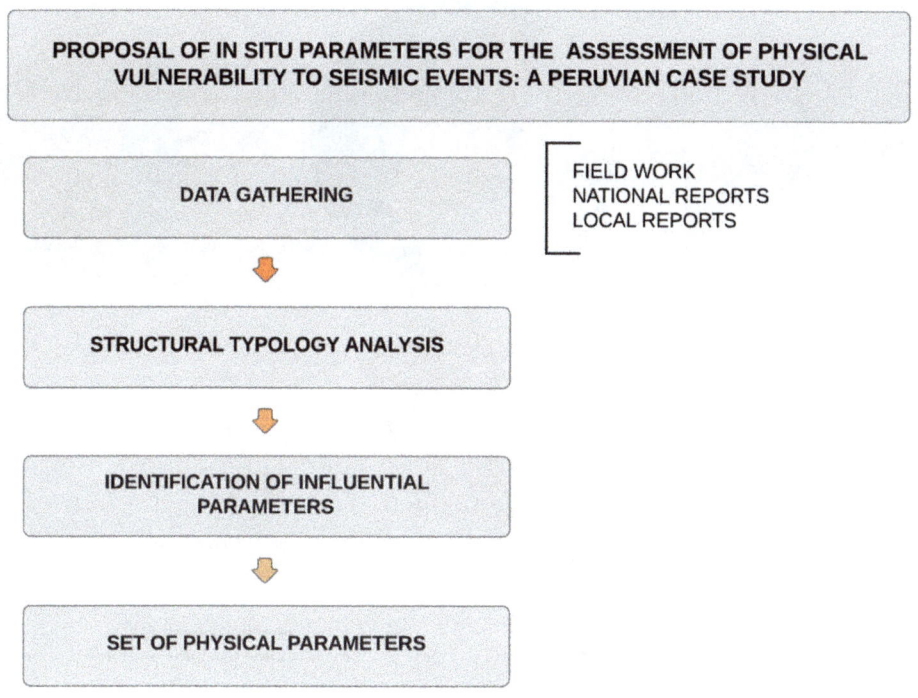

Figure 1: Methodological framework.

2.1 Description of the study area

The district of Los Olivos (Lima, Peru) has a territorial extension of 18.25 km², located at the coordinates 11°58′13″ of South latitude and 77°04′26″ of West longitude [25]. According to data registered from the last national census, the population is 325,884 inhabitants, of which 51.87% were women and 48.13% were men; likewise, registering a total of 82,848 dwellings [26]. Despite being a recently created district in Lima, it has optimal levels of employability, health and education, exceeding the percentage average compared to other districts [27]. The seismic hazard that Los Olivos will have to face is governed by its location on the coast of Peru, where different earthquakes occurred in the last 5 centuries, highlighting – among them – those that occurred in 1746, 1966 and 1974 with magnitudes of 9 Mw, 7.5 Mw and 8 Mw, respectively [2]. Therefore, the A.H. Laura Caller will be used as a sample to identify the parameters of physical vulnerability. Fig. 2 shows the location of the case study.

2.2 Determination of the structural typology

The damage that a building can suffer after a seismic movement is linked to the type of material used, the building design and the location where its execution was carried out, since the damage manifested will never be the same if we analyze and compare a masonry construction (regardless of whether it is with or without confinement) with one of adobe, considering both as correctly made constructions [1]. Now, if in these a structural component had been omitted or a construction stage was performed poorly, it would not be so easy to

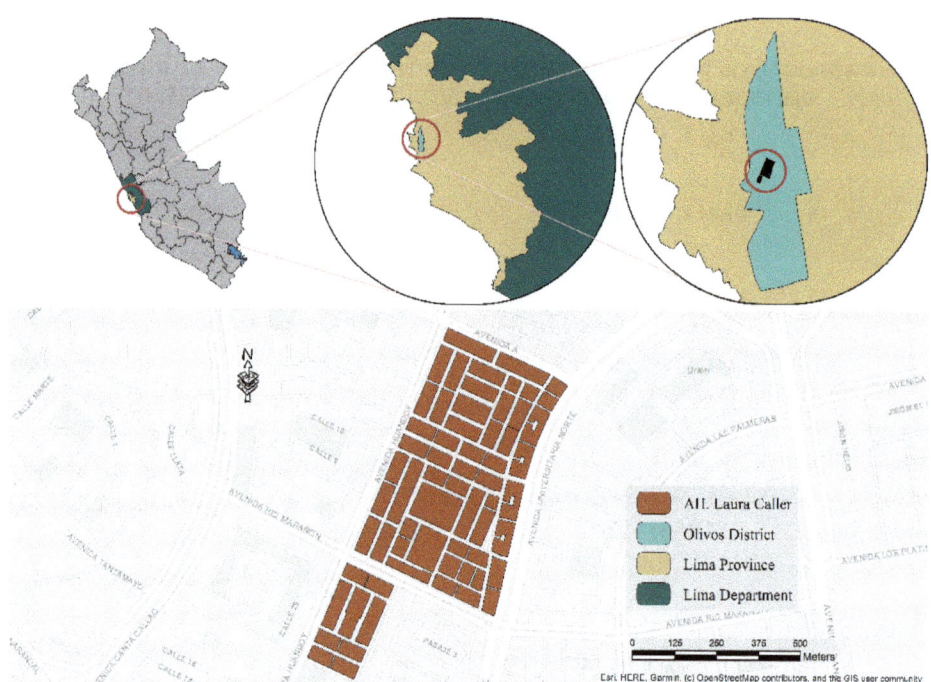

Figure 2: Location map of the A.H. Laura Caller.

indicate which construction would be most affected. Considering the above, the identification of the structural typology of the study area was developed through recognition, first of the study area, through the cadastre of the area in dwg format (AutoCAD program), listing the blocks within the chosen sector and establishing the number of dwellings that comprise it. This first tour was complemented with a virtual tour (through Google Maps). Then, after viewing each home and parameters such as: predominant material, number of floors, current wear and tear, and stage of the construction process, it was possible to define 4 different types. Table 2, in the results section, shows the structural typology of the dwellings analyzed in the case study.

2.3 Identification of the most influential parameters

The compilation of the parameters that contributed as input in different methodologies was previously used, grouping them into three categories, which will be corroborated by existing research and related to the study area. The first component is the location and condition of the soil, since – according to the type of material and the properties that are present – can increase the seismic force that influences the construction [28]. The second component will be the type of structural system, since, according to the material used, the conditions or construction standards that should have been considered based on it are established [29]. The third and last component is the number of floors, which is conditioned to the first two components [30]. With this preliminary information, a set of parameters related to the area of interest is outlined. For this, concepts for each one are established in order to avoid duplication of measurement and scope of the same parameter. In order to simplify and achieve an adequate understanding, it was decided to group in five categories i.e., structural construction system, irregularities and condition of the soil, floors and roofs, state of conservation and non-structural elements and as the fifth and last category, structural interaction. Table 1 shows the preliminary classification grouped into the categories shown above.

Table 1: Classification of parameters in relation to categories.

Category	Item	Parameter	Previous studies
Structural construction system	P1	Resilient system organization	[21], [23]
	P2	Quality of the resilience system	[11], [23]
	P3	Conventional resilience	[23], [29]
	P4	Maximum walls distance	[11], [15], [19]
Irregularities and soil condition	P5	Added position and interaction	[31], [32]
	P6	Plant configuration	[4], [11]
	P7	Height regularity	[5], [17]
Slabs and roofs	P8	Slab facade openings	[4], [31]
	P9	Horizontal diaphragms	[17], [29]
	P10	Roofing system	[17], [29]
Conservation state and other elements	P11	Fragility and state of conservation	[4], [13]
	P12	Non-structural elements	[23], [31]
Structural interaction	P13	Height difference	[5], [16], [33]
	P14	Position within a block	[5], [10], [16]
	P15	Number of staggered floors	[5], [16]
	P16	Structural heterogeneity	[16], [17]
	P17	Difference of the opening area percentage	[15], [16]

On the other hand, considering that it is desired to achieve an adequate differentiation and classification of the level of physical vulnerability of the dwellings, the parameters were analyzed and selected based on two criteria based on the preliminary collection of parameters and the structural typology of the study area. These selection criteria consist in avoiding that the selected parameter is present or absent in the same way or measure in all the dwellings and also in considering a parameter even when it is little present.

3 RESULTS

Regarding the first component related to the location and condition of the soil, the Japanese Peruvian Center for Seismic Research and Disaster Mitigation (CISMID) determined that the Los Olivos district found that the soils present favorable characteristics for the foundation of buildings [34]. In addition, local reports maintain that the topography is flat throughout the territory except for two areas with a slope greater than 45% [25]. In accordance with the second component, the analysis of the structural typology resulted in that 25.3% of the houses analyzed are made of adobe, which had covered with corrugated sheets or mats, likewise, of 73.2% of the houses analyzed are made of masonry, with no roof, indicating that the construction was paralyzed in the last level, and in case of having them, these were either of reinforced concrete or sheets of corrugated iron. Table 2 shows in detail the configuration of the 4 types of structural typology identified in the study area. Finally, in relation to the third and last component, which is the number of flats, in the area the existence of one-story homes was recorded by 4%, two-story homes by 11%, and three-story homes by 17%. Those with four floors were 25% and for those with 5 floors 42%, which are randomly distributed in the sector. Considering the information described for the 3 components with respect to the study area, the parameters that would be most useful and help to identify and classify the level of physical vulnerability of the dwellings are presented in Table 3.

Table 2: Structural typology of the study area.

Type	Characteristics
A	One-story adobe buildings, without any type of confinement, with a light roof such as calamine, simply leaned-on, showing wear at the base due to humidity.
B	Masonry buildings completed, finding in the study area up to 5 floors.
C	Masonry buildings of up to 5 levels, with a corrugated sheet or mat roof, placed and nailed on circular timbers placed throughout the roofing area (ranging from ends between two parallel perimeter walls) at random distances, independent of the presence of the perimeter beam.
D	Unfinished masonry buildings of 1 to 4 levels, observing the state of abandonment due to the presence of oxidation of the protruding or exposed steels, absence of columns in some of the cases or presence of walls at an intermediate level, indicating that the continuation of said construction will be carried out according to the economic disposition at home.

4 DISCUSSION

From the typological analysis carried out, it is important to consider that approximately 25% of the analyzed houses were built based on adobe (type A), generating a disadvantage in terms of resilience compared to the other types identified. Part of this housing conformation can be explained through social or economic factors [35]. Regarding the categories of preliminary parameters, the structural construction system is one of the most important because they contemplate common failures within a construction – whether it is self-built or not –, since it is not the same to evaluate a construction with the presence of a basement with those that present only higher levels. On the other hand, one of the most important parameters is the type of soil, which can cause very serious problems such as what happened in 2017 in Tehuantepec, Mexico [32]. However, it is not representative for the study area, since in small sectors such as the one taken into consideration, where the geotechnical and typological composition do not vary, still being one of the optimal ones, it would not be a differential of physical vulnerability between the dwellings. The district has two small sectors with a slope greater than 45%, nevertheless. Likewise, the roof parameter is one of the most important and neglected ones in the study area, proof of this is the existence of type D and C dwellings, which will not present homogeneous displacement in all their structures, damaging the elements and adjacent buildings.

Table 3: Final parameter set.

Parameter	Contribution	Optimal condition
Resilient system organization	Avoid diagonal fracturing of the walls through the presence of confinement.	Confinement of the walls on its four sides.
Quality of the resilient system	Confinement according to the material to avoid the detachment of any element.	Respecting the verticality and thickness of the mortar, which must be homogeneous.
Conventional resilience	Capacity to support external forces, whether seismic or produced by the environment.	The density of the walls between the horizontal force should be less than 1.
Maximum walls distance	Achieve confinement to the greatest extent possible.	For masonry, the cavity of the wall must not exceed twice the height of the wall; in the case of adobe, it depends on the thickness of the wall used.
Height regularity	The transmission of the seismic force occurs from the upper floor to the lower one cumulatively.	Ideally, each home has homogeneous levels or variation that does not modify the resistance dramatically.
Slab facade openings	The resistance of the slab will be weakened by the presence of internal voids from where the different failures will occur.	Ideally, it should not have voids and, if they do, it should be confined around the perimeter of the void.
Horizontal diaphragms	Achieve displacement homogeneity in full seismic activity.	Embedment in the beams and correctly supported by columns.
Roofing system	Material and quality.	It should not show bending or damage.
Fragility and state of conservation	The damage that the construction may present causing the resistance itself to be weakened.	It must not have cracks or wear in the structural elements.
Non-structural elements	Elements that, although they do not provide resistance to the construction system, can generate damage to the inhabitants.	Must be correctly placed, avoiding detachment.
Height difference	The mezzanine of the adjacent building would hit the wall, causing it to have to withstand a greater horizontal force.	All must be at the same level, since the additional force would fall between the slabs that will be in charge of distributing said force.
Position within a block	Location of the lot within the block, being able to be located on a corner or between two lots.	Be located in the middle of two lots, as the movement can be expected to be homogeneous.
Number of staggered floors	Regarding the heights of both the house to be analyzed and that of its adjacent ones.	Find yourself on the same level.
Difference of the opening area percentage	Proportion of voids in the walls between adjoining buildings.	In general terms, it is necessary to try to have the lowest percentage of voids, since this weakens the resistance of the wall.

5 CONCLUSIONS

In summary, a methodology has been presented to determine a set of parameters based on the typology of the study area, which conditions the assessment of the physical vulnerability of a dwelling. This proposed set of parameters is easy to understand and identify. With a view to later creating or adapting a methodology to identify the level of fiscal vulnerability of dwellings with a typology equal to that found in Los Olivos, the parameters that were presented can be used as input, leaving to each individual criteria, the degrees of vulnerability and the partial or total use of the selected indicators, achieving that the elaborated instrument can be easily implemented by people with basic construction knowledge, thereby achieving a reduction in time and costs, unlike the implementation of complex methodology where they require highly trained staff or other application techniques (i.e., machine learning).

ACKNOWLEDGEMENTS

The authors would like to thank Rubén Varillas and the Department of Civil Engineering of the Universidad Tecnológica del Perú for their valuable support in the process of publication and dissemination of this manuscript.

REFERENCES

[1]   Ordaz, A., Hernández, J. & Garatachia, J., Cartographic approach to structural vulnerability to earthquakes using a qualitative methodology: Application to the city of Toluca. *Cuad Geográficos*, **59**, pp. 178–198, 2020. DOI: 10.30827/cuadgeo.v59i2.9340.

[2]   sMorales-soto, N. & Zavala, C., Earthquakes in the Central Coast of Peru: Could it be lima the scene of a future disaster? *Revista Peruana de Medicina Experimental y Salud Pública*, **25**, pp. 217–224, 2008.

[3]   Bambarén, C. & Alatrista, M del S., Estimation of the socioeconomic impact of the Pisco earthquake on the Peruvian health sector. *Revista Medica Herediana*, **20**, pp. 89, 2012. DOI: 10.20453/rmh.v20i2.988.

[4]   Khan, S., Qureshi, M., Rana, I. & Maqsoom, A., An empirical relationship between seismic risk perception and physical vulnerability: A case study of Malakand, Pakistan. *International Journal of Disaster Risk Reduction*, **41**, pp. 1–9, 2019. DOI: 10.1016/j.ijdrr.2019.101317.

[5]   Guardiola-Víllora, A. & Basset-Salom, L., Seismic risk scenarios for the Eixample district of the city of Valencia. *Revista Internacional de Métodos Numéricos para Cálculo y Diseño en Ingeniería*, **31**, pp. 81–90, 2015. DOI: 10.1016/j.rimni.2014.01.002.

[6]   Ocola, L., Hazard, vulnerability, risk and the possibility of seismic disasters in Peru. *Rev Geofísica*, **61**, pp. 60–125, 2005.

[7]   Birkmann, J., Assesing vulnerability in the context of multiple-stressors. *Vulnerability Assessment and Adaptation Planning*, pp. 1–25, 2010.

[8]   Wisner, B., Blaikie, P., Cannon, T. & Davis, I., At risk. **43**, 2003.

[9]   Izquierdo-Horna, L. & Kahhat, R., Methodological framework to integrate social and physical vulnerability in the prevention of seismic risk. *Risk Analysis*, **121**, pp. 69–79, 2018. DOI: 10.2495/RISK180061.

[10] Gent, K., Astroza, M. & Giulano, G., *IX Chilean Congress of Seismology and Earthquake Engineering IX Conference*. Calibration GNDT Vulnerability Index to Chil. Build. Confin. Mason. Struct., pp. 1–15, 2005.

[11] López, O., Coronel, G. & Rojas, R., Prioritization indices for seismic risk management in existing buildings. *Revista de la Facultad de Ingeniería Universidad Central de Venezuela*, **29**, pp. 107–126, 2014.

[12] Maldonado, E. & Chio, G., Estimation of seismic vulnerability functions in earthen buildings. *Ing y Desarro*, **25**, pp. 180–99, 2009.

[13] Zora, F. & Acevedo, A., Seismic vulnerability index of schools of the metropolitan area of Medellin, Colombia. *Environmental Impact Assessment Review*, **16,** pp. 195–207, 2019. DOI: 10.24050/reia.v16i32.1035.

[14] Benedetti, D. & Petrini, V., Sulla vulnerabilita sismica di edifici in muratura: un metodo di valutazione. A method for evaluating the seismic vulnerability of masonry buildings, 1984.

[15] Ortega, J., Vasconcelos, G., Rodrigues, H. & Correia, M., A vulnerability index formulation for the seismic vulnerability assessment of vernacular architecture. *Engineering Structures*, **197**, pp. 1–20, 2019. DOI: 10.1016/j.engstruct.2019.109381.

[16] Formisano, A., Florio, G., Landolfo, R. & Mazzolani, F., Numerical calibration of an easy method for seismic behaviour assessment on large scale of masonry building aggregates. *Advances in Engineering Software*, **80**, pp. 116–38, 2015. DOI: 10.1016/j.advengsoft.2014.09.013.

[17] Kassem, M., Mohamed, F., & Noroozinejad, E., Development of seismic vulnerability index methodology for reinforced concrete buildings based on nonlinear parametric analyses. *MethodsX*, **6**: pp. 199–211, 2019. DOI: 10.1016/j.mex.2019.01.006.

[18] Chieffo, N., Clementi, F., Formisano, A. & Lenci, S., Comparative fragility methods for seismic assessment of masonry buildings located in Muccia (Italy). *Journal of Building Engineering*, **25**, pp. 1–14, 2019. DOI: 10.1016/j.jobe.2019.100813.

[19] Mosoarca, M., Onescu, I., Onescu, E., Azap, B., Chieffo, N. & Szitar-Sirbu, M., Seismic vulnerability assessment for the historical areas of the Timisoara city, Romania. *Engineering Failure Analysis*, **101**, pp. 86–112, 2019. DOI: 10.1016/j.engfailanal.2019.03.013.

[20] Mosoarca, M., Onescu, I., Onescu, E. & Anastasiadis, A., Seismic vulnerability assessment methodology for historic masonry buildings in the near-field areas. *Engineering Failure Analysis*, **115**, pp. 1–20, 2020. DOI: 10.1016/j.engfailanal.2020.104662.

[21] Kappes, M.S., Papathoma-Köhle, M. & Keiler M., Assessing physical vulnerability for multi-hazards using an indicator-based methodology. *Applied Geography*, **32**, pp. 577–590, 2012. DOI: 10.1016/j.apgeog.2011.07.002.

[22] Vargas, Y., Pujades, L., Barbat, A. & Hurtado, J., Probabilistic assessment of capacity, fragility and seismic damage of reinforced concrete buildings. *Revista Internacional de Métodos Numéricos para Cálculo y Diseño en Ingeniería*, **29**, pp. 63–78, 2013. DOI: 10.1016/j.rimni.2013.04.003.

[23] Maldonado, E., Gómez, I. & Chio, G., Application of fuzzy sets in the evaluation of seismic vulnerability parameters of masonry buildings. *Ing y Desarro*, **22**, pp. 1–22, 2007.

[24] Calvi, G.M., Pinho, R., Magenes, G., Bommer, J.J., Restrepo-Vélez, L.F. & Crowley, H., Development of seismic vulnerability assessment methodologies over the past 30 years. *ISET Journal of Earthquake Technology*, **43**, pp. 75–104, 2006.

[25] Municipalidad de Los Olivos, *Earthquake Disaster Risk Reduction and Prevention Plan District of Los Olivos*, Lima, 2019.

[26] INEI, *National Census 2017*, 2017.

[27]   Lázaro, I., Los Olivos: A district on the road to prosperity. *Pensam Crítico*, **9**, pp. 61–76, 2014. DOI: 10.15381/pc.v9i0.9023.

[28]   Carrillo, J., Evaluation of the seismic vulnerability of structures using a performance design. *DYNA*, **155**, pp. 91–102, 2008.

[29]   Chica, A. & Fuertes, A., Approach to the structural analysis and seismic protection of heritage buildings based on the characterization of their distinctive technical features. *Revista Ingeniería de Construcción*, **33**, pp. 315–326, 2018. DOI: 10.4067/s0718-50732018000300315.

[30]   Iglesias, S., Irigaray, C. & Chacón, J., Analysis of seismic risk in urban areas using geographic information systems. Application to the City of Granada. *Cuad Geográficos*, **39**, pp. 147–166, 2006.

[31]   Maldonado, E., Chio, G., Gómez, I., Seismic vulnerability index for masonry buildings based on expert opinion. *Ingeniería y Universidad*, **11**, pp. 149–168, 2007.

[32]   Guzmán, J., Williams, F., Riquer, G., Vargas, A. & Leyva, R., Soil liquefaction failures induced by the September 7, 2017 earthquake in the City of Coatzacoalcos, Veracruz, Mexico. *Revista Ingeniería Sísmica*, pp. 82–106, 2020. DOI: 10.18867/ris.102.526.

[33]   San Bartolomé, Á. & Quiun, D., Seismic design of confined masonry buildings. *Ciencia*, **13**, pp. 163–87, 2010.

[34]   CISMID, *Seismic Microzonation of the District of Los Olivos*, 2014.

[35]   Birkmann, J., Measuring vulnerability to promote disaster-resilient societies: Conceptual frameworks and definition. **14**, 2009. DOI: 10.1002/aehe.3640140303.

STRENGTH BALANCE OF STEEL DAMPER COLUMNS AND SURROUNDING BEAMS IN REINFORCED CONCRETE FRAMES

KENJI FUJII & MIZUKI KATO
Department of Architecture, Chiba Institute of Technology, Japan

ABSTRACT

In earthquake-prone countries, energy dissipated devices (dampers) have recently been widely used in building structures. Their main purpose is to mitigate damage to beams and columns during strong seismic events. The dampers act as energy absorbing members. In this study, the strength balance of steel damper columns and surrounding beams contained in reinforced concrete frames is investigated. First, the damper column strength ratio is defined on the basis of the rigid-perfectly plastic mechanical model. Next, a nonlinear dynamic analysis of various frame models was performed to investigate the influence of the damper column strength ratio on the nonlinear seismic response of reinforced concrete frames containing steel damper columns. The results of the analysis indicate that that the proper strength balance of the steel damper columns and surrounding beams is important in maximizing the energy dissipation into the damper columns. The beam-end section connected to the damper columns needs sufficient strength to avoid premature yielding prior to any energy dissipation. To discuss the strength balance of steel damper columns and surrounding beams, the damper column strength ratio is a possible index.

Keywords: steel damper column, reinforced concrete frame, energy dissipation, damper column strength ratio.

1 INTRODUCTION

The steel damper column is one energy dissipating device (damper) used to control seismic damage in building structures [1], [2]. Fig. 1 illustrates a steel damper column and its application in a reinforced concrete (RC) frame structure. In the steel damper column (Fig. 1(a)), low-yield-strength steel is used in shear panel dampers, which absorb the hysteresis energy [1], [2].

Steel damper columns are installed in the mid-span of RC beams (Fig. 1(b)). The deformation of a shear damper panel (Fig. 1(a)) is influenced not only by the deformation of the roll-formed H-section column (elastic column), but also the deformation of the RC beams connected to the damper column. The main purpose of these dampers is to minimize the damage to beams and columns caused by strong seismic displacements. Using dampers enables the energy to be absorbed. Because the shear damper panel begins to absorb hysteresis energy when it yields, the effectiveness of the damper column depends strongly on the strength balance of the damper columns and their surrounding RC members. In addition, the yield strength of the damper panel increases through strain hardening [1]–[3]. Therefore, strain hardening should be considered when evaluating strength balance. On the basis of these discussions, two questions arise:

- What kind of index is appropriate to evaluate quantitatively the strength balance of steel damper columns and surrounding RC beams?
- What criterion, based on the strength balance, holds for steel damper columns to be effective?

WIT Transactions on The Built Environment, Vol 202, © 2021 WIT Press
www.witpress.com, ISSN 1743-3509 (on-line)
doi:10.2495/ERES210031

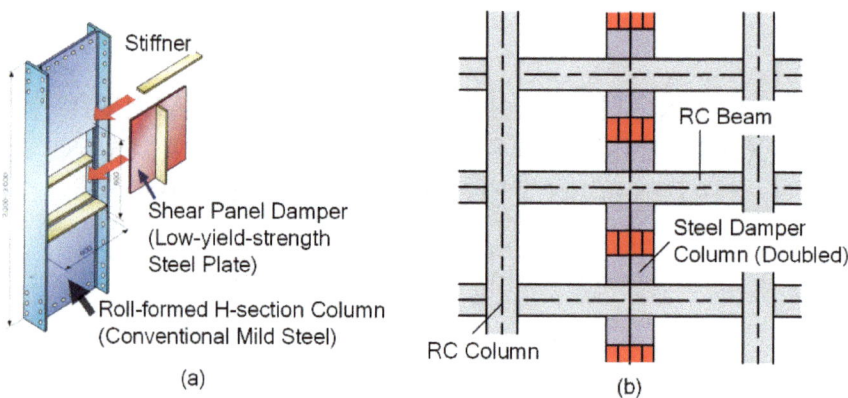

Figure 1:   Steel damper column and its application to reinforced concrete (RC) frame structures. (a) Steel damper column [2]; and (b) RC frame with steel damper columns.

In this study, the strength balance of steel damper columns and surrounding beams in reinforced concrete frames with damper columns is investigated. First, the damper column strength ratio is defined on the basis of the rigid-perfectly plastic mechanical model. Next, a nonlinear dynamic analysis of frame models is conducted to investigate the influence of the damper column strength ratio on the nonlinear seismic response of RC frames with steel damper columns.

2 DEFINITION OF DAMPER COLUMN STRENGTH RATIO

Consider a RC frame with a damper column (Fig. 2). Here, H_i denotes the height of level i above ground level, and h_i the height of the ith story. For simplicity in discussions, three assumptions are made:

- All RC beams undergo rigid-perfectly plastic behaviour, and all RC columns are infinitely rigid and strong.
- All shear panels undergo rigid-perfectly plastic behaviour, and all elastic columns are infinitely rigid and strong.

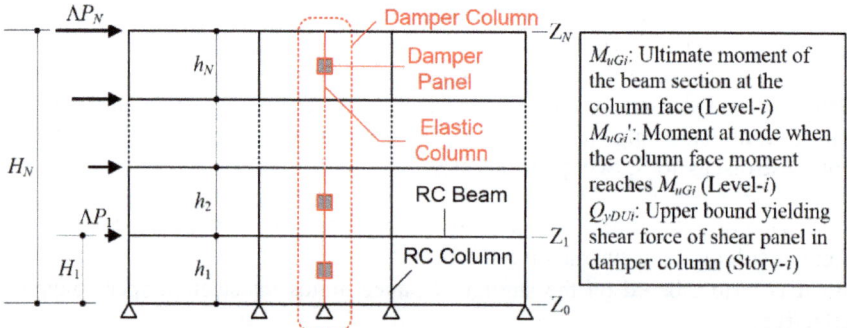

Figure 2: RC frame with damper column.

- In the calculation of internal virtual work, the work done by each plastic hinge at a node is assumed to be the product of the moment $M_{uGi}{}'$ and the plastic rotation.

On the basis of these assumptions, we consider two collapse mechanisms (Fig. 3); here, P_i denotes the assumed horizontal force at level i, θ_p the plastic rotation. Assuming all RC columns are infinitely rigid and strong, both mechanisms belong to the "whole collapse" class; moreover, for mechanism 1, all shear damper panels yield whereas, for mechanism 2, no shear damper panels yield. Let Λ_j denote the collapse load factor for mechanism j; the external virtual work is calculated as

$$\delta W_{Oj} = \Lambda_j \left(\sum_{i=1}^{N} P_i H_i \right) \theta_p .$$ (1)

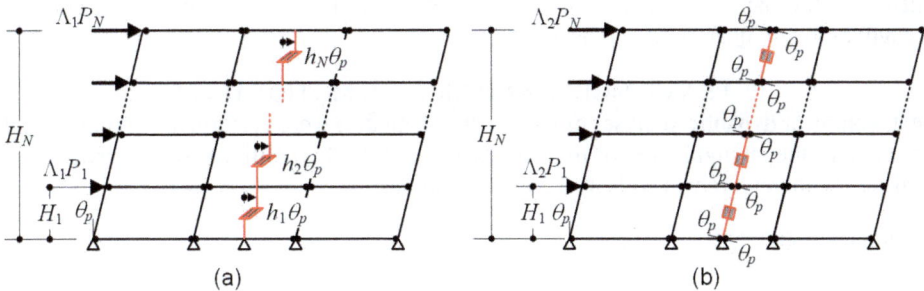

Figure 3:   Collapse mechanism of RC frame with damper column. (a) Mechanism 1 (damper panel yields); and (b) Mechanism 2 (damper panel does NOT yield).

Next, the internal virtual work of mechanism j, δW_{Ij}, is assumed to be the sum of the internal virtual work done by (a) the damper columns and the beam-end sections connected to the damper columns, δW_{IDj}, and (b) the other RC beam sections, δW_{IFj}. That is,

$$\delta W_{Ij} = \delta W_{IDj} + \delta W_{IFj} .$$ (2)

In collapse mechanism 1, the internal virtual work δW_{ID1} is calculated by considering the virtual work done by the damper panel,

$$\delta W_{ID1} = \left(\sum_{i=1}^{N} Q_{yDUi} h_i \right) \theta_p .$$ (3)

Similarly, in collapse mechanism 2, the internal virtual work δW_{ID2} is calculated considering the virtual work done by the beam-end sections connected to damper column,

$$\delta W_{ID2} = \left\{ \sum_{i=0}^{N} \left(M_{uGLi}{}' + M_{uGRi}{}' \right) \right\} \theta_p .$$ (4)

Following the principle of virtual work, the collapse load factor for mechanism j, Λ_j, is obtained by equating δW_{Oj}, in eqn (1) and δW_{Ij}, in eqn (2). The true collapse load factor Λ is then the minimum of Λ_1 and Λ_2, and hence the true collapse mechanism of the given frame corresponds to Λ.

Here, for both mechanisms, the internal virtual work done by the other RC beam sections, denoted δW_{IF1} and δW_{IF2}, are assumed that to be the same. In this instance, the relationship between the magnitudes of Λ_1 and Λ_2 is determined from that between δW_{ID1} and δW_{ID2}. The strength ratio, α, for the damper column is then defined as

$$\alpha \equiv \frac{\delta W_{ID1}}{\delta W_{ID2}} = \sum_{i=1}^{N} Q_{yDUi} h_i \left/ \left\{ \sum_{i=0}^{N} \left(M_{uGLi}{}' + M_{uGRi}{}' \right) \right\} \right. . \tag{5}$$

If α is smaller (larger) than unity, collapse mechanism 1 (2) occurs. In discussing the strength balance of the damper column and surrounding RC beams, we chose for this study the strength ratio as a quantitative indicator.

3 FRAME MODEL AND GROUND MOTION DATA

Next, nonlinear dynamic analyses of various frame models were performed to investigate the influence of the strength ratio of the damper column on the nonlinear seismic response of reinforced concrete frames with steel damper columns.

3.1 Frame model data

3.1.1 Simplified frame model
Fig. 4 illustrates one of the simplified frame models considered. This frame model is based on the ten-story RC frame model presented in Mukoyama et al. [4]; the lower part of the frame Y2 (frame with damper columns, range: stories 2 to 5) is extracted for a simplified frame model. The floor masses of levels 1 to 3 are assumed equal, with $m_1 = m_2 = m_3 = 270$ t, whereas the mass of level 4 is assumed to be $m_4 = 1620$ t, which takes into account the mass of the upper floors. Young's modulus and the shear modulus of the concrete are assumed to be $E_C = 2.52 \times 10^4$ MPa and $G_C = 1.08 \times 10^4$ MPa, respectively. Fig. 5 presents the envelope of the force-deformation relationship of the nonlinear flexural spring of RC members. The same modelling applied in Mukoyama et al. [4] is used in this study except for the RC columns; only stiffness degradation due to cracking is considered and no flexural yielding is considered. Fig. 6 shows the hysteresis rule of the members. The Muto hysteresis model [5] with one modification (Fig. 6(a)) is used to model the flexural spring in the RC members, whereas the hysteresis model proposed by Ono and Kaneko [6] (Fig. 6(b)) is used to model the shear behaviour of the damper columns taking into consideration the strain-hardening behaviour.

Tables 1 and 2 list the properties of RC beams and columns used in the simplified model. In both tables, b and D denote the width and depth of the sections, respectively, I, A_S and A_N the moment of inertia, and the sectional area for shear and axial deformation, respectively, that takes into account the difference in concrete strength for each level. In addition, M_c and M_y denote the flexural cracking and yielding moments, respectively, and α_y denotes the secant stiffness degradation ratio at the yield point (Fig. 5). Extracted from a 10-story frame structure, our simplified frame model comprises beams for levels Z_0 and Z_4 corresponding to the boundary; therefore, the stiffness and strength of these beams are assumed to be 1/2 that of the original model.

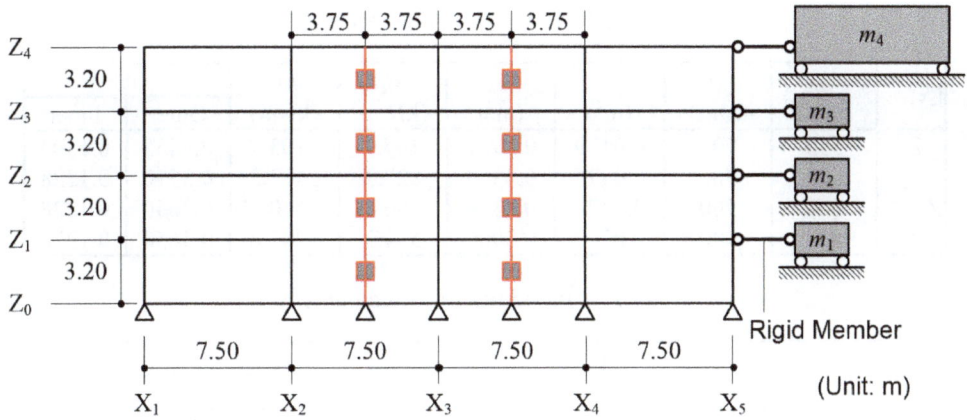

Figure 4: Simplified frame model.

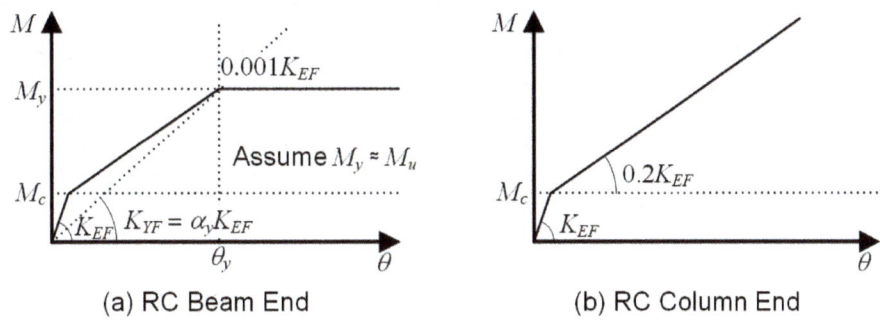

(a) RC Beam End

(b) RC Column End

Figure 5: Envelope of the force–deformation relationship of nonlinear flexural spring.

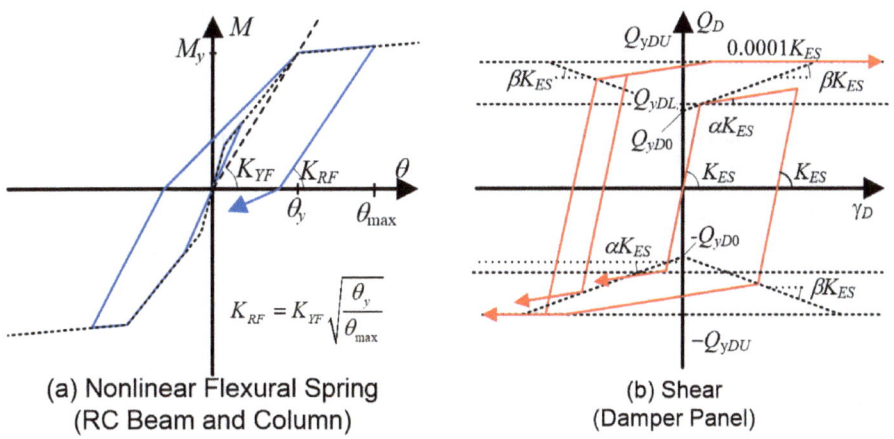

(a) Nonlinear Flexural Spring
(RC Beam and Column)

(b) Shear
(Damper Panel)

Figure 6: Hysteresis rules.

WIT Transactions on The Built Environment, Vol 202, © 2021 WIT Press
www.witpress.com, ISSN 1743-3509 (on-line)

Table 1: Properties of RC beams.

Lv.	b (mm)	D (mm)	I (m⁴)	A_S (m²)	M_c (kNm)	M_y (kNm)	α_y Outer	Inner
Z_4	550	900	0.0179	0.1929	367.9	1103.7	0.3455	0.1303
Z_3	600	900	0.0376	0.4359	827.7	2483.2	0.3510	0.1228
Z_1, Z_2	600	900	0.0376	0.4359	886.8	2660.4	0.3640	0.1228
Z_0	800	900	0.0243	0.3000	594.2	1782.6	0.3650	0.1228

Table 2: Properties of RC columns.

Level	b (mm)	D (mm)	I (m⁴)	A_S (m²)	A_N (m²)	M_c (kNm) Outer	Inner
4	900	900	0.0547	0.6750	0.8100	696.5	995.2
3	900	900	0.0547	0.6750	0.8100	746.3	1094.8
2	900	900	0.0547	0.6750	0.8100	796.1	1194.3
1	900	900	0.0547	0.6750	0.8100	845.9	1293.9

Table 3 lists the properties of a steel damper column; Q_{yDL} and Q_{yDU} denote respectively the initial and the upper bound yield strength of the damper panel (Fig. 6(b)). In addition, t_p denotes the thickness of the damper panel, A_{SDp} the sectional area for shear deformation of damper panel, and H_{D0} the height of damper panel. Young's modulus and the shear modulus for steel are set to $E_S = 2.05 \times 10^5$ MPa and $G_S = 7.88 \times 10^4$ MPa, respectively. The initial normal yield stress of the steel used for damper panels is set to 205 MPa whereas the nominal yield stress after appreciable cyclic loading is set to 300 MPa.

Table 3: Properties of steel damper column.

Level	Q_{yDL} (kN)	Q_{yDU} (kN)	t_p (mm)	A_{SDp} (m²)	H_{D0} (m)	Section of elastic column (mm × mm × mm × mm)
4	1251	1831	9	0.0106	0.600	H-600×250×16×32×2
1 to 3	1511	2211	9	0.0128	0.700	H-700×300×16×28×2

The damping matrix is assumed to be proportional to the instantaneous stiffness matrix without a damper column. The damping ratio of the first elastic mode of the model without a damper column is assumed to be 0.05; its natural period without damper is 0.725 s, whereas that with damper is 0.477 s.

3.1.2 Analysis parameters

In this analysis, the structural parameter is the strength balance of the damper column and surround RC beams. The assumptions regarding yield strength of the beam sections connected to damper columns are described below.

Fig. 7 depicts the bending moment at the beam-end section connected to the damper columns. At the beam-end sections A₁ and B₂, the bending moment is assumed to be the yielding moment M_{yGi}. The equilibrium of the moment at node C is

$$M_{GLi}{}' + M_{GRi}{}' = \frac{1}{2}\left(Q_{yDUi}h_i + Q_{yDUi+1}h_{i+1}\right).\tag{6}$$

Assuming moment M_{GLi}' equals M_{GRi}', the moment at the beam-end sections B_1 and A_2 is

$$M_{GLi} = M_{GRi} = \frac{1}{1 + dL_d/L_0}\left\{\frac{1}{4}\left(Q_{yDUi}h_i + Q_{yDUi+1}h_{i+1}\right) - \frac{dL_d}{L_0}M_{yGi}\right\}. \qquad (7)$$

Note that for $i = 0$, the value of $Q_{yDUi}h_i$ is set to zero. Similarly, the value of $Q_{yDUi+1}h_{i+1}$ is set to zero for $i = 4$.

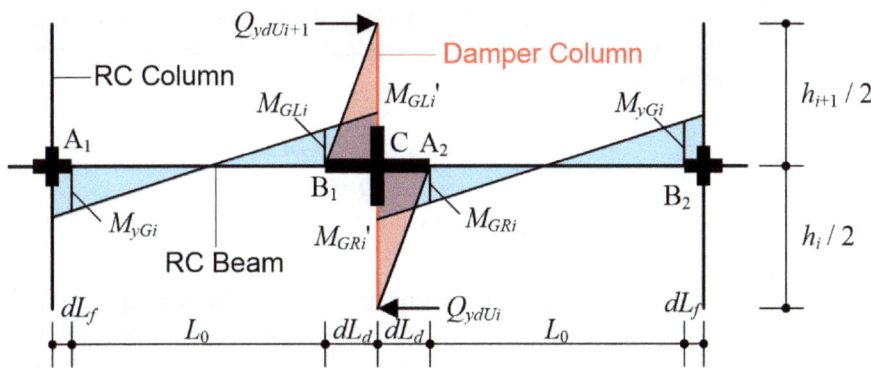

Figure 7: Bending moment at the beam-end sections connected to damper columns.

Given the above, the yield strength M_y at the beam-end sections connected to damper columns are determined using beam-end strength ratio α_G ($=M_y/M_{GL}$); for $\alpha_G = 1$, M_y equals to M_{GL}.

Fig. 8 shows the assumed moment–deformation angle relationship of a flexural spring at the beam-end section connected to the damper column. The cracking moment M_c and the secant stiffness degradation ratio at yielding α_y is set independent of α_G for simplicity. Therefore, the tangent stiffness degradation ratio after crack α_1 is assumed to increase as α_G increases. Parameter α_G is set from 0.7 to 1.5 with step intervals of 0.1. Table 4 lists the beam-end strength ratio α_G and corresponding damper column strength ratio α. For the calculation of α, the ultimate moment of the beam section at the column face is assumed to be their yielding moment. As listed, the damper column strength ratio α decreases as α_G increases.

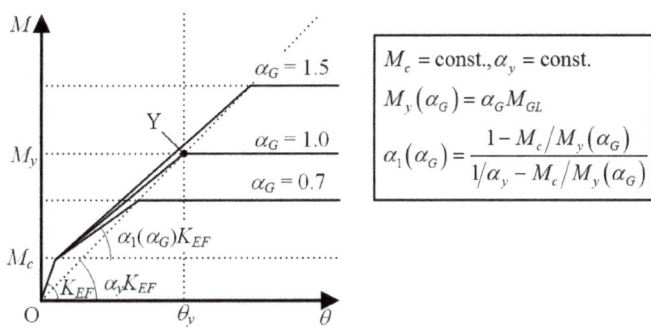

Figure 8:   Assumed moment–deformation angle relationship of flexural spring at beam-end section connected to damper column.

Table 4: Beam-end strength ratio α_G and corresponding damper column strength ratio α.

α_G	0.7	0.8	0.9	1.0	1.1	1.2	1.3	1.4	1.5
α	1.320	1.193	1.088	1.000	0.925	0.861	0.805	0.756	0.712

3.2 Ground motion data

Ten artificial ground motions are generated. The target elastic spectrum with 5% critical damping (pseudo-acceleration spectrum $_pS_A(T, 0.05)$, with T denoting the natural period of the system) was determined from the Building Standard Law of Japan [7] applying type-2 soil conditions. Specifically,

$$_pS_A(T,0.05)=\begin{cases}4.8+45T \quad \text{m/s}^2 & T \le 0.16 \text{ s}\\ 12.0 & 0.16 \text{ s} \le T \le 0.864 \text{ s}\\ 12.0(0.864/T) & T > 0.864 \text{ s}\end{cases}. \quad (8)$$

The phase angle is obtained as a uniform random value and the Jenning-type envelope function $e(t)$ proposed by the Building Centre of Japan [8]. Fig. 9 shows the pseudo-acceleration spectrum and the total energy spectrum [9] of the artificial ground motions used in this study. The ground accelerations are scaled using a multiplicative factor λ that is set between 0.5 and 1.0.

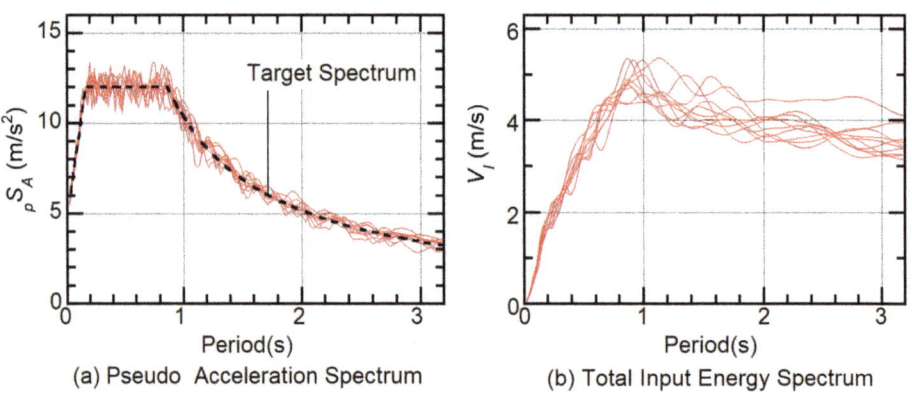

(a) Pseudo Acceleration Spectrum   (b) Total Input Energy Spectrum

Figure 9: Elastic response spectra of artificial ground motions used in this study.

4 ANALYSIS RESULTS AND DISCUSSIONS

4.1 Analysis results

Averages of results obtained from the nonlinear dynamic analysis of the ten artificial ground motions were considered. Fig. 10 shows the relationship between the maximum of the whole deformation angle R^*_{max}, defined as the peak horizontal displacement at level Z_4 divided by the total height H_4 and the damper column strength ratio α. The whole deformation angle

R^*_{max} increases as α increases, the variation being more noticeable for λ = 1.0 (Fig. 10(b)) than with λ = 0.5 (Fig. 10(a)).

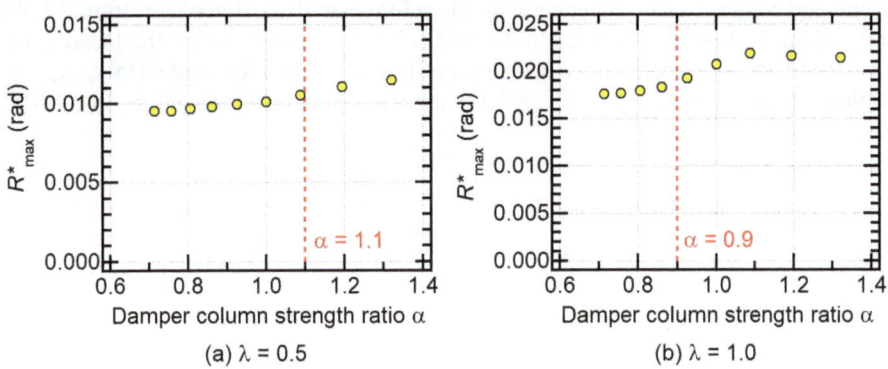

(a) λ = 0.5   (b) λ = 1.0

Figure 10: Whole deformation angle.

Fig. 11 shows the relationship between the total input energy E_I, cumulative strain energy (whole frame: E_S, RC frame: E_{Sf}, steel damper columns: E_{Sd}) and the damper column strength ratio α. The variation in E_I and E_S is not noticeable. However, E_{Sf} increases whereas E_{Sd} decreases, as α increases. For λ = 0.5 (Fig. 11(a)), E_{Sf} exceeds E_{Sd} when α is larger than 1.1, whereas for λ = 1.0 (Fig. 11(b)), E_{Sf} exceeds E_{Sd} when α is larger than 0.9. This implies that the effectiveness of steel damper columns is strongly related to the damper column strength ratio α; the steel damper column is more efficient at energy dissipation if α is small.

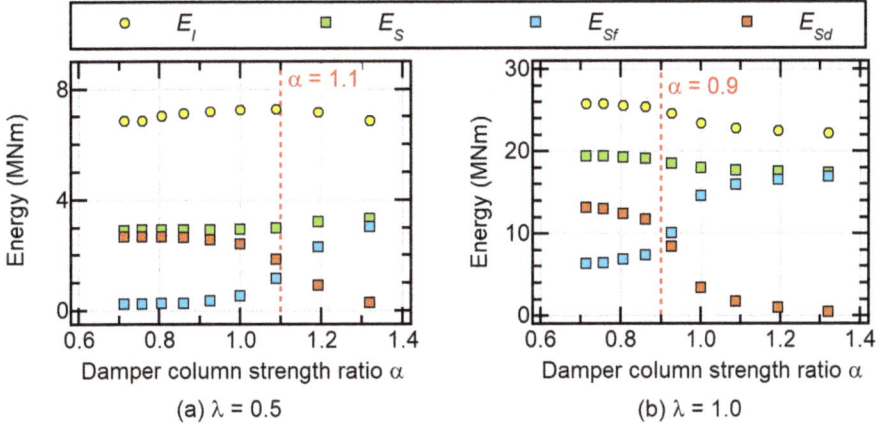

(a) λ = 0.5   (b) λ = 1.0

Figure 11: Total input energy and cumulative strain energy.

In summary, the damper column strength ratio α is suitable in discussing the strength balance of the damper columns and surrounding RC beams; for a better effectiveness of steel damper columns, α should be low in value. Based on our numerical analysis results, α should not exceed 0.9 when the expected whole deformation angle is close to 0.02 rad.

4.2 Discussion

As described in Section 2, the difference in the two collapse mechanisms is the yielding of the beam-end connected to damper column. Therefore, our discussion begins from the plastic rotation of the beam-end connected to damper columns. Fig. 12 depicts the location of the beam-end investigated herein and a definition of the plastic rotation angle. Here, θ_{pmax} is the peak plastic rotation angle; when θ_{max} is less than θ_y, θ_{pmax} is zero (no yielding occurs).

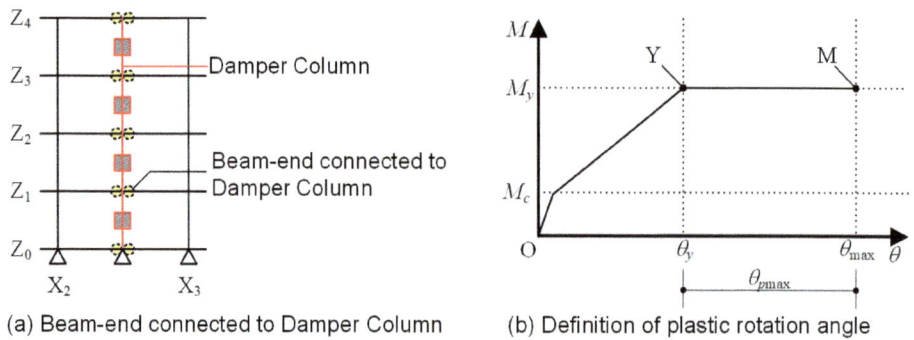

(a) Beam-end connected to Damper Column   (b) Definition of plastic rotation angle

Figure 12: Plastic rotation at beam-end connected to damper column.

Fig. 13 shows the relationship between the peak plastic rotation at beam-ends connected to damper columns θ_{pmax} at each level and the damper column strength ratio α. As expected, θ_{pmax} increases as α increases. For $\lambda = 0.5$ (Fig. 13(a)), θ_{pmax} is almost zero when α is less than 0.9. A similar observation holds for $\lambda = 1.0$ (Fig. 13(b)). This implies that the behaviour of the simplified frame model is similar to collapse mechanism 1 (damper panel yield) when α is small.

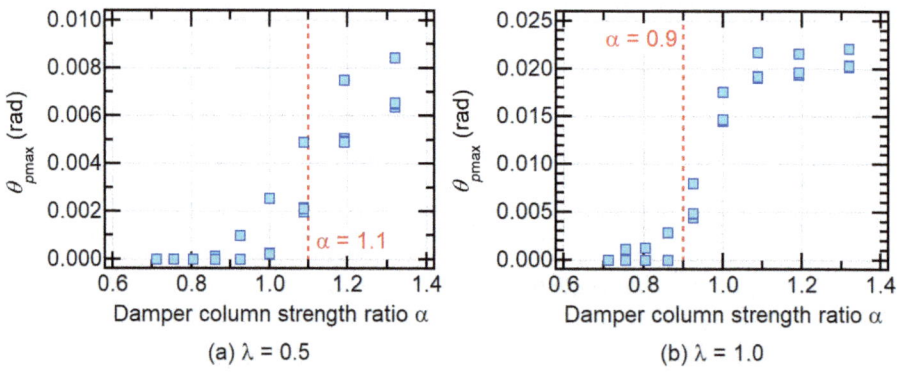

(a) $\lambda = 0.5$ (b) $\lambda = 1.0$

Figure 13: Peak plastic rotation angle at beam-ends connected to damper column.

Next, focusing on the behaviour of the damper panel, Fig. 14 presents the relationship between the peak shear strain of the damper panel in each damper column γ_{Dmaxi} and the damper column strength ratio α. As expected, γ_{Dmaxi} is large and stable when α is small. For

$\lambda = 0.5$ (see Fig. 14(a)), the peak shear strain γ_{Dmaxi} is stable if α is less than 1.1, whereas if α is larger than 1.1 the strain decreases rapidly as α increases. A similar observation holds if $\lambda = 1.0$ (Fig. 14(b)). The peak shear strain γ_{Dmaxi} decreases rapidly as α increases if α is larger than 0.9. Note that the peak shear strain γ_{Dmaxi} reaches 5% and more if α is less than 0.9. This implies that the condition of the damper panel is close to the ultimate stage [2]. Therefore, $\alpha = 0.9$ seems suitable in eliciting full performance from the damper columns.

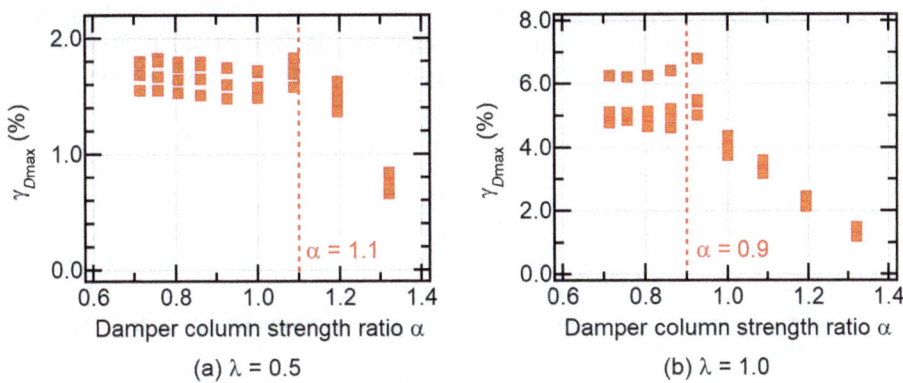

(a) $\lambda = 0.5$ (b) $\lambda = 1.0$

Figure 14: Peak shear strain of damper panel.

In summary, the effectiveness of steel damper columns is pronounced if α is less than 0.9. The mechanism underlying the behaviour of the RC frame with a damper column is collapse mechanism 1 (Fig. 3(a)) if α is less than 0.9. In this instance, the deformation of damper panels is pronounced, and the energy absorption of the damper panels maximized. Note that in theory (see Section 2), the boundary between the two collapse mechanisms occurs at $\alpha = 1$. However, the nonlinear dynamic analysis results indicate that a reasonable criterion should be 0.9. This arises from the variation in moment distribution in the damper column arising from the influence of higher mode. Therefore, some safety factor is needed for the yielding moment at the beam-end section connected to the damper column. Based on the results obtained, the recommended value for the safety factor is $\alpha_G = 1.2$ ($\alpha = 0.861$, Table 4).

5 CONCLUSIONS

The strength balance of steel damper columns and surrounding beams in reinforced concrete frame with a damper column was investigated. The main contribution and results of this study are the following:

- The damper column strength ratio α proposed, eqn (5), is a potential index to discuss the strength balance of steel damper columns and surrounding beams.
- Based on the numerical analysis results presented, the damper column strength ratio α should not exceed 0.9 when the expected whole deformation angle is close to 0.02 rad.

ACKNOWLEDGEMENTS

The authors thank the financial support from JFE Civil Engineering and Construction Corp. Valuable comments from Chizuru Irie and Mitsuhide Yoshinaga, JFE Civil Engineering and Construction Corp. are also appreciated. The original frame model data for the reinforced concrete building was provided from Riho Mukoyama, a graduate student of Chiba Institute of Technology. We thank Richard Haase, PhD, from Edanz Group (https://en-author-services.edanz.com/ac) for editing a draft of this manuscript.

REFERENCES

[1]   Katayama, T., Ito, S., Kamura, H., Ueki, T. & Okamoto, H., Experimental study on hysteretic damper with low yield strength steel under dynamic loading. *Proceedings of the 12th World Conference on Earthquake Engineering*, Paper No. 1020, 2000.

[2]   JFE Civil Engineering & Construction Corp., JFE no Seishin-mabashira, JFE no Seishin-panel (Vibration control column and panel product by JFE). 2019. www.jfe-civil.com/pdf/catalog/vibration_control_column.pdf. Accessed on: 31 Mar. 2021 (in Japanese).

[3]   Nakashima, M., Strain-hardening behaviour of shear panels made of low-yield steel 1: Test. *Journal of Structural Engineering*, **121**(12), pp. 1742–1749, 1995.

[4]   Mukoyama, R., Fujii, K., Irie, C., Tobari, R., Yoshinaga, M. & Miyagawa, K., Displacement-controlled seismic design method of reinforced concrete frame with steel damper column. *Proceedings of the 17th World Conference on Earthquake Engineering*, Paper No. 2g-0143, 2020.

[5]   Muto, K., Hisada, T., Tsugawa, T. & Bessho, S., Earthquake resistant design of a 20 story reinforced concrete buildings. *Proceedings of the Fifth World Conference on Earthquake Engineering*, pp. 1960–1969, 1974.

[6]   Ono, Y. & Kaneko, H., Constitutive rules of the steel damper and source code for the analysis program. *Passive Control Symposium 2001*, Structural Engineering Research Center, Tokyo Institute of Technology, pp. 163–170, 2001 (in Japanese).

[7]   BCJ, *The Building Standard Law of Japan on CD-ROM*. The Building Center of Japan, 2016.

[8]   Otani, S., Japanese seismic design of high-rise reinforced concrete buildings – An example of performance-based design code and state of practices. *Proceedings of the 13th World Conference on Earthquake Engineering*, Paper No. 5010, Vancouver, Canada, 2004.

[9]   Akiyama, H., *Earthquake-Resistant Limit-State Design for Buildings*, University of Tokyo Press: Tokyo, Japan, 1985.

[10] Akiyama, H., *Earthquake-Resistant Design Method for Buildings Based on Energy Balance*, Gihodo Shuppan, 230 p., 1999 (in Japanese).

PROPOSAL OF TERRITORIAL PARAMETERS FOR SEISMIC HAZARD ASSESSMENT IN PISCO, PERU

LUIS IZQUIERDO-HORNA & PATRICIA ARANIBAR
Universidad Tecnológica del Perú, Perú

ABSTRACT
During the last few decades, Peru has experienced numerous natural phenomena (e.g., earthquakes, tsunamis, debris flows, etc.) that have caused considerable economic and human losses. However, the most prevalent phenomenon in Peru is earthquakes due to its geographic location (e.g., Ancash 1970 (7.9 Mw), Pisco 2007 (7.9 Mw), Arequipa 2013 (7.1 Mw), etc.). Therefore, in this research, we focus on improving the techniques for identifying the level of seismic hazard through a set of parameters specific to the territorial conditions. Thus, due to the antecedents presented in the city of Pisco, it has been chosen as a case study. In this sense, a systematic review of parameters and techniques used to determine the level of seismic hazard of a given sector was carried out. Then, the availability of information on the parameters previously identified was evaluated in order to obtain that the most representative variables of the territory are: soil type, slope and soil liquefaction. We consider that this set of parameters is compatible with machine learning techniques or any other method established by the corresponding authorities related to risk and disaster management. Finally, these parameters are applicable and reproducible at any territorial scale.
Keywords: hazard, earthquakes, risk and disaster management, Peru.

1 INTRODUCTION

The occurrence of a seismic event is a worldwide problem in developing communities since they usually generate large economic and human losses [1]. For this reason, at present, numerous investigations have been carried out in order to identify sectors with a high seismic hazard in order – in this way – to reduce the impacts generated on people and their livelihoods. According to the Pan American Health Organization [2], Peru is one of the countries with the highest seismic activity in the world due to its geographical location. In other words, it is part of the Pacific ring of fire, a sector that experiences approximately 80% of this type of event in the world. According to the National Institute of Civil Defense (INDECI) [3], the collision of tectonic plates has caused devastating earthquakes like the one of 1970 in Yungay, Ancash (7.8 Mw earthquake). This earthquake caused the death of 66,795 people and completely buried the city of Yungay (80% of the houses were destroyed) [4]. Another catastrophic event was the 1996 Nazca earthquake (6.4 Mw earthquake), which left a balance of 94,047 people and 12,700 homes affected. The population was isolated by landslides and the most affected areas were Pisco, Palpa and Ica [3].

The need to know the seismic hazard of a certain sector generated great curiosity to determine which are the most optimal parameters to be used in various techniques that help with this purpose. In this sense, in Syria a numerical classification scheme was developed to identify parameters that allow the assessment of seismic hazard in a GIS environment. These indicators were based on: historical seismicity, tectonic characteristics, topography and slope [5]. Likewise, in Indonesia the seismic hazard was estimated based on territorial parameters such as: slope, curvature, elevation, lithology, amplification factor, distance from faults, fault density, distance from the epicenter, among others. These parameters were used in a machine learning environment (i.e., neural network) and hierarchical analysis [6]. Along the same lines, geological, geodetic, geotechnical and geophysical parameters were used in Tabriz, Iran to be able to build a seismic microzoning map to indirectly assess the current level of

WIT Transactions on The Built Environment, Vol 202, © 2021 WIT Press
www.witpress.com, ISSN 1743-3509 (on-line)
doi:10.2495/ERES210041

seismic hazard in the sector [7]. In addition, parameters related to demographic, environmental and physical criteria were also used for this purpose in a neural network approach. The territorial parameters that stand out the most in this research are slope, elevation and geology [8]. Similarly, failure parameters such as impact, fall, trail, slip, location, maximum and minimum depth can also be used to indirectly assess the level of seismic hazard [9].

On the other hand, at the national level, territorial parameters obtained through satellite images (i.e., digital elevation models, topography, etc.) are also used to estimate the level of hazard [10]. Similarly, Matsuoka et al. [11] propose the use of parameters obtained through satellite images accompanied by field data and censuses to indirectly estimate the seismic hazard in Peru through a GIS environment. In the same way, Lazarus [12] proposes the use of surveys with technical sheets and satellite photographs to map the seismic scene of the place. Similarly, the Japanese Peruvian Center for Seismic Research and Disaster Mitigation (CISMID) [13] proposes visual evaluation sheets (building material, number of floors, predominant structural system, etc.) complemented with photographs contrasted to a GIS database for the mapping of seismic microzoning and seismic hazard analysis. Thus, various studies focused on seismic zoning activities used political-administrative, physical-geographic, socio-demographic, economic-productive and functional criteria [14]. Finally, according to Matsuzaki et al. [15] it is possible to obtain parameters based on data from damage survey, seismic movement simulation, ground conditions and damage rate for each zone.

Consequently, it can be observed that there is no consensus on which set of parameters is the most optimal to determine the level of seismic hazard. However, common aspects of these projects lie in the so-called territorial conditions. The latter due to the possibility of implementing various techniques (i.e., hierarchical analysis, machine learning, GIS environments, etc.) and the availability of information. Therefore, this study shows an extensive bibliographic review accompanied by an analysis of the study area to be able to determine a set of – applicable and reproducible – parameters that allow an adequate and coherent evaluation of the level of seismic hazard at different territorial scales.

The rest of the article is organized as follows: After this first introductory section, Section 2 presents the research methods, including the collection and processing of information, the definition, selection and analysis of criteria to choose the set of parameters that allow assessing the level of seismic hazard; Section 3 reports the results obtained, while, in Section 4, we present an in-depth discussion of them. Finally, the conclusions are shown in Section 5.

2 METHODOLOGY

The objective of this work is focused on determining a set of territorial parameters that allow an adequate identification of the level of seismic hazard. For the methodological implementation, the city of Pisco was chosen as a case study. Fig. 1 shows the methodological framework designed to fulfill this purpose.

2.1 Data gathering

This project focuses on a comprehensive review of the literature on how territorial parameters can help identify the level of seismic hazard of a given area. At the local level, the most relevant research corresponds to that carried out in 2001 by INDECI [14] whose objective was to determine the hazard levels of Pisco categorized into 4 levels (i.e., very high, high,

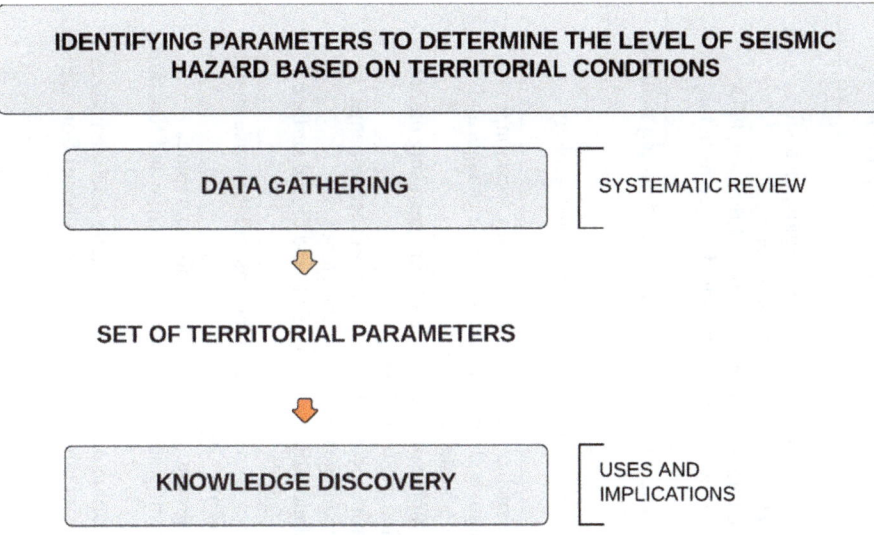

Figure 1: Methodological framework.

medium and low) using territory parameters such as the water table, soil type and slope. However, these parameters and hazard classification cease to have effect when an event reoccurs and site considerations change [16]. At the international level, Table 1 shows the summary of the most relevant revised bibliography for determining the level of hazard.

2.2 Proposal of territorial parameters

This section explains the process for selecting the territorial parameters that were considered in this proposal. To obtain the final proposal of the territorial parameters set, different bibliographic sources were studied to subsequently select the most appropriate literature with the most influential parameters. Following that, territorial parameters were proposed for the Pisco area, considering criteria such as accessibility and information availability. For this, the research developed by Birkmann was taken as a main reference [18] which mentions that for a parameter to be influential it must be reproducible, robust, reliable and easy to measure. In this sense, the result of this adaptation led to the determination of a set of parameters that will be presented and discussed in later sections.

2.3 Description of the case study

The province of Pisco is located to the north of the province of Ica and to the south of the province of Chincha. Pisco has an area of 3,978 km² and a population of 125,879 inhabitants according to the National Institute of Statistics and Informatics (INEI) [19]. In this sense, the district of Pisco has been selected as a case study not only because of its location in the coastal zone near the convergent edge of the plates, but also because it is continuously affected by earthquakes of different intensity, reinforcing the idea of that Pisco is one of the cities most affected by seismic movements [16]. Fig. 2 shows the location of the case study and Table 2 shows the history suffered, the 2007 earthquake being the event that has caused the most

Table 1: Systematic review of parameters used to identify the level of seismic hazard. Based on [5]–[9], [11], [14]–[17].

Author	Investigatory project	Parameters
Ahmad et al. [5]	Seismic hazard assessment of Syria using seismicity, DEM, slope, active faults and GIS	Historical seismicity data of the region, tectonic characteristics, topography and slope
Jena et al. [6]	Integrated model for earthquake risk assessment using neural network and analytic hierarchy process: Aceh province, Indonesia	Slope, curvature, elevation, aspect, lithology, PGA density, magnitude density, and distance from the epicenter
Karimzadeh et al. [7]	A GIS-based seismic hazard, building vulnerability and human loss assessment for the earthquake scenario in Tabriz	Geological, geodetic, geotechnical and geophysical of the sector
Yariyan et al. [8]	Earthquake risk assessment using an integrated Fuzzy Analytic Hierarchy Process with Artificial Neural Networks based on GIS: A case study of Sanandaj in Iran	Distance to a fault, slope, elevation and geology
Funning et al. [9]	The 1998 Aiguille, Bolivia earthquake: A seismically active fault revealed with InSAR	Failure such as impact, fall, trail, slip, location, maximum and minimum depth, static change, and gradients in displacement
Maruyama et al. [15]	Evaluation of building damage and tsunami inundation based on satellite images and GIS data following the 2010 Chile earthquake	Topographic maps and satellite images using digital elevation model (DEM)
Matsuoka et al. [11]	Development of building inventory data and earthquake damage estimation in Lima, Peru for future earthquakes	Data from local and national reports accompanied by satellite images (elevation and slope)
Matsuoka et al. [17]	Extraction of urban information for seismic hazard and risk assessment in Lima, Peru using satellite imagery	Digital elevation model (DEM) and digital surface model (DSM)
Matsuzaki et al. [15]	Evaluation of Seismic Vulnerability of Buildings Based on Damage Survey Data from the 2007 Pisco, Peru Earthquake	Base rock movement simulations and response analysis in terms of PGA and PGV
INDECI [14]	Hazard map, land use plan and mitigation measures for the city of Pisco	Political-administrative, physical-geographic, sociodemographic, economic-productive and functional studies
INDECI [16]	Pisco and San Andrés hazard map	Topographic maps and panchromatic satellite images and profile information in pits and trenches from CESEL and INDECI

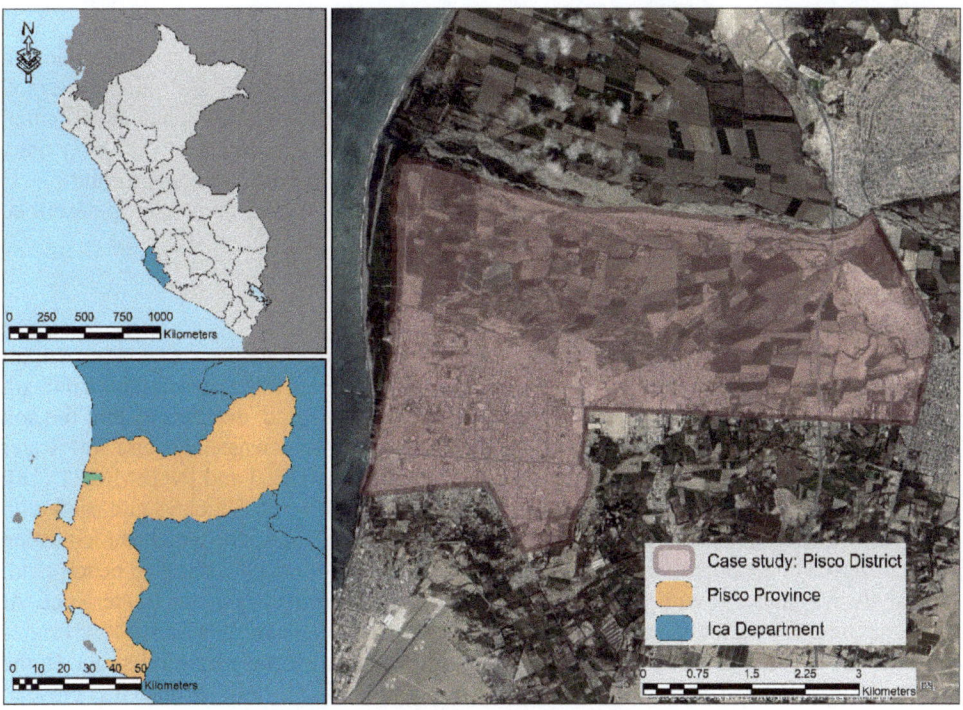

Figure 2: Case study: Pisco District.

Table 2: Record of seismic events in Pisco. Adapted from [16].

Date	Intensity	Location	Date	Intensity	Location
05/12/1664	VIII	Pisco	01/15/1960	IV	Ica
03/30/1813	VIII	Ica	01/27/1961	VI	Ica
11/21/1901	IV	Ica	09/28/1968	VI	Pisco
02/23/1907	V	Ica	09/28/1978	V	Ica
09/11/1914	IV	Ica	10/20/2006	VII	Ica
09/20/1915	V	Ica	10/26/2006	VI	Ica
10/07/1920	V	Ica	08/15/2007	VII	Pisco
10/11/1922	IV	Ica	10/28/2011	VII	Ica
08/24/1942	IX	Ica	01/30/2012	VI	Ica
09/29/1946	VII	Pisco	05/06/2012	VI	Ica
12/09/1950	VII	Pisco	03/15/2014	VI	Ica
04/04/1951	IV	Ica	10/19/2016	V	Ica
01/15/1960	VII	Ica	01/16/2021	V	Ica

damage [20]. Among the main losses generated are 59,971 victims, 383 deaths, 50,522 homes affected and multiple damages to critical infrastructure [2]. The most relevant feature of this event was its long duration [21] revealing the need for an updated study of hazards conditioned by the susceptibility of the territory [22].

3 RESULTS

After processing and analyzing the information obtained from the study of the relevant literature (Table 1), we propose as territorial parameters the type of soil, slope and soil liquefaction. Also, it is important to mention that these parameters show adequate potential to study the seismic hazard present in the sector and that not only can they vary over time, but also according to the study area since each area of interest has different geomorphological characteristics among other peculiarities [6]. Finally, each proposed parameter is described below.

3.1 Soil type

The city of Pisco is made up of sedimentary rocks and quaternary deposits. Recent accumulated quaternary deposits consist of thick conglomerates interspersed with sand, silt and clay [23]. According to Walsh [24], after a soil study in Pisco, determined that the soil in the area exhibits a sandy texture (loamy sand with a loose consistency) and low water retention capacity. According to the research carried out by INDECI and INGEMMET [16], the area is composed of white diatomite interspersed with silt – whitish clays and sand with particles of different sizes, but with a very small amount of fines, for which the cohesion tends to zero. On the other hand, these deposits are made up of poorly classified beach sand; almost always saturated with water and, therefore, have poor mechanical characteristics. At the same time, this sector is made up of landfill and rubble. Likewise, the CISMID [25] states that the soil of Pisco is made up of loose silty sand with semi-loose silty sand strata. In the same way, Pisco shows a problem of chemical aggressivity in the soil, because it is close to the coast. Fig. 3 shows the distribution of land types in the sector of interest.

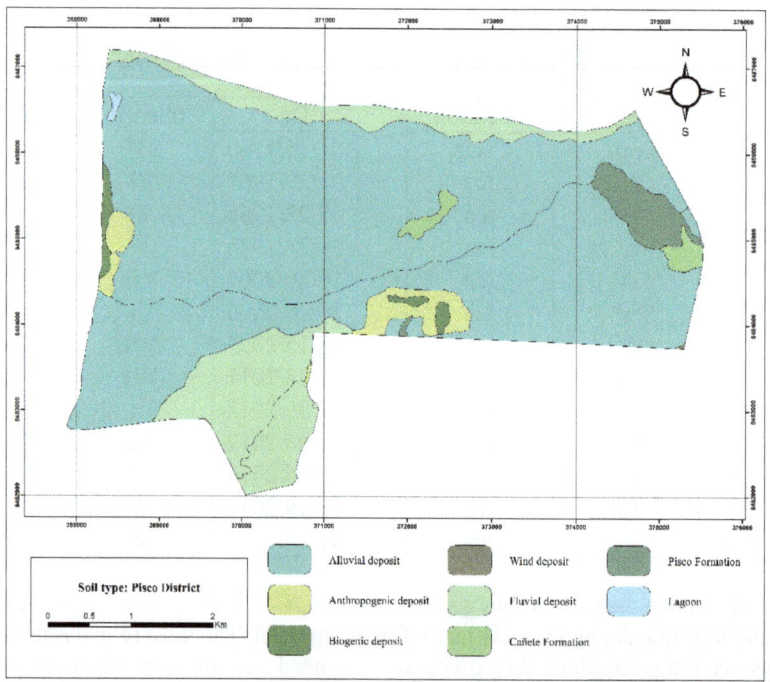

Figure 3: Soil type map: Pisco District.

3.2 Slope

The study area is made up of a marine plain on which wind materials have been deposited. The surface presents slopes of less than 9% with steep sectors that reach slopes of up to 35%. The relief is flat, moderately inclined [24]; that is, the elevation profile of the area represents a sector with a slight slope. Fig. 4 shows the mentioned characteristics of the sector.

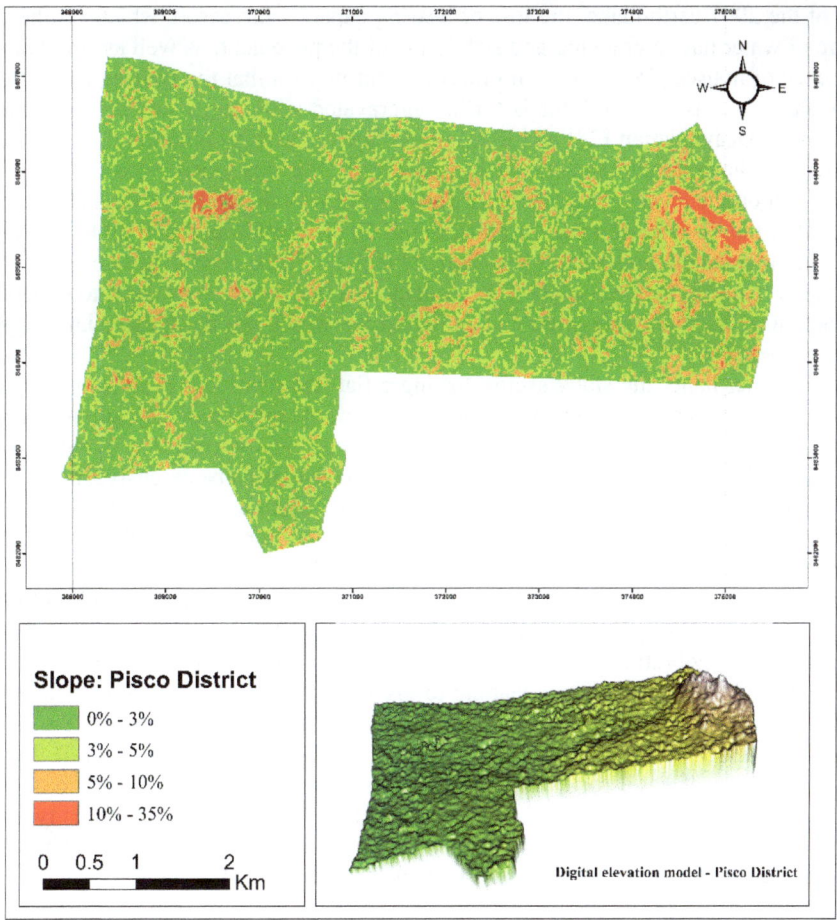

Figure 4: Distribution of slopes: Pisco District.

3.3 Soil liquefaction

Soil liquefaction is a phenomenon in which soils, due to water saturation (i.e., water table) and particularly soils such as sand or gravel, lose their resistance and flow as a result of the stresses caused in them due to seismic events; that is, during a seismic event, the pore water pressure increases in the subsoil and the effective effort is reduced until causing the pore water pressure to equal the total effort and the effective effort becomes zero ($\sigma' = 0$). The latter implies that the soil loses resistance and behaves like a fluid [26]. On the other hand, although there are almost no bibliographic references of the geotechnical characteristics of the sandy soil in the Pisco area, we can deduce analytically how the reduction of the bearing capacity (q_{adm}) of the land was, which can be determined from the load capacity (q_u) with Terzaghi's general expression [27].

4 DISCUSSION

For all the above, the sandy soil of Pisco always had a high liquefaction potential and, as a result of the 2007 earthquake, the loss of bearing capacity increased, which resulted in the outcrop of water due to cracking and slab joints of the pavement, as well as small turns and subsidence of houses [28]. Low or no fines content implies that the bearing capacity of the natural sandy soil is low, and this is further aggravated when the sandy soil is submerged. Carrillo and Alcayhuamán [28] and Rodríguez-Marek et al. [29] mention that, in Pisco, the sand deposits have an almost superficial water table. For this reason, this parameter has a great impact on the level of seismic hazard associated with the sector [30], [31]. Although it is true, after the Pisco earthquake in 2007, the E030 standard was modified to reinforce the houses in the area and make them less vulnerable [32]. There is no evidence of any type of soil improvement in the area. Therefore, considering that Pisco is exposed to experiencing liquefaction of its soils, it is important to determine which are the areas with the greatest predisposition.

On the other hand, the study sector, having a flat and relatively even slope, is prone to experiencing a tsunami in the event of a seismic event of considerable magnitude because it is located on the convergent edge and close to the sea. The implication of this parameter implies that, in semi-flat terrain, the penetration of the wave can be extensive; however, as it progresses, the wave height decreases due to the existing friction forces. Therefore, considering the implications that this may have on the conditions and livelihoods of residents, it is important to take it into account in a seismic hazard assessment [14].

Finally, it is important to mention that seismic waves travel through different types of soils. In the case of Pisco, the predominant soil is sandy (soft soil), which amplifies the waves and causes strong vibrations, causing the soil to experience liquefaction effects, differential settlements and resonance effects, leading to instability, collapse, and damage to buildings [33]. Likewise, the type of soil in the study area made up of sands and water tables causes the phenomenon of soil liquefaction to occur in the event of a seismic event. If the soil of Pisco were made of solid rock, this phenomenon would not occur and the consequences would be less lethal [34]. Likewise, in view of the fact that the cohesion of this type of soil tends to zero and, in turn, this parameter is a function of shear resistance, which would represent the absence of soil rigidity and large deformations [35].

5 CONCLUSION

This research determined a set of territorial parameters that influence the level of seismic hazard in a sector of interest through a vast bibliographic review and data from the study area. These parameters are soil liquefaction, soil type, and slope. Thus, it was identified that the Pisco area presents a semi-uniform relief with a type of sandy soil with very little amounts of fines, which generates a high susceptibility to soil liquefaction. Therefore, the effective stress would increase and, despite the rigidity of the structure, would cause subsidence and differential settlements, generating economic and human losses. On the other hand, considering that the type of soil will remain constant with minimum cohesion under normal conditions and that the slope will have a minimum percentage of variation due to erosion and human action, it would be important to work on establishing soil improvement strategies (i.e., injection by compaction) to reduce the liquefaction potential of the sector. Finally, this research is presented as a starting point for future research works that will implement and validate the parameters analyzed in this proposal.

ACKNOWLEDGEMENTS

This project was funded by Universidad Tecnológica del Perú, within the framework of the "Research Projects I+D+i 2021 – 1" agreement. The authors would like to thank José Zevallos and Yustin Yepez for their valuable comments on the previous version of this manuscript.

REFERENCES

[1]   Tavera, H., *Seismic Risk. Peru, A Highly Seism Country*, 2019.
[2]   OPS, Earthquake in Pisco, Peru – Two years after the quake, chronicle and lessons learned in the health sector. Lima, 2010.
[3]   INDECI, Earthquakes in Peru over time. 2006.
[4]   INDECI, Hazard map, mitigation program and disaster mitigation measures for the cities of Yungay and Ranrahirca. 2005.
[5]   Ahmad, R.A., Singh, R.P. & Adris, A., Seismic Hazard Assessment of Syria using seismicity, DEM, slope, active faults and GIS. *Remote Sensing Applications: Society and Environment*, **6**, pp. 59–70, 2017. DOI: 10.1016/j.rsase.2017.04.003.
[6]   Jena, R. et al., Integrated model for earthquake risk assessment using neural network and analytic hierarchy process: Aceh province, Indonesia. *Geoscience Frontiers*, **11**, pp. 613–634, 2020. DOI: 10.1016/j.gsf.2019.07.006.
[7]   Karimzadeh, S., Miyajima, M., Hassanzadeh, R., Amiraslanzadeh, R. & Kamel, B., A GIS-based seismic hazard, building vulnerability and human loss assessment for the earthquake scenario in Tabriz. *Soil Dynamics and Earthquake Engineering*, **66**, pp. 263–280, 2014. DOI: 10.1016/j.soildyn.2014.06.026.
[8]   Yariyan, P., Zabihi, H., Wolf, I.D., Karami, M. & Amiriyan, S., Earthquake risk assessment using an integrated fuzzy analytic hierarchy process with artificial neural networks based on GIS: A case study of Sanandaj in Iran. *International Journal of Disaster Risk Reduction*, **50**, 2020. DOI: 10.1016/j.ijdrr.2020.101705.
[9]   Funning, G.J., Barke, R.M.D., Lamb, S.H., Minaya, E., Parsons, B. & Wright, T.J., The 1998 Aiquile, Bolivia earthquake: A seismically active fault revealed with InSAR. *Earth and Planetary Science Letters*, **232**, pp. 39–49, 2005. DOI: 10.1016/j.epsl.2005.01.013.
[10] Maruyama, Y., Yamazaki, F., Matsuzaki, S., Miura, H. & Estrada, M., Evaluation of building damage and tsunami inundation based on satellite images and GIS data following the 2010 Chile earthquake. *Earthquake Spectra*, **28**, pp. 165–178, 2012. DOI: 10.1193/1.4000023.
[11] Matsuoka, M. et al., Development of building inventory data and earthquake damage estimation in lima, peru for future earthquakes. *Journal of Disaster Research*, **9**, pp. 1032–1041, 2014. DOI: 10.20965/jdr.2014.p1032.
[12] Lazaro Barreto, M.R., *Analysis of Hazards and Vulnerabilities for Disaster Risk Management, using the Geographic Information System (GIS) in the Town of Acopampa – Carhuaz, Ancash*, Universidad Nacional Santiago Antunez de Mayolo, 2015.
[13] CISMID, *Risk Analysis in Urban Areas of the Punta Negra District*, vol. 2, 2017.
[14] INDECI, UNICA, *Hazard Map, Land Use Plan and Proposed Mitigation Measures for the Effects of Natural Disasters in the City of Pisco*, 2001.

[15] Matsuzaki, S., Pulido, N., Maruyama, Y., Estrada, M., Zavala, C. & Yamazaki, F., Evaluation of seismic vulnerability of buildings based on damage survey data from the 2007 Pisco, Peru earthquake. *Journal of Disaster Research*, **9**, pp. 1050–1058, 2014. DOI: 10.20965/jdr.2014.p1050.

[16] INDECI, INGEMMET, CESEL, CONIDA, Hazard map of Pisco and San Andres information for reconstruction – Earthquake August 15, 2007. 2008.

[17] Matsuoka, M., Miura, H., Midorikawa, S. & Estrada, M., Extraction of urban information for seismic hazard and risk assessment in Lima, Peru using satellite imagery. *Journal of Disaster Research*, **8**, pp. 328–345, 2013. DOI: 10.20965/jdr.2013.p0328.

[18] Birkmann, J., Measuring vulnerability to promote disaster-resilient societies: Conceptual frameworks and definitions. *Institute of Environmental Human Security Journal*, **5**, pp. 7–54, 2006.

[19] INEI. Ica: Final results. 2018.

[20] Tavera, H. & Bernal, I., The Pisco (Peru) earthquake of 15 August 2007. *Seismological Research Letters*, **79**, pp. 510–515, 2008. DOI: 10.1785/gssrl.79.4.510.

[21] Tavera, H., Bernal, I., Strasser, F.O., Arango-Gaviria, M.C., Alarcón, J.E., Bommer, J.J., Ground motions observed during the 15 August 2007 Pisco, Peru, earthquake. *Bulletin of Earthquake Engineering*, **7**, pp. 71–111, 2009. DOI: 10.1007/s10518-008-9083-4.

[22] Elnashai, A.S. et al., The Pisco-Chincha earthquake of August 15, 2007: Seismological, geotechnical, and structural assessments. 2008.

[23] Parra, D., Vasquez, D. & Alva, J., Geotechnical microzonation of Pisco. *CISMID UNI*, **1**, p. 19, 2015.

[24] Walsh., Environmental and social impact study "Peru Nitrates Project", 2006.

[25] CISMID, MVCS, Seismic microzonation study for the coastal areas of Pisco, San Clemente, Túpac Amaru, San Andrés and Paracas for the Coastal Areas of Pisco, San Clemente, San Andrés and Paracas Lima – March 2012. 2012.

[26] Braja M.D., Geotechnical engineering, 2001.

[27] Terzaghi, K., Bearing capacity. *Theoretical Soil Mechanics*, John Wiley & Sons, Ltd., pp. 118–143, 1943. DOI: 10.1002/9780470172766.ch8.

[28] Carrillo, A. & Alcayhuamán, L., Liquefaction of soils during the Pisco-Peru earthquake-2007. Universidad Ricardo Palma, vol. 9, 2007.

[29] Rodríguez-Marek, A. et al., Geotechnical aspects of the Pisco, Peru earthquake of August 15, 2007. *Rev Int Disastr Nat*, **7**, 2007.

[30] Graf, E., Compaction grouting technique and observations. *Journal of the Soil Mechanics and Foundations Division,* pp. 1151–1158, 1969.

[31] El-Kelesh, A.M., Matsui, T., Tokida, K., Field investigation into effectiveness of compaction grouting. *Journal of Geotechnical and Geoenvironmental Engineering*, **138**, pp. 451–460, 2012. DOI: 10.1061/(asce)gt.1943-5606.0000540.

[32] MVCS, National building regulations. Lima, 2018.

[33] Bernal, I., The effects of earthquakes on soil types 2020. https://www.gob.pe/institucion/igp/noticias/309366-los-efectos-de-los-sismos-en-los-tipos-de-suelo. Accessed on: 8 May 2021.

[34] Tavera, H., Evaluation of hazards associated with earthquakes and secondary effects in Peru, 2014.

[35] Socualaya Cardenas, K., Soil characterization to obtain ballast coefficient. San Agustin de Cajas District, 2017.

STRUCTURAL MONITORING FOR SEISMIC DAMAGE EVALUATION: A CASE STUDY

STEFANO ANASTASIA, PEDRO POVEDA-MARTÍNEZ,
BENJAMÍN TORRES-GORRIZ & SALVADOR IVORRA-CHORRO
Departamento de Ingeniería Civil, Universidad de Alicante, Spain

ABSTRACT
Seismic events are one of the phenomena with greater influence on the structural condition of bridges and viaducts. For this reason, its design and construction must be carried out under dynamic criteria to guarantee its resistance in a wide range of scenarios. Into the Spanish territory, these requirements are included in the Earthquake-Resistant Construction Standard – NCSE02. However, a large number of structures were built before the appearance of seismic regulations and thus, its indications were not taken into account. In these cases, it is crucial to monitor the behaviour of the structure in order to assess possible damage due to dynamic action. This work presents a case study, focusing on the Santa Ana viaduct, constructed in the Quisi ravine, in Benissa, Alicante (Spain). The rivet structure was built in the early 20th century, being unusual finding similar case studies in the literature. The viaduct, which is in the final stage of its useful life, serves as a bridge for a tram line between neighbouring towns. This work aims at studying the viaduct's structural modal shapes and damping factor to establish its possible interaction with a seism. To this end, the viaduct was monitored with eighteen accelerometers distributed along its length. Through an acquisition system, the vibrations suffered by the structure were automatically registered after the passage of each train. The signals were subsequently processed using different operational modal analysis techniques. This allows not only to obtain the modal parameters associated with the structure, but also its temporary evolution and therefore, predictively determine the possible appearance of structural damage.
Keywords: *structural damage, operational modal analysis, preservation, conservation, predictive analysis.*

1 INTRODUCTION
During the past decade, the European territory suffered a series of seismic events, including a number of catastrophic ones, which caused, among other things, the collapse of viaducts and bridges leading to the loss of numerous human lives. These events had a strong impact on Civil Engineering by drawing the attention of many experts both technical and scientific. Their efforts are focused on keeping the "health status" of the structures, a discipline known as Structural Health Monitoring (SHM) [1] and for which it is necessary to establish the behaviour of the structure in order to determine its apparent stability.

The current international scenario, where many efforts have been focused in bridges, has led to a deeper understanding of the problem. In this context, SHM normally encompasses all those techniques intended to monitor structures and it results applicable not only in buildings, but also in civil structures, especially in bridges and viaducts [2]. The purpose of monitoring civil structures is to identify sudden or progressive damages by controlling its behaviour in operating conditions or even during particular environmental conditions [3]. It should be stressed that modern constructions may include elements capable of dissipating the energy transmitted by a seism to the system (seismic isolators). In this way, the fundamental frequency of the structural vibration is reduced to a value below the earthquake frequencies. However, previous constructions lack this type of elements and therefore, the damage suffered by the structure can be irreversible and gradually lead to its collapse. For this reason, measuring the damage suffered by the structure after a seismic event is of crucial importance to ensure its integrity.

WIT Transactions on The Built Environment, Vol 202, © 2021 WIT Press
www.witpress.com, ISSN 1743-3509 (on-line)
doi:10.2495/ERES210051

The damage identification procedure is usually based on modal analysis, a fundamental tool for SHM [4]. The method requires to register the dynamic responses of the structure, which generally involves the use of acceleration transducers. These measurements allow to evaluate the modal parameters of the structure, which variation highlights the damage suffered by the construction. A clear example is the damping factor, which can be classified as a "damage index" and it is based on the vibration response of the structure. In order to obtain the structural condition, modal parameters should be obtained before and after an excitation event. The results allow to verify if the variation of the modal parameters corresponds to a critical damage level for the structure, which in our case is represented by a bridge [5]. Therefore, since the damage is always accompanied by a reduction in rigidity, the SHM based on output-only modal analysis will consist of identifying changes in modal frequencies. Generally, the SHM using vibration analysis comprises five steps: (i) sensing; (ii) location; (iii) classification; (iv) evaluation; and (v) forecast. Modal analysis involves the determination of natural frequencies, damping ratios and modal shapes of a structure from its dynamic response.

The SHM based on modal analysis is already used by many technicians, and it is especially important in bridges. The damage suffered by the structure in different locations and components actually leads to different frequency changes. However, it remains difficult to determine the location of damages simply by observing changes in modal frequencies, since modal shape is the only parameter related to position. To address this problem, many studies have focused on searching indices that can identify both the damage and its position from modal shapes. Some clear examples are the Modal Curvature Index [6], the Modal Assurance Criterion (MAC) [7] and Coordinate Modal Assurance Criterion (COMAC) [8].

This work aims to illustrate some practical aspects of damage detection methods. The research activity, conducted by the Department of Civil Engineering of the University of Alicante, is part of this complex and articulated panorama. It refers to the structural monitoring of the QUISI railway bridge to control possible damages produced by earthquakes or the deterioration of the structure. To do this, the natural modal frequencies of the structure are automatically obtained with the passage of each train. The data provided by the system, corresponding to the signal acquired by different accelerometers fixed on the bridge, are analysed in time and frequency. Different operational modal analysis techniques are applied: (i) in the non-parametric field, FDD; and (ii) in the parametric field, SSI-Cov. The results of this study will allow to identify any anomaly in the modal shapes and therefore, determine the presence and location of any damage produced in the structure.

2 MODAL PARAMETERS IDENTIFICATION

The data provided by the measurement system was analysed in time and frequency domains. First, the maximum acceleration of each sensor was obtained for all recorded events. Secondly, an Operational Modal Analysis (OMA) was performed by using several techniques. The purpose of this analysis was to determine the modal frequencies of the structure in order to detect significant changes in its behaviour. In addition to this, the damping factor was calculated.

2.1 Spectral analysis

Fig. 1 illustrates the signals obtained after a railway crossing. The peak values recorded by the accelerometers were determined for each event. This value, which gives an idea of the stresses suffered by the structure, was stored together with some temporal information (date, time …) to conform a historical register.

WIT Transactions on The Built Environment, Vol 202, © 2021 WIT Press
www.witpress.com, ISSN 1743-3509 (on-line)

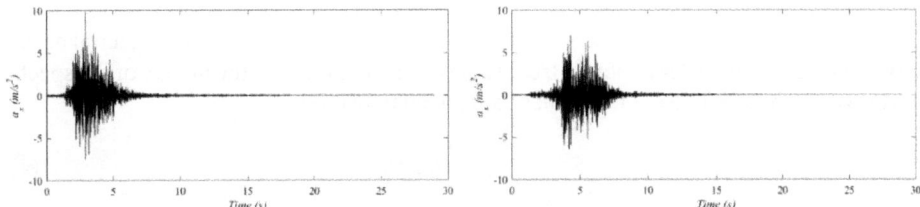

Figure 1: Acceleration recorded when a railway passes at different positions on the viaduct.

A first approximation to determine the modal behaviour of the structure can be obtained by means of a spectral analysis of the signals. However, to avoid frequency components belonging to the excitation source (railway), only the last part of the vibration signal was selected. The time window applied to each register was established from 3 seconds after the peak of maximum amplitude to the end of the signal. The power spectral density (PSD) corresponding to each signal was obtained by using the Welch method (periodogram). To do this, the signals were divided into small segments. Then, thanks to the application of a time window, the Fast Fourier Transform (FFT) was calculated. Finally, all blocks were averaged and PSD was estimated according to the following expression:

$$\varepsilon\{P^\wedge{}_W(f)\} = \frac{1}{P}\sum_{p=0}^{P-1} \quad \varepsilon\left\{P \quad \overset{(p)}{_{xx}}(f)\right\},\tag{1}$$

where:

$$P_{xx}(f) = \sum_{k=-\infty}^{\infty} \quad r_{xx}(k)e^{-j2\pi fk} = \left|\sum_{n=-\infty}^{\infty} \quad x[n]e^{-j2\pi fn}\right|^2\tag{2}$$

From the frequency analysis, the structural modes were manually extracted. Since the excitation signal was unknown, the results might contain spurious frequencies that do not belong to the structure. For this reason, it was necessary to implement more advanced methodologies that were able to provide greater accuracy.

2.2 Modal behaviour of the structure

It is extremely useful to identify the natural frequencies of the structure automatically readily. This would allow to establish a continuous monitoring of the structure and thus, an early diagnosis of the damage. In this sense, there are several OMA techniques that efficiently provide modal frequencies and damping factors from the relationship between the vibration corresponding to different points of the structure.

In this study, the Santa Ana viaduct was analysed using two specific OMA-algorithms: the Frequency Domain Decomposition method and the Covariance-driven Stochastic Subspace Identification-Cov [9], [10]. Both methods start with a previous processing where the correlation matrix of the signals is determined:

$$R_j = \frac{1}{n_t}\sum_{k=0}^{n_t-1} \quad y_k y_{k+j}^T.\tag{3}$$

WIT Transactions on The Built Environment, Vol 202, © 2021 WIT Press
www.witpress.com, ISSN 1743-3509 (on-line)

2.2.1 Frequency domain decomposition (FDD) method

The FDD method assumes orthogonal modes. They are estimated through the spectral density of vibration signals. In order to obtain frequencies and modal forms, the matrix of the spectral densities is decomposed through singular value decomposition as:

$$S^{\wedge}_{yy}(\omega_j) = U_j S_j U_j^H, \tag{4}$$

where U_j corresponds to an orthogonal matrix with singular vectors $S^{\wedge}_{yy}(\omega_j)$, and S_j is a diagonal matrix of singular values. Singular vectors will be related to the modal shapes of the structure.

2.2.2 Covariance-driven stochastic subspace identification, SSI-Cov

The SSI-Cov [11] method is based on the identification of the model through state variables (State Space Model). Modal parameters are identified through the only output system response data. Their covariance function, on which this non-parametric method is based, can be seen as the free dynamic response of the structure. The estimation of the model takes place through the resolution of the Toeplitz matrix (composed of various covariance functions). The matrix is composed and defined according to the following expression:

$$T_{1|j_b} = \begin{bmatrix} R_{j_b} & R_{j_b-1} & \cdots & R_1 & R_{j_b+1} & R_{j_b} & \cdots & R_2 & \cdots & \cdots & \cdots & \cdots & R_{2j_b-1} & R_{2j_b+2} & \cdots & R_{j_b} \end{bmatrix}. \tag{5}$$

The decomposition of this matrix into singular values (SVD) allows to estimate the modal frequencies of the structure.

2.3 Automated operation modal analysis

The automation of the OMA algorithm aims to obtain a series of snapshots containing the dynamic characteristics of the structure. Therefore, it allows to establish a time evolution of the modal behaviour of the bridge, which is extremely useful to detect possible damages from a deviation. The system involves two basic steps: (i) automatic calculation of modal frequencies (FDD, SSI-Cov) and (ii) the automated tracking of modal properties over time. It must be noted that the FDD parametric identification technique is significant because of its simplicity, and its rigorous physical meaning. Methods used in this work were developed in frequency domain, starting from the estimation of an output spectrum or half spectrum matrices from the measured dynamic responses. The results of this model allowed to produce a series of graphs consisting of a succession of frequency values and damping factors.

3 BRIDGE AND MONITORING SYSTEM

3.1 Structure under analysis

The Santa Ana viaduct, in the Quisi ravine, was built between 1913 and 1915. It has a six-span structure bridged by Pratt-type metal lattices over piles of metal profiles. The viaduct is currently used as a crossing for the tram line 9 between the towns of Benidorm and Denia, in Alicante (Spain) (Fig. 2).

The structure is approximately 170 m long and consists of six spans of various lengths: $21.48 + 21.12 + 42.00 + 42.00 + 21.12 + 21.48$ m. The two central spans present a continuous structural scheme, while the four lateral ones are isostatic (Fig. 3).

Figure 2: Front view of the Santa Ana viaduct.

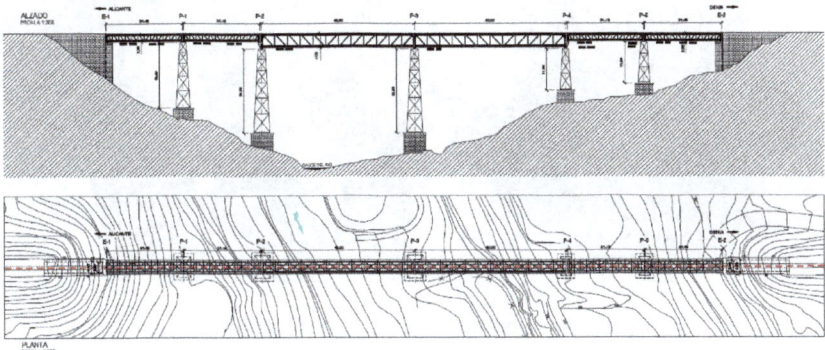

Figure 3: Ground and elevation plan of the Santa Ana viaduct.

3.2 Monitoring system

The monitoring of the structure was carried out through a number of accelerometers located along its length. Moreover, the system included all the necessary equipment for the automatic and continuous acquisition of the signals.

3.2.1 Vibration signals acquisition system

The signals provided by the transducers installed along the structure were acquired by means of a HBK QUANTUMX MX 1601B interrogator (Fig. 4). The system was formed by 16 channels, allowing the supply of accelerometers through constant current or IEPE and a sample frequency of 20 kHz.

Figure 4: Acquisition system QUANTUMX MX1601B.

3.2.2 Acceleration transducers

The monitoring system had a total of 18 accelerometers: 12 uniaxial (B&K 4507-B-006) and 6 triaxial (B&K 4506-B-003); both with a sensitivity of 490 mV/g and a working range between 0.3 and 2000 Hz (Fig. 5). The transducer characteristics guarantee a dynamic range of ±5 g, avoiding the saturation due to the pass of a train. The sensors, stainless steel, were hermetically sealed and their connections were water-resistant. Communication between accelerometers and data acquisition system was carried out through coaxial cable.

Figure 5: Accelerometers installed in Quisi viaduct.

Accelerometers were distributed along the structure, including two uniaxial sensors and one triaxial sensor in each span as shown in Fig. 6 (symmetrical installation for spans 4, 5 and 6).

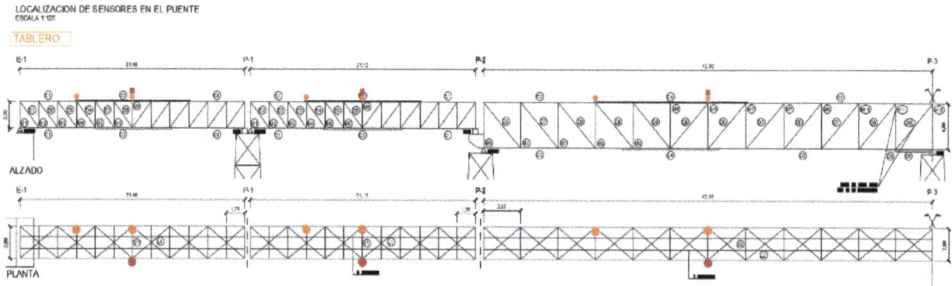

Figure 6: Uniaxial (orange) and triaxial (red) accelerometers location diagram in the first three spans.

3.2.3 Monitoring system configuration

The measurement system was configured to register the vibration of the viaduct with the passage of each railway. To do this, two photoelectric cells were located at both sides of the viaduct. When the passage of a convoy was detected, the acquisition system was activated, recording the signal provided by the accelerometers for a period of 30 seconds. The data was stored on a remotely accessible server, allowing its subsequent processing and analysis. The acquisition corresponding to the X component was carried out using a sample frequency of 200 Hz. In the other cases, the frequency was established to 50 Hz. Table 1 summarises the most relevant characteristics of the signals, including the position and direction of each accelerometer.

Table 1: Measurement position and direction.

Span	Direction	Position*
	X	0.25
	X	0.5
Span 1/6	X	0.5
	Y	0.5
	Z	0.5
	X	0.25
	X	0.5
Span 2/5	X	0.5
	Y	0.5
	Z	0.5
	X	0.4
	X	0.6
Span 3/4	X	0.5
	Y	0.5
	Z	0.5

*Relative position with respect to the span length.

For every train passage, the time signal provided by each accelerometer included into the metal structure was stored in a comma separated value (CSV) file. In addition, a column corresponding to a temperature probe, as well as the necessary data to determine the direction of each train were included. The files generated by the system were automatically stored on a server using Synology Drive technology. The system allows the continuous and simultaneous access of multiple clients remotely for the management of stored information and its subsequent analysis.

From a numerical model of the viaduct, the first modal frequency of each span was obtained (Table 2). These values allowed to verify the proper functioning of the modal analysis algorithms used in this work.

Table 2: Theoretical first modal frequency for each span.

Span	Theoretical first modal frequency (Hz)
1	9.40
2	9.35
3	4.73
4	4.73
5	9.35
6	9.40

4 ANALYSIS AND RESULTS

The PSD analysis clearly showed the most relevant modal frequencies of each span. As shown in Figs 7 and 8, corresponding to the frequency behaviour of span 2 and 3, the experimental results confirmed those obtained theoretically for the structure. In the first case, a frequency of 8.85 Hz was derived from the signals (9.40 Hz by numerical methods, 5% deviation). For span 3, the first mode was located at 4.57 Hz (theoretical results: 4.73 Hz, 3% deviation).

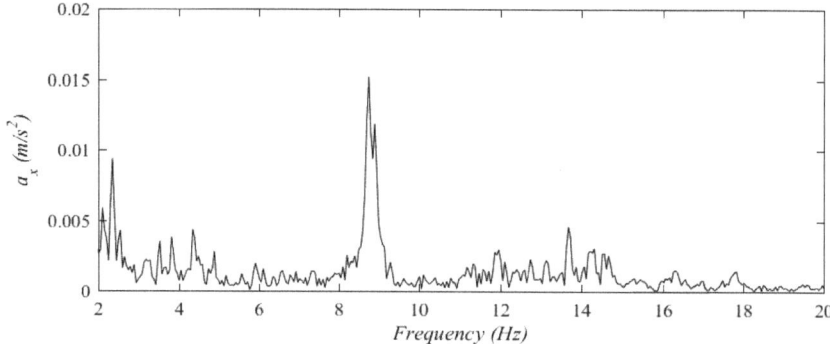

Figure 7: Frequency spectrum corresponding to the central point of spam 2 (X component).

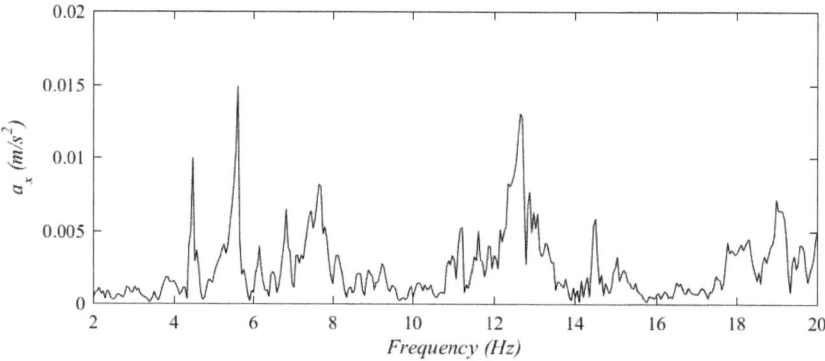

Figure 8: Frequency spectrum corresponding to the central point of spam 3 (X component).

To automatically obtain the modal frequencies of the structure, two different OMA algorithms were used: SSI-Cov and FDD. In the first case, the results provided by the model allowed to produce a series of graphs (Fig. 9) consisting of a succession of poles or peaks, which can be identified along the vertical alignments in the diagram. By means of these figures, it was possible to observe the stability of the poles which had been obtained for the different natural frequencies of the state space model.

Using the OMA methods above, it was possible to measure the modal frequencies of the structure for each event. This information allowed to visualise its evolution over time and therefore, to detect important deviations due to structural damages. Figs 10–12 show the evolution of the vibration modes over a period of nearly 10 days.

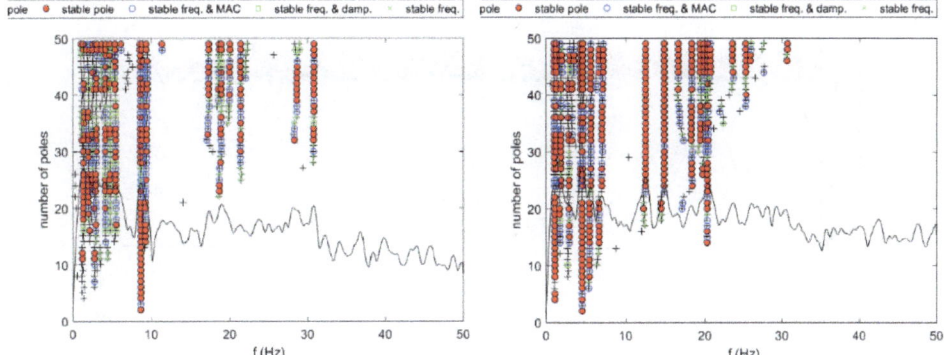

Figure 9: Modal frequencies obtained by means of SSI-Cov method.

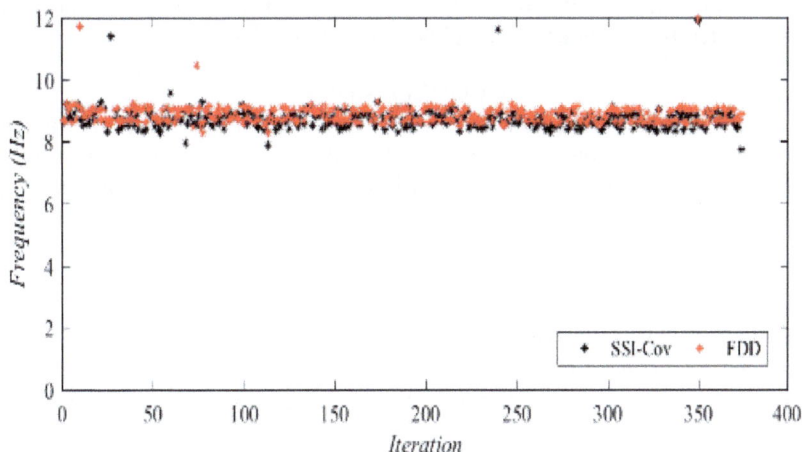

Figure 10: Evolution of the first modal frequency for span 2.

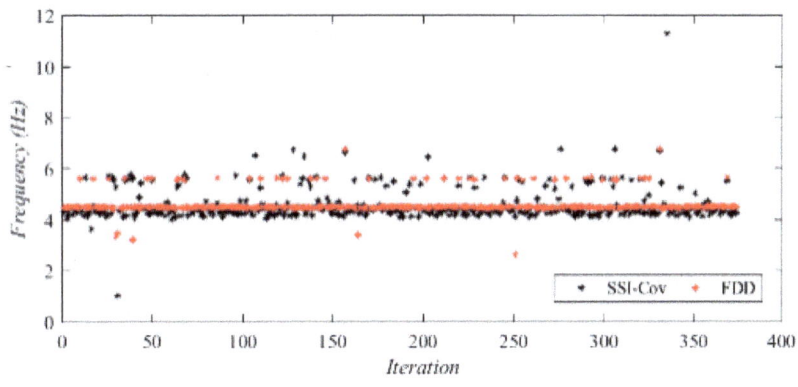

Figure 11: Evolution of the first modal frequency for span 4.

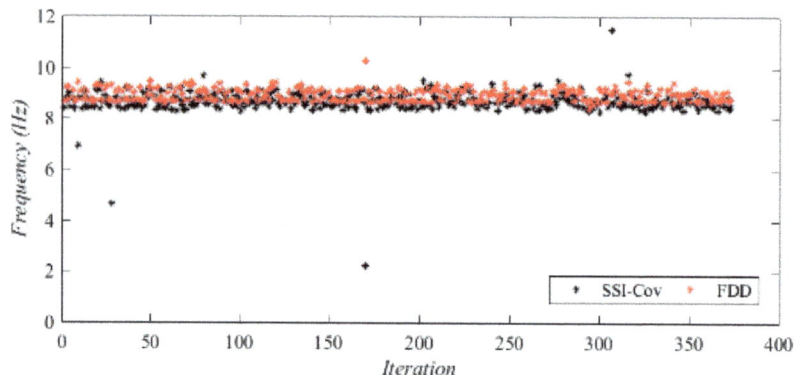

Figure 12: Evolution of the first modal frequency for span 6.

The modal properties derived from SSI-Cov were consistent with those obtained by the FDD method in terms of frequency, damping and shape modal.

5 CONCLUSIONS

The illustrated procedure for automated identification and modal tracking was based on parametric SSI-Cov and non-parametric FDD methods. The procedure was tested and verified with the data recorded on the historic Quisi railway bridge. For this purpose, a permanent monitoring system consisting of accelerometer sensors was designed. The results included in this work correspond to a preliminary study and show a first approximation to the structural monitoring and damage detection. It is extremely important for the application of SHM to gather accurate information of the results in order to obtain a faithful interpretation of the dynamic behaviour of the structure. The preliminary results showed a great correlation between the algorithms used in the analysis and the theoretical behaviour of the structure. However, a better optimisation of the algorithms should be carried out in order to avoid outliers.

To summarize, the proposed methodology is promising for the interpretation of the data provided by a continuous dynamic monitoring system. The developed procedure should take into account some factors that affect the modal identification, such as environmental conditions and background noise.

ACKNOWLEDGEMENTS

The research leading to these results has received funding from the Research Fund for Coal and Steel under grant agreement No. 800687 (DESDEMONA EU project). Authors are particularly grateful to ECISA for its major support.

REFERENCES

[1]   Magalhães, F., Cunha, A. & Caetano, E., Vibration based structural health monitoring of an arch bridge: From automated OMA to damage detection. *Mechanical Systems and Signal Processing*, pp. 212–228, 2012.

[2]   Chen, Z., Zhou, X., Wang, X., Dong, L. & Qian, Y., Deployment of a smart structural health monitoring system for long-span arch bridges: A review and a case study. *Sensors MDPI*, 2017.

[3]   Brownjohn, J.M.W., Magalhaes, F., Caetano, E. & Cunha, A., Ambient vibration re-
       testing and operational modal analysis of the Humber Bridge. *Engineering Structures*,
       32, pp. 2003–2018, 2010.

[4]   Benedettinia, F. & Gentile, C., Operational modal testing and FE model tuning of a
       cable-stayed bridge. *Engineering Structures*, pp. 2063–2073, 2011.

[5]   Peeters, B. & Ventura, C.E., Comparative study of modal analysis techniques for
       bridge dynamic characteristics. *Mechanical Systems and Signal Processing*, pp. 965–
       988, 2003.

[6]   Pandey, A.K., Biswas, M. & Samman, M.M., Damage detection from changes in
       curvature mode shapes. *Journal of Sound and Vibration*, pp. 321–332, 1991.

[7]   Wanga, Y.M. & Elhag, T.M.S., A comparison of neural network, evidential reasoning
       and multiple regression analysis in modelling bridge risks. *Expert Systems with
       Applications*, pp. 336–348, 2007.

[8]   Lieven, N.A.J. & Ewins, D.J., Spatial correlation of mode shapes, the coordinate modal
       assurance criterion (COMAC). *Proceedings of the 6th International Modal Analysis
       Conference*, Kissimmee, Florida, USA, pp. 690–695, 1988.

[9]   Brebbia, C.A., Telles, J.C.F. & Wrobel, L.C., Boundary integral formulation for
       inelastic problems. *Boundary Element Techniques*, Berlin and New York, 1984.

[10]   Bratanow, T. & De Grande, G., Numerical analysis of normal stresses in non-
       Newtonian boundary layer flow. *Engineering Analysis*, pp. 20–25, 1985.

[11]   Ranieri, C. & Fabbrocino, G., *Operational Modal Analysis of Civil Engineering
       Structures*, Springer: New York, 2014.

EXAMINING THE ADEQUACY OF SEPARATION GAPS BETWEEN ADJACENT BUILDINGS UNDER NEAR-FIELD AND FAR-FIELD EARTHQUAKES

YAZAN JARADAT, HARRY FAR & ALI SALEH
School of Civil and Environmental Engineering, Faculty of Engineering and Information Technology,
University of Technology Sydney (UTS), Australia

ABSTRACT
Earthquake-induced pounding is a phenomenon that has been observed in almost every major earthquake since the 1960s. Pounding between adjacent buildings occurs due to insufficient separation and with different dynamic properties. This usually causes local damage, and in some extreme cases, total collapse of structures. Building codes in seismically active zones recommended a minimum separation gap between adjacent buildings to avoid pounding during severe earthquakes. AS1170.4-2007 is an Australian standard that requires 1% of the building height as a minimum separation gap between buildings to preclude pounding. This article presents experimental and numerical results to examine the adequacy of this specification to avoid seismic pounding between steel-frame structures under near-field and far-field earthquakes. It is found that AS1170.4-2007 is inadequate if the shorter building is used to estimate the required separation between adjacent structures under both near-field and far-field earthquakes. The code specification is adequate if the taller building is used to estimate the required separation between adjacent structures under far-field earthquakes only. The results are also compared with corresponding results obtained using the ABS and SRSS methods.
Keywords: adjacent building, separation gap, steel frame, seismic code, SAP2000, shake table.

1 INTRODUCTION

Investigations have shown that collisions between adjacent buildings during earthquakes may cause severe structural damage. These collisions, generally called 'structural pounding', occur due to several factors such as insufficient separation between adjacent buildings and different dynamic characteristics, which cause an out-of-phase vibration [1].

The phenomenon of structural pounding has been observed in many earthquakes. Analysis of damage statistics show that pounding was present in over 40% of the 330 collapsed or severely damaged buildings during the 1985 Mexico earthquake, and for at least 15% of them pounding was the main cause for collapse [2]. Pounding damage was also observed in the San Francisco Bay area during the 1989 Loma Prieta earthquake; out of the 500 buildings surveyed, more than 200 showed damage caused by pounding [3]. Also, in the 1994 Northridge (US) earthquake and the 1995 Kobe (Japan) earthquake, damage due to pounding was observed [4].

Structural pounding is a complex phenomenon involving plastic deformations at contact points, local cracking or crushing, and fracturing due to impact and friction [5]. According to several researchers, the main reason for seismic pounding is insufficient separation gap between adjacent buildings [1], [6].

The effectiveness of different techniques to mitigate pounding have been presented in several studies; for example (1) the use of collision shear walls and bracing systems [7]; (2) the installation of soft material layers, such as rubber, at certain locations on adjacent buildings where pounding is expected [8]; and (3) connecting adjacent structures with links (such as spring links, dashpot links or viscoelastic links) to produce in-phase vibrations [6], [9].

WIT Transactions on The Built Environment, Vol 202, © 2021 WIT Press
www.witpress.com, ISSN 1743-3509 (on-line)
doi:10.2495/ERES210061

However, the above methods have advantages and disadvantages. For example, using shear walls decreases the top displacements and number of impacts while increasing the maximum impact force; filling gaps with rubber pads may reduce peak impact force while increasing the number of poundings; and joining two structures is beneficial to the flexible adjacent structure while increasing the response of the stiffer building. Therefore, and based on the definition of pounding, providing a sufficiently large gap between adjacent buildings is the most common natural way to prevent structural pounding. It is concluded that the most effective method is to increase the separation distance to completely preclude pounding.

Building codes in zones of active seismicity around the world have recognised the destructive effects that pounding may induce. The approach commonly adopted in building codes has been to avoid contact interactions between structures by providing sufficient separation between them. In some cases, codes depend on the maximum displacements of each building only, while in other cases a small proportion of maximum displacement is used. In other cases, the separation distance depends on the building height, while still others adopt a combination of maximum displacement and building height. Furthermore, some codes depend on the type of soil and seismic action. Below are examples of various international codes. According to the 2009 edition of the International Building Code [10], minimum separations are given by:

$$S = U_a + U_b,$$ (1)

$$S = \sqrt{U_a{}^2 + U_b{}^2},$$ (2)

where S is separation distance and U_a, U_b are the maximum displacement responses of the adjacent structures "a" and "b", respectively, at the location where pounding is expected to occur (i.e. at the level coinciding with the roof level of the shorter building) [11]. Eqns (1) and (2) are usually referred to as the absolute sum ABS and SRSS rules (square root of sum of squares), respectively. The Uniform Building Code [12] also follows the same codal provisions. The National Building Code of Canada (NBCC) [13] requires the separation between adjacent structures or across construction joints to equal or exceed the value of

$$S = \sqrt{U_a + U_b},$$ (3)

in which U_a, U_b are the maximum displacement responses of the adjacent structures "a" and "b", respectively, at the location where pounding is expected to occur. Additionally, Eurocode 8 [14] proposed stricter requirements, according to which the minimum separation distance S (in metres) between two adjacent buildings should be less than (a) $S = 0.05 + 0.005h$ and (b) $S = 0.4q(u_a + u_b)$, where h, q and u_a, u_b are the building height (in metres), the behaviour factor, and the top floor displacement of the two buildings due to seismic design forces, respectively.

The Australian Earthquake Standard (AS1170.4-2007) [15] states that pounding needs to be considered for structures over 15 m for design category II or III. Clauses 5.4.5 and 5.5.5 for design category II and III states "this clause is deemed to be satisfied if the setback from a boundary is more than 1% of the structure height". Hao [16] rephrased the previous statement as that the required separation is equal to 1% of the adjacent building height, expressed in eqn (4).

$$S = 0.01H,$$ (4)

where H is the building height.

The minimum required separation between adjacent structures to avoid pounding has been investigated by many authors. In previous studies, building structures are idealised as single-degree-of-freedom (SDOF) oscillators [17], [18] or multiple-degree-of-freedom (MDOF) oscillators [1], [19], [20] and structural responses are considered as either linear elastic or non-linear inelastic. All the above-mentioned researchers carried out functional research regarding earthquake-induced pounding between adjacent structures and the required separation distance to preclude pounding. These studies showed that an earthquake's impact depends on its epicentre. They distinguished two major types of earthquake motion: spatially varying ground motion and uniform motion. Spatially varying motion generates larger relative responses between adjacent structures than uniform motion; its effect is especially important in stiff low-rise adjacent structures.

This article examines the adequacy of the minimum separation gap prescribed by AS1170.4, to find out whether or not the minimum separation gap of 1% between two adjacent steel-frame structures is enough to preclude pounding under earthquake ground motion.

2 EXPERIMENT

2.1 Description of test frames

The experiment tested 1/30 scale single-bay moment resisting steel-frame models – as 15-storey, 10-storey and 5-storey structures – on an MTS 354.20 multi-axial simulation table of size 2.2 m × 2.2 m at the University of Technology, Sydney. This table is capable of testing samples of 2 tonnes at 5 g accelerations, 1,000 mm/s velocity and up to ±200 mm stroke. The three frame structures were designed individually, at their reduced scale, according to AS/NZS 3678–2011 (Structural Steel). The general arrangement of the three test frames, placed on the shake table, is shown in Fig. 1. The overall floor plan dimensions of all models are 0.4 m × 0.4 m. The height of the 15-storey frame, 10-storey frame and 5-storey frame are 1.5 m, 1.0 m and 0.5 m, respectively. Columns and floors of the three models are made of rectangular flat steel sections of 40 mm × 2 mm and 400 mm × 5 mm, respectively. More details are available in [21], [22].

2.2 Preliminary system identification tests

The dynamic characteristics of each steel frame were identified by conducting several preliminary tests: a free vibration test, stiffness test and sine sweep test. In the free vibration test, the experiment aimed to measure the fundamental period and damping of the structures. Experimentally, damping can be estimated by various methods, including one that uses the width of the peak value of the frequency response function of the structure [23]. In the stiffness test, the experiment aimed to measure the stiffness parameter for the frame structure. Finally, the purpose of a sine sweep test is to determine the natural frequency and modes of vibration, especially mode 1, mode 2 and mode 3, since a free vibration test cannot measure these modes. Tables 1 and 2 illustrate the dynamic characteristics of the experimental and numerical results for the 15-storey, 10-storey and 5-storey models. Results are closely similar in natural period and stiffness.

Figure 1: Test frames on shake table.

2.3 Adopted earthquake acceleration records

Four scaled earthquake acceleration records – El Centro 1940 (Fig. 2(a)), Hachinohe 1968 (Fig. 2(b)), Kobe 1995 (Fig. 2(c)) and Northridge 1994 (Fig. 2(d)) – were adopted for the shake table tests [22]. The first two earthquakes involved far-field ground motion, and the second two involved near-field motion. These earthquake records have been chosen by the International Association for Structural Control and Monitoring for benchmark seismic studies [24].

Structures behave differently during near-field compared to far-field earthquakes, and this behaviour should be considered in the design process of structures [25]. Several researchers reported the importance of the near-field ground motion characteristics on the elastic and inelastic dynamic behaviour of structures [26], [27].

Table 1: Experimental dynamic characteristics of the structural models.

	Experimental					
	Free vibration		Sine sweep test			Stiffness kN/mm
	Natural frequency Hz	Damping %	Mode 1 Hz	Mode 2 Hz	Mode 3 Hz	
5-storey	6.53	0.467	7.05	21.15	36.83	0.0275
10-storey	3.54	0.431	3.61	11.26	18.70	0.0144
15-storey	2.27	0.503	2.33	7.11	11.76	0.0081

Table 2: Numerical dynamic characteristics of the structural models.

	Numerical			
	Modal load analysis			Stiffness kN/mm
	Mode 1 Hz	Mode 2 Hz	Mode 3 Hz	
5-storey	6.76	20.31	33.24	0.0278
10-storey	3.53	10.57	17.56	0.0149
15-storey	2.29	6.87	11.44	0.0090

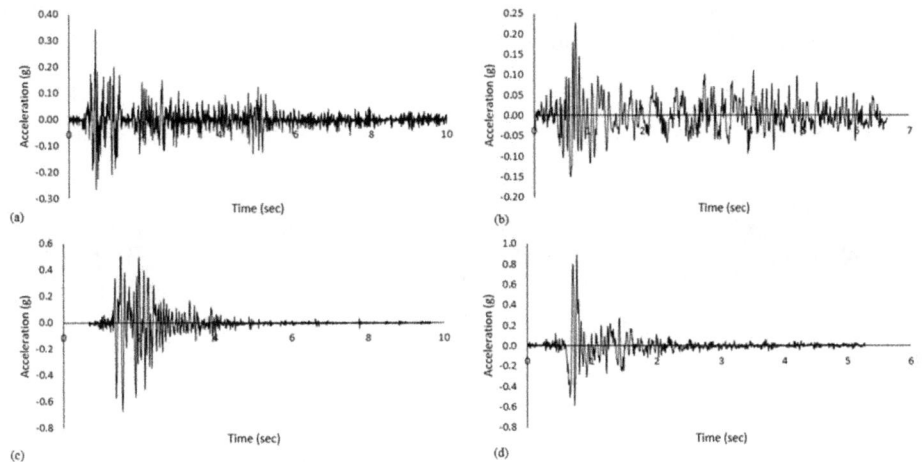

Figure 2:   Scaled earthquake ground motion. (a) El Centro earthquake 1940; (b) Hachinohe earthquake 1968; (c) Kobe earthquake 1995; and (d) Northridge earthquake 1994.

Shake table tests were then performed on the 15-storey, 10-storey and 5-storey structural models, which were directly fixed on top of the shake table, to obtain seismic responses of the structural models to be used for numerical verification. After securing the structural models on the shake table as shown in Fig. 1, accelerometers were installed on each model. The arrangement of the accelerometers was as follows: for the 15-storey structural model, one on the fifth floor, one on the tenth floor and one on the fifteenth floor; for the 10-storey structural model, one on the fifth floor and one on the tenth floor; for the 5-storey structural model, one on the fifth floor only. In addition, a further accelerometer was mounted on the shake table. The acceleration time history recorded by this accelerometer was used as input ground motion in the numerical analyses, to eliminate reproduction errors. The models also had a significantly large separation gap to avoid pounding.

3 DISCUSSION

The software used for numerical investigation as a three-dimensional frame is known as SAP2000 version 20 [28]. It is appropriate for the output of time-history analysis with a non-linear gap element, and specifically, to model pounding conditions and obtain structural responses such as displacement and acceleration. Ritz vector was selected as the mode type, in the modal load case [29]. The selected maximum numbers of modes were 99 and 99%

target dynamic participation ratios. Non-linear dynamic analysis or fast non-linear analysis (FNA) were performed under the influence of the four scaled earthquake acceleration records with 9,000 time steps at 0.001 s step size, 5,500 time steps at 0.001 s step size, 10,000 time steps at 0.001 s step size, and 6,000 time steps with 0.001 s step size of the El Centro, Hachinohe, Kobe and Northridge earthquakes, respectively.

The numerical predictions and experimental values of the absolute acceleration time histories for the three structural frame models are presented and compared. It becomes apparent that the trend and the values of the experimental shake table test results are in good agreement and consistent with the numerical predictions. Numerical displacement was used, since the purpose of this experiment is to determine the minimum separation distance between adjacent structures to avoid pounding. In this article, only one structural frame model is shown (in Fig. 3). This is due to page limitations.

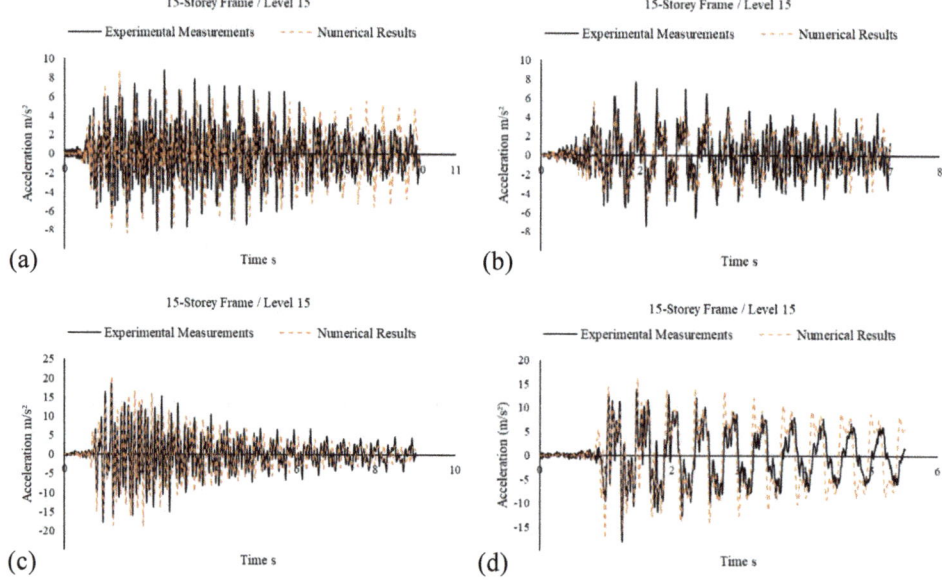

Figure 3: Experimental and numerical absolute acceleration time histories for 15-storey frame (top floor) under (a) Scaled El Centro earthquake; (b) Scaled Hachinohe earthquake; (c) Scaled Kobe earthquake; and (d) Scaled Northridge earthquake.

Top floor numerical relative displacement time histories for the 15-storey, 10-storey and 5-storey frames under scaled earthquake ground motion are discussed. Only the 15-storey frame is shown in Fig. 4. Relative displacement time histories under the four scaled seismic excitations are compared. For the top floor of the 15-storey frame. It can be seen that the highest relative displacement caused under the Northridge earthquake was 40.9 mm. This displacement was 19.7 mm, 14.7 mm and 16.9 mm for Kobe, El Centro and Hachinohe, respectively. Top floor relative displacement time histories of the 10-storey frame under the four scaled seismic excitations showed the highest relative displacement under the Northridge earthquake, at 31.5 mm, and 26.3 mm, 7.3 mm and 3.8 mm for Kobe, El Centro and Hachinohe, respectively. Relative displacement time histories for the top floor of the

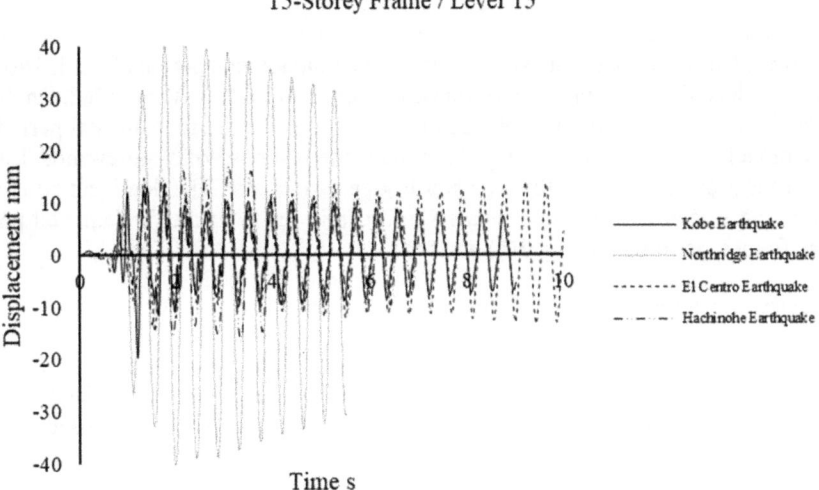

Figure 4:   Top floor numerical relative displacement time histories of the 15-storey frame, under four scaled earthquakes.

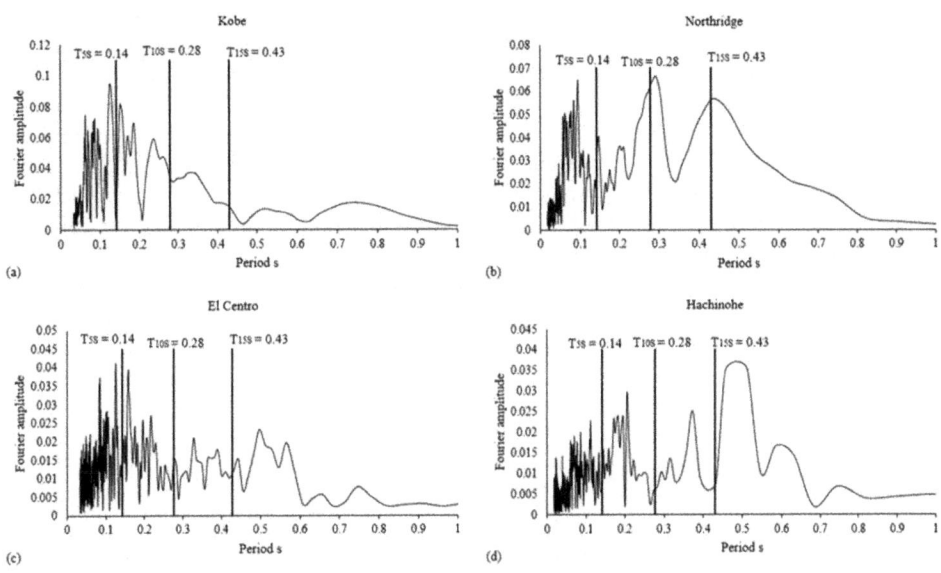

Figure 5:   Fundamental periods of the 5-storey, 10-storey and 15-storey frames, and of Fourier spectrum of ground motion of (a) Scaled Kobe earthquake; (b) Scaled Northridge earthquake; (c) Scaled El Centro earthquake; and (d) scaled Hachinohe earthquake.

5-storey frame, by contrast, showed the highest relative displacement under the Kobe earthquake, at 15.4 mm, and 9 mm, 5.3 mm and 3.5 mm for Northridge, El Centro and Hachinohe, respectively.

Building response increases when the characteristic period of ground motion is close to a building's fundamental period [25], [30]. It is apparent from Fig. 5(b) that the dominant periods of the Northridge earthquake are close to the fundamental period of the 15-storey and 10-storey models. Consequently, the response of these two models is amplified. On the other hand, the dominant period of the Kobe earthquake is close to the fundamental period of the 5-storey model, as shown in Fig. 5(a). Therefore, the response of 5-storey model is much higher for the Kobe earthquake than the Northridge earthquake. The dominant period of the El Centro and Hachinohe earthquakes are significantly far from the fundamental period of the three frames, as shown in Fig. 5(c) and 5(d).

3.1 Required separation distance to avoid structural pounding

Relative displacement is considered to be the most important parameter in structural pounding problems. If $u_a(t)$ and $u_b(t)$ are the displacement time histories of adjacent buildings "a" and "b" at the potential pounding location (see Fig. 6) then the minimum separation gap required to prevent structural pounding S is given by Lin and Weng [31]:

$$S = max[u_a(t) - u_b(t)], \tag{5}$$

where *max* is the maximum value of the entire range of the relative displacement time history. Structural pounding may occur once the gap between potential pounding locations is less than S. In other words, pounding will occur when $u_a(t) - u_b(t) - S > 0$. The minimum separation distance to avoid pounding, $u_a(t) - u_b(t) - S \leq 0$, between the 15-storey and 10-storey frames, the 15-storey and 5-storey frames, and the 10-storey and 5-storey frames under the aforementioned scaled earthquakes are presented in Fig. 7. The minimum separation distances to preclude pounding for the 15-storey frame adjacent to the 10-storey frame are 16 mm, 15 mm, 32 mm and 66 mm, under the influence of the four scaled earthquake accelerations of El Centro, Hachinohe, Kobe and Northridge, respectively. Furthermore, the minimum separation distance to avoid pounding between the 15-storey frame and the 5-storey frame are 10 mm, 10 mm, 15 mm and 20 mm for El Centro, Hachinohe Kobe and Northridge, respectively. Finally, the minimum separation distance to avoid pounding between the 10-storey frame and the 5-storey frame are 10 mm, 5 mm, 27 mm and 27 mm for El Centro, Hachinohe, Kobe and Northridge, respectively.

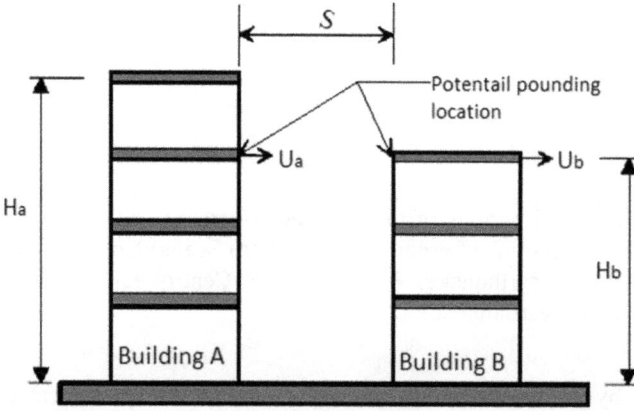

Figure 6: Potential pounding location between adjacent buildings with different heights.

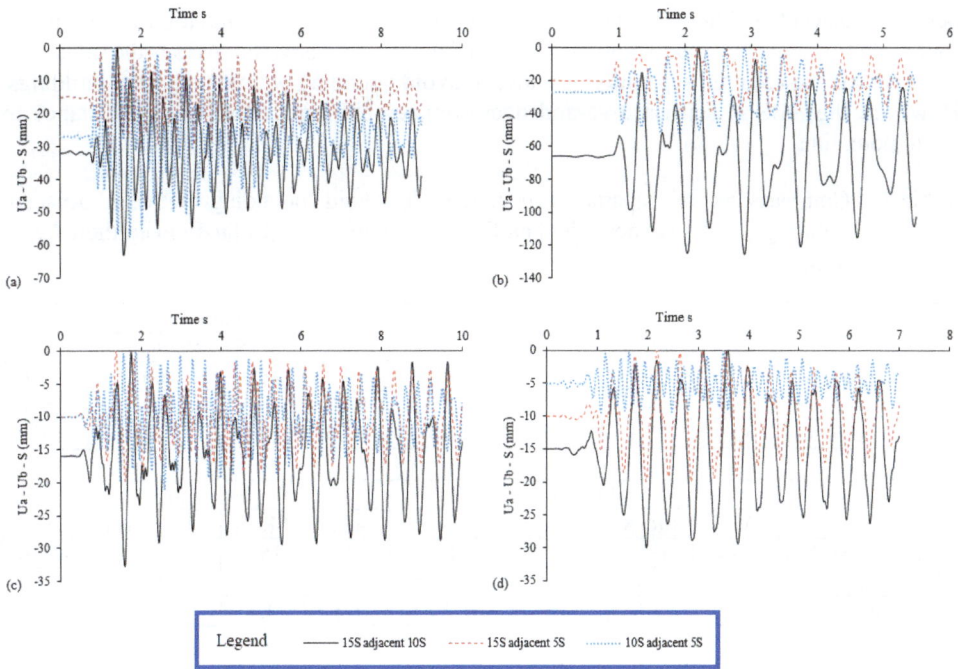

Figure 7:   Minimum separation distance to avoid pounding under (a) Scaled Kobe earthquake; (b) Scaled Northridge earthquake; (c) Scaled El Centro earthquake; and (d) Scaled Hachinohe earthquake.

3.2 Code-specified separation distance

The Australian Standards Earthquake Design Code AS1170.4-2007 requires that any building of earthquake design category II greater than 15 m in height, or any building of earthquake design category III, must be separated from adjacent structures or set back from an adjacent building boundary to avoid pounding. The minimum setback distance stipulated by the code is 1% of the structure height (see clauses 5.4.5 and 5.5.5 of AS1170.4-2007) [15]. By reading the code that a second building is adjacent to a first building, then the required separation gap is naturally assumed as the setback distance required for the second building. If both buildings have the same height, then it is clear that the required separation gap is 1% of the structure height.

However, if adjacent buildings have different heights then uncertainty naturally arises as to which structure height to use – the first building, already in place, or the second due for construction adjacent to it. Separation distance could be determined by either height [16].

The calculated required separation distances between two buildings using 1% height of either the taller or the shorter building are given in Table 3; these results will be compared with the numerical simulation results to evaluate the adequacy of code specifications. As previously discussed, building codes recommend separation gaps between adjacent buildings in order to avoid collision during earthquakes. ABS, SRSS and AS1170.4-2007 give equations to measure separation distance based on maximum lateral displacement (see eqns (1) and (2)) and height (see eqn (4)). Table 3 compares the code-required separation distance between adjacent buildings to avoid pounding and the numerical simulation results under four scaled earthquake ground motions. The tabulated results show that ABS is the safest rule to determine the required separation distance to avoid pounding between adjacent buildings. However, the ABS is conservative and overestimates the required separation distance to avoid pounding.

Table 3: Comparisons of separation distances to avoid pounding between adjacent buildings, under (a) near-field earthquake, in mm; and (b) far-field earthquake, in mm.

(a)	Kobe				Northridge			
	ABS	SRSS	Numerical	AS	ABS	SRSS	Numerical	AS
15S adjacent 10S	38.5	29.0	32	15/10	68.9	48.9	66	15/10
15S adjacent 5S	25.1	18.2	15	15/5	30.7	23.5	20	15/5
10S adjacent 5S	32.9	23.3	27	10/5	31.5	24.2	27	10/5

(b)	El Centro				Hachinohe			
	ABS	SRSS	Numerical	AS	ABS	SRSS	Numerical	AS
15S adjacent 10S	18.4	13.3	16	15/10	19.1	15.7	15	15/10
15S adjacent 5S	12.3	8.8	10	15/5	13.4	10.5	10	15/5
10S adjacent 5S	10.1	7.1	10	10/5	6.1	4.4	5	10/5

Results from the SRSS rule are significantly and reasonably accurate (but not always conservative) and vary to conservative as well (but not as conservative as the ABS rule). These finding were also reported by [1], [11].

As illustrated in Table 3, separation distance calculated using 1% of the taller adjacent structure always *underestimates* the required separation distance to avoid pounding under near-field earthquakes except in one case; and in this case, the code-required separation distance of 15 mm (1% of the 15-storey frame height) is adequate to preclude pounding between 15-storey and 5-storey frames under a scaled Kobe earthquake. On the other hand, separation distance calculated using 1% of the taller adjacent structure *overestimates* the required separation distance to avoid pounding under far-field earthquakes in most cases. However, if the separation distance is calculated as 1% of the shorter structure, this underestimates the required separation distance to avoid pounding under near-field and far-field earthquakes except in one case; and in this case, the code-required separation distance of 5 mm (1% of the 5-storey frame height) is adequate to preclude pounding between 10-storey and 5-storey frames under a scaled Hachinohe earthquake.

These results indicate that the code specification is inadequate if the shorter building is used to estimate the required separation between adjacent structures under near-field and far-field earthquakes. By contrast, the code specification is adequate if the taller building is used to estimate the required separation between adjacent structures under far-field earthquakes only.

4 CONCLUSION

This article presents experimental and numerical investigations of the minimum separation distance to avoid pounding between adjacent buildings. The aim was to evaluate the adequacy of the minimum separation gap prescribed by AS1170.4 to avoid pounding of adjacent steel-frame structures.

Shake table tests under the influence of four scaled earthquake acceleration records have been performed on the scale models, and the experimental measurements in terms of absolute acceleration time histories were determined. Afterwards, the numerical models of the constructed structural models were created, and fully non-linear time history dynamic analyses were performed under the influence of the four scaled earthquake acceleration records. Then, absolute acceleration time histories were determined and compared with the experimental measurements, which show good agreement. Subsequently, the numerical prediction in terms of relative displacement time histories were determined.

The results of this study conclude that the recommendations of the ABS are the safest method; however, ABS overestimates the required separation distance between adjacent buildings to avoid pounding. SRSS is significantly reasonably accurate, and not always conservative nor very conservative. It appears that the SRSS code-specified peak design displacement is the most appropriate method sufficient to avoid pounding or to minimise its effects. The required separation distance defined in AS1170.4-2007 is inadequate to avoid pounding when the shorter adjacent building is used under earthquake ground motion. The required separation distance defined in AS1170.4-2007 is adequate to avoid pounding only if the taller adjacent building is used under far-field earthquake ground motion.

The characteristics of ground motions whether far-field or near-field have a significant influence on the required separation gap to avoid pounding between adjacent buildings. It can be concluded that the minimum separation gap prescribed by AS1170.4 was based only on far-field earthquake ground motion, since Australia is a seismically inactive region. Appropriate gap distance for adjacent structures should be determined with regard to the characteristics of an expected earthquake along with the properties of structures.

REFERENCES

[1] Jeng, V., Kasai, K. & Jagiasi, A. (eds), The separation to avoid seismic pounding. *Proceedings of 10th World Conference on Earthquake Engineering*, 1992.

[2] Rosenblueth, E. & Meli, R., The 1985 Mexico earthquake. *Concrete International*, **8**(5), pp. 23–34, 1986.

[3] Kasai, K. & Maison, B.F., Building pounding damage during the 1989 Loma Prieta earthquake. *Engineering Structures*, **19**(3), pp. 195–207, 1997.

[4] Anagnostopoulos, S. (ed), Building pounding re-examined: How serious a problem is it. *11th World Conference on Earthquake Engineering*, Pergamon, Elsevier Science Oxford, UK, 1996.

[5] Jankowski, R. (ed), Comparison of numerical models of impact force for simulation of earthquake-induced structural pounding. *International Conference on Computational Science*, Springer, 2008.

[6] Jankowski, R. & Mahmoud, S., Linking of adjacent three-storey buildings for mitigation of structural pounding during earthquakes. *Bulletin of Earthquake Engineering*, **14**(11), pp. 3075–3097, 2016.

[7] Barros, R.C. & Khatami, S.M. (eds), Seismic response effect of shear walls in reducing pounding risk of reinforced concrete buildings subjected to near-fault ground motions. *Proceedings of the 15th World Conference on Earthquake Engineering*, Lisbon, Portugal, 2012.

[8] Raheem, S.E.A., Mitigation measures for seismic pounding effects on adjacent buildings responses. *4th ECCOMAS Thematic Conference*, Kos Island, Greece, 12–14 Jun. 2013.

[9] Richardson, A., Walsh, K. & Abdullah, M., Closed-form design equations for controlling vibrations in connected structures. *Journal of Earthquake Engineering*, **17**(5), pp. 699–719, 2013.

[10] IBC, *International Building Code*, International Code Council: Country Club Hills, Illinois, 2009.

[11] Lopez-Garcia, D. & Soong, T.T., Assessment of the separation necessary to prevent seismic pounding between linear structural systems. *Probabilistic Engineering Mechanics*, **24**(2), pp. 210–223, 2009.

[12] Council, B.S.S. & Council, A.T., NEHRP guidelines for the seismic rehabilitation of buildings. Federal Emergency Management Agency, 1997.

[13] NBCC, *National Building Code of Canada*, National Research Council: Ottawa, 2010.

[14] Eurocode8, *Design of Structures for Earthquake Resistance – Part 1: General Rules, Seismic Actions and Rules for Buildings*, European Committee for Standardization: Brussels, 2005.

[15] AS1170.4-2007, *Structural Design Actions Part 4: Earthquake Actions in Australia*, Standards Australia, 2007.

[16] Hao, H., Analysis of seismic pounding between adjacent buildings. *Australian Journal of Structural Engineering*, **16**(3), pp. 208–225, 2015.

[17] Anagnostopoulos, S.A., Pounding of buildings in series during earthquakes. *Earthquake Engineering & Structural Dynamics*, **16**(3), pp. 443–456, 1988.

[18] Kasai, K., Jagiasi, A.R. & Jeng, V., Inelastic vibration phase theory for seismic pounding mitigation. *Journal of Structural Engineering*, **122**(10), pp. 1136–1146, 1996.

[19] Abdel Raheem, S.E., Mitigation measures for earthquake induced pounding effects on seismic performance of adjacent buildings. *Bulletin of Earthquake Engineering*, **12**(4), pp. 1705–1724, 2014.

[20] Favvata, M.J., Minimum required separation gap for adjacent RC frames with potential inter-story seismic pounding. *Engineering Structures*, **152**, pp. 643–659, 2017.

[21] Tabatabaiefar, S.H.R., Fatahi, B. & Samali, B., Numerical and experimental investigations on seismic response of building frames under influence of soil-structure interaction. *Advances in Structural Engineering*, **17**(1), pp. 109–130, 2014.

[22] Tabatabaiefar, H.R. & Mansoury, B., Detail design, building and commissioning of tall building structural models for experimental shaking table tests. *The Structural Design of Tall and Special Buildings*, **25**(8), pp. 357–374, 2016.

[23] Chopra, A., *Dynamics of Structures*, 3rd edn, Prentice Hall: New Jersey, 2007.

[24] Karamodin, A. & Kazemi, H., Semi-active control of structures using neuro-predictive algorithm for MR dampers. *Structural Control and Health Monitoring: The Official Journal of the International Association for Structural Control and Monitoring and of the European Association for the Control of Structures*, **17**(3), pp. 237–253, 2010.

[25] Yaghmaei-Sabegh, S. & Jalali-Milani, N., Pounding force response spectrum for near-field and far-field earthquakes. *Scientia Iranica*, **19**(5), pp. 1236–1250, 2012.

[26] Hatzigeorgiou, G.D., Damping modification factors for SDOF systems subjected to near-fault, far-fault and artificial earthquakes. *Earthquake Engineering & Structural Dynamics*, **39**(11), pp. 1239–1258, 2010.

[27] Yaghmaei-Sabegh, S. & Tsang, H., An updated study on near-fault ground motions of the 1978 Tabas, Iran, earthquake (Mw = 7.4). *Scientia Iranica*, **18**(4), pp. 895–905, 2011.

[28] Sap, C., Integrated software for structural analysis and design. *Analysis Reference Manual*, 2000.

[29] Jaradat, Y. & Far, H., Optimum stiffness values for impact element models to determine pounding forces between adjacent buildings. *Structural Engineering and Mechanics*, 2020.

[30] Abdel Raheem, S.E., Seismic pounding between adjacent building structures. *Electronic Journal of Structural Engineering*, **6**, pp. 66–74, 2006.

[31] Lin, J.H. & Weng, C.C., A study on seismic pounding probability of buildings in Taipei metropolitan area. *Journal of the Chinese Institute of Engineers*, **25**(2), pp. 123–135, 2002.

Author index

Anastasia S. PII-47
Aranibar P. PII-37

Far H. PII-59
Favvata M. J. PII-3
Flenga M. G. PII-3
Fujii K. PII-25

Galván A. PII-15

Ivorra-Chorro S. PII-47

Izquierdo-Horna L. PII-15, PII-37

Jaradat Y. PII-59

Kato M. PII-25

Poveda-Martínez P. PII-47

Saleh A. PII-59

Torres-Gorriz B. PII-47

Structural Studies, Repairs and Maintenance of Heritage Architecture XVI

Edited by: P. DE WILDE, Free University of Brussels, Belgium

Originating from the 16th edition of the Conference on Studies, Repairs and Maintenance of Heritage Architecture, this volume brings together latest contributions from scientists, architects, engineers and restoration experts dealing with different aspects of heritage buildings, including the preservation of architectural heritage.

The importance of retaining the built cultural heritage cannot be overemphasised. Rapid development and the inappropriate conservation techniques are threatening many built cultural heritage unique sites in different parts of the world.

This current volume covers a wide range of topics related to the historical aspects and the reuse of heritage buildings, as well as technical issues on the structural integrity of different types of buildings, such as those constructed with materials as varied as iron and steel, concrete, masonry, wood or earth. Material characterisation techniques are also addressed, including non-destructive tests via computer simulation. Modern computer simulation can provide accurate results demonstrating the stress state of the building and possible failure mechanisms affecting its stability.

The included papers focus on such topics as: Heritage and tourism; Heritage architecture and historical aspects; Management and assessment of heritage buildings; Modern (19th/20th century) heritage; Re-use of heritage buildings; Adaptability and accessibility; Social, cultural and economic aspects; Material characterization; Learning from the past; Industrial heritage; Heritage masonry structures.

WIT Transactions on The Built Environment, vol. 191
ISBN: 978-1-78466-359-9   eISBN: 978-1-78466-360-5

Published 2019 / 534pp

Lightning Source UK Ltd.
Milton Keynes UK
UKHW051009200821
389166UK00003B/53